WITHDRAWN

ENGINEERING APPLICATIONS
OF FRACTURE ANALYSIS

INTERNATIONAL SERIES ON THE
STRENGTH AND FRACTURE OF MATERIALS AND STRUCTURE
General Editor: D M R Taplin, D Sc, D Phil, F I M

OTHER TITLES IN THE SERIES

EASTERLING	Mechanisms of Deformation and Fracture
HAASEN, GEROLD & KOSTORZ	Strength of Metals and Alloys (ICSMA 5) (3 Volumes)
MILLER & SMITH	Mechanical Behaviour of Materials (ICM 3) (3 Volumes)
PIGGOTT	Load Bearing Fibre Composites
SMITH	Fracture Mechanics - Current Status, Future Prospects
TAPLIN	Advances in Research on the Strength and Fracture of Materials (ICF 4) (6 Volumes)

Related Pergamon Journals

Acta Metallurgica
Canadian Metallurgical Quarterly
Corrosion Science
Engineering Fracture Mechanics
Fatigue of Engineering Materials and Structures
Materials Research Bulletin
Metals Forum
Physics of Metals and Metallography
Scripta Metallurgica

ENGINEERING APPLICATIONS OF FRACTURE ANALYSIS

Proceedings of the First National Conference on Fracture held in Johannesburg,
South Africa, 7-9 November 1979

Edited by

G. G. GARRETT

Professor of Physical and Fabrication Metallurgy, University of the Witwatersrand,
Johannesburg, Republic of South Africa

and

D. L. MARRIOTT

Associate Professor, Department of Industrial and Mechanical Engineering,
University of Illinois at Urbana - Champaign, USA

PERGAMON PRESS
OXFORD · NEW YORK · TORONTO · SYDNEY · PARIS · FRANKFURT

U.K.	Pergamon Press Ltd., Headington Hill Hall, Oxford OX3 0BW, England
U.S.A.	Pergamon Press Inc., Maxwell House, Fairview Park, Elmsford, New York 10523, U.S.A.
CANADA	Pergamon of Canada, Suite 104, 150 Consumers Road, Willowdale, Ontario M2J 1P9, Canada
AUSTRALIA	Pergamon Press (Aust.) Pty. Ltd., P.O. Box 544, Potts Point, N.S.W. 2011, Australia
FRANCE	Pergamon Press SARL, 24 rue des Ecoles, 75240 Paris, Cedex 05, France
FEDERAL REPUBLIC OF GERMANY	Pergamon Press GmbH, 6242 Kronberg-Taunus, Hammerweg 6, Federal Republic of Germany

Copyright © 1980 Pergamon Press Ltd.

All Rights Reserved. No part of this publication may be reproduced, stored in a retrieval system or transmitted in any form or by any means: electronic, electrostatic, magnetic tape, mechanical, photocopying, recording or otherwise, without permission in writing from the publishers.

First edition 1980

British Library Cataloguing in Publication Data

National Conference on Fracture, *1st,
Johannesburg, 1979*
Engineering Applications of Fracture Analysis. -
(International series on the strength and fracture
of materials and structures).
1. Metals - Fracture - Congresses
I. Title II. Garrett, G G
III. Marriott, D L IV. Series
620.1'6'6   TA460   80-41074
ISBN 0-08-025437-3

In order to make this volume available as economically and as rapidly as possible the authors' typescripts have been reproduced in their original forms. This method has its typographical limitations but it is hoped that they in no way distract the reader.

TA
409
N37
1980

Printed in Great Britain by A. Wheaton and Co. Ltd., Exeter

ORGANISING COMMITTEE

Prof. G.G. Garrett   Department of Metallurgy
   (Chairman)   University of the Witwatersrand

Dr S. Luyckx   Department of Physics
   (Hon. Secretary)   University of the Witwatersrand

Dr D. Chandler   Department of Mechanical Engineering
                            University of the Witwatersrand

J. Heyman   Iscor Research Laboratories, Pretoria

J.J. Marais   Department of Mechanical Engineering
                            University of Pretoria

Dr D.L. Marriott   Licensing Branch, Atomic Energy Board,
                            Pretoria

Prof. L.O. Nicolaysen   Bernard Price Institute of Geophysics,
                            University of the Witwatersrand

B. Protheroe   Chamber of Mines Research Laboratories
                            Johannesburg

Prof. F.P.A. Robinson   Department of Metallurgy
                            University of the Witwatersrand

C. Smallbone   South African Institute of Welding

V. Abery (Mrs)   Department of Metallurgy
   (Conference Secretary)   University of the Witwatersrand

SPONSORS

Institution of Metallurgists (S.A. Branch)

S.A. Institute of Physics

S.A. Institute of Welding

CONTENTS

Nomenclature	xi
Conversion Units	xv
Introduction F.R.N. NABARRO	xvii
The Brittle Fracture Story J.D. HARRISON	xix

SECTION 1: SOME PROBLEMS OF FRACTURE

Fractures in Springs of Steel Strip G. PERSSON	3
Metallurgical Failures in the Mining Industry R.N. TAYLOR and W.J. VAN DEN BERG	19
Progress in Extending the Life of Steel Components Exposed to High Temperature Environments D. DAVIDSON	33
Unusual Fractures in the Mining Industry D.D. HOWAT	45

SECTION 2: UNDERSTANDING FRACTURE

The Strength and Fracture of Two-Phase Alloys – A Comparison of Two Alloy Systems J. GURLAND	63
Failure by Fatigue G.G. GARRETT	79
On the Microstructural Control of the Fracture Processes Involved in Wear C.J. HEATHCOCK, C. ALLEN, B.E. PROTHEROE and A. BALL	95
The General Characteristics and Evaluation of Stress Corrosion Cracking D. TWIGG	103
Factors Controlling HAZ and Weld Metal Toughness in C-Mn Steels R.E. DOLBY	117

SECTION 3A: SOLVING FRACTURE PROBLEMS - THE TOOLS AVAILABLE

Design and the Prevention of Metallurgical Failures
    P.J. ECCLESTON   137

Fracture Mechanics and the Assessment of
Structural Reliability
    G.G. GARRETT   187

Defect Assessment by Means of Non-Destructive Testing
    J.J. MARAIS   203

Estimation of Risk of Failure of Components due to
Fast Fracture
    D.L. MARRIOTT   219

Fractography: A Tool for Failure Analysis
    S.B. LUYCKX   231

SECTION 3B: SOLVING FRACTURE PROBLEMS - SOME CASE STUDIES

The COD Approach and its Application to Welded Structures
    J.D. HARRISON, M.G. DAWES, G.L. ARCHER and M.S. KAMATH   249

Simplified Stress Intensity Evaluation of a Nuclear Reactor
Pressure Vessel under a given Accident Loading
    W. VAN DER WALT   269

Avoiding Fracture in Pressure Vessels
    J.R. CAMPBELL   283

Fracture Toughness Considerations in the Design and
Use of High Strength Components
    G.T. VAN ROOYEN   295

Cracking in Weldments of Structural Steels
    R.E. DOLBY   313

SECTION 4 : ADVANCES IN FRACTURE

Factors Influencing the Impact Toughness of Multipass
Submerged-Arc Welds in Micro-Alloyed Steel
    J.I.J. FICK   327

A Study of Crack Arrest Related to Nuclear Plant Integrity
    D.L. MARRIOTT, R.P.G. ANDERSON and G.G. GARRETT   337

Stress Corrosion and Corrosion Fatigue in Light
Water Reactor Environments
    D. de G. JONES   353

A Review of the Use of Isoparametric Finite
Elements for Fracture Mechanics
    F.J. HEYMANN   371

Fracture and Plastic Deformation
    P.J. JACKSON and O.L. de LANGE   389

The Significance of Rock Fracturing in the
Design and Support of Mine Excavations
    H. WAGNER and N. WISEMAN   399

Rock Fracturing Processes in Deep Mines
    N.C. GAY   409

The Deformation and Fracture of Quartz
    G. GLOVER and A. BALL   419

Author Index   429

NOMENCLATURE

A	Area of cross-section of a specimen
A_o	Area of cross-section of a specimen at the start of testing
A_f	Area of cross-section of a specimen at fracture
a	Crack length - one half the total length of an internal crack or depth of a surface crack
a_o	Original crack length - one-half of total length of an internal crack at the start of a fracture toughness test, or depth of a surface crack at the start of a fracture toughness test
a_p	Measured crack length - one-half the effective total length of an internal crack or effective depth of a surface crack as measured by physical methods
a_e	Effective crack length - one-half the effective total length of an internal crack or effective depth of a surface crack (adjusted for the influences of a crack-tip plastic zone)
$\Delta a, \Delta a_p, \ldots$	Crack growth increment
da/dN	Rate of fatigue crack propagation
B	Test piece thickness
b	Atomic interval (Burgers vector magnitude)
d	Average grain diameter
D_L	Lattice diffusion rate
D_B	Grain boundary diffusion rate
D_S	Surface diffusion rate
E	Young's modulus of elasticity
exp	Exponential base of natural logarithms
G	Strain energy release rate with crack extension per unit length of crack border or crack extension force
$G_I\ G_{II}\ G_{III}$	Crack extension forces for various modes of crack opening
h	Planck's constant
I	Moment of inertia
J	Path-independent integral characterising elastic/plastic deformation field intensity at crack tip; also, energy release rate for non-linear elastic material

Nomenclature

K	Stress-intensity factor - a measure of the stress-field intensity near the tip of a perfect crack in a linear elastic solid
K_c	Fracture toughness - the largest value of the stress-intensity factor that exists prior to the onset of rapid fracture
K_{max}	Maximum stress-intensity factor
K_{min}	Minimum stress-intensity factor
K_{th}	Threshold stress intensity factor below which fatigue crack growth will not occur
K_I	Opening mode stress-intensity factor
K_{IC}	Plane-strain fracture toughness
K_{Ii}	Elastic stress-intensity factor at the start of a sustained-load flaw-growth test
K_{ISCC}	Plane-strain K_I threshold above which sustained-load flaw-growth occurs
K_{II}	Edge-sliding mode stress-itensity factor
K_{III}	Tearing mode stress-intensity factor
$\dot{K}$	Rate of change of stress-intensity factor with time
ΔK	Stress intensity range
k	Boltzmann constant
k_y	Parameter that determines grain-size dependence of yield strength
l_o	Gauge length
$\ln$	Natural logarithm
$\log$	Common logarithm
m	Strain-rate sensitivity exponent
N_f	Number of cycles to failure
n	Strain hardening exponent
P	Force
P_{max}	Maximum force
Q	Activation energy

Q_a	Activation energy for crack growth
Q_c	Activation energy for creep
Q_d	Activation energy for self diffusion
T	Temperature
T_M	Absolute melting temperature
T_D	Ductile-brittle transformation temperature
t	Time
t_o	Time at the onset of a test
t_f	Fracture time
U	Potential energy
γ_s	True surface energy
γ_B	Grain boundary surface energy
γ_p	Effective surface energy of plastic layer
δ	Value of crack opening displacement
δ_c	Critical crack opening displacement, being one of the following: (1) Crack opening displacement at fracture (2) Crack opening displacement at first instability or discontinuity (3) Crack opening displacement at which an amount of crack growth commences
δ_m	Crack opening displacement at first attainment of maximum force
δ_z	Thickness of grain boundary layer
ε	Normal strain
ε_e	Normal strain, elastic
ε_p	Normal strain, plastic
ε_T	Normal strain, total
ε_{max}	Normal strain at maximum tensile load
ε_E	Engineering normal strain

ε_f	Normal strain, critical value at failure
ε_i	Principal strains (i = 1, 2, 3)
ε_{pi}	Principal strains, plastic
$\varepsilon_x\ \varepsilon_y\ \varepsilon_z$	Cartesian strain components
ε_{ij}	Strain tensor
$\dot{\varepsilon}$	Strain rate
$\dot{\varepsilon}_e$	Strain rate, elastic
$\dot{\varepsilon}_p$	Strain rate, plastic
$\dot{\varepsilon}_o$	Strain rate, initial value
$\Delta\varepsilon$	Strain range
$\Delta\varepsilon_p$	Plastic strain range
ν	Poisson's ratio
σ	Normal stress
σ_y	Yield stress under uniaxial tension
$\sigma_1\ \sigma_2\ \sigma_3$	Principal normal stresses
σ_e	Fatigue strength, endurance limit
σ_f	Fracture stress
σ_{max}	Maximum stress
$\sigma_x\ \sigma_y\ \sigma_z$	Cartesian components of normal stress
$\dot{\sigma}$	Stress rate
$\Delta\sigma$	Stress range
τ	Shear stress
τ_o	Critical shear stress
$\tau_1\ \tau_2\ \tau_3$	Principal shear stresses
τ_{max}	Shear stress, maximum value
Ω	Atomic volume

CONVERSION UNITS

To convert from	to	multiply by
inch	metre (m)	2.54×10^{-2}
pound force	newton (N)	4.448
kilogram force	newton (N)	9.807
kilogram force/metre^2	pascal (Pa)	9.807
pound mass	kilogram mass (kg)	4.536×10^{-1}
ksi	pascal (Pa)	6.895×10^{6}
ksi $\sqrt{\text{in}}$	MN m$^{-3/2}$	1.099
torr	pascal (Pa)	1.333×10^{2}
bar	pascal (Pa)	1×10^{5}
angstrom	metre (m)	1×10^{-10}
calorie	joule (J)	4.184
foot-pound	joule (J)	1.356
degree Celsius	kelvin (K)	$T_K = T_C + 273.15$

IMPORTANT MULTIPLES

Multiplication factor	Prefix	Symbol
10^{-12}	pico	p
10^{-9}	nano	n
10^{-6}	micro	μ
10^{-3}	milli	m
10^{3}	kilo	k
10^{6}	mega	M
10^{9}	giga	G

INTRODUCTION

F. R. N. Nabarro

*Deputy Vice-Chancellor, University of the Witwatersrand,
Johannesburg, R.S.A.*

A welcome custom is growing in South Africa of bringing together people involved in the scientific and the technical aspects of a subject to exchange knowledge in a common conference. This meeting, Fracture '79, arranged by the Institute of Metallurgists (South African Branch), the South African Institute of Physics and the South African Institute of Welding, is a case in point. The contributions range from theoretical analyses of the influence of dislocation stresses in initiating fracture and of the strengths of two-phase alloys of very different structures through formal applications of fracture mechanics to the practical design of steel components exposed to high-temperature environments. There is a special emphasis on problems relating directly to South African industry, such as failures in rock drilling equipment, unwanted rock fractures in mines and safety problems in nuclear reactors. Techniques for the study of fractures have their place: fractography, ultrasonic testing and analysis of inclusions. Most of the work has been done in South Africa, and the papers give a good general picture of fracture studies in the country, but there are also valuable contributions from the United States of America and Sweden.

While this Conference forms one of a series, each meeting in this series has owed its success largely to the enthusiasm of one man. In this case, the drive has come from Professor Geoff Garrett of the Department of Metallurgy of the University of the Witwatersrand. He has shown the appropriate qualities of toughness, resilience, heat resistance, and lack of susceptibility to fatigue and corrosion by stress or fretting. Writing as I do before the event, I can only hope that the meeting will have the success it deserves. If it succeeds in advancing our knowledge of fracture processes, it will serve an important purpose. If it succeeds only in the more modest task of spreading existing knowledge among practical metallurgists and engineers, it will justify many times over the cost and effort that have gone into its organisation.

THE BRITTLE FRACTURE STORY

J. D. Harrison

The Welding Institute, Abington, Cambridge, England

INTRODUCTION

First it must be acknowledged that the title of this paper is the same as that of a book by Mrs Tipper, who carried out a lot of the early research on brittle fracture at Cambridge University in the 1950s.

A short story concerning Dr Geoff Egan, who is well known in the fracture field, illustrates the type of problem that may confront those at the forefront of

Fig.1 The World Concord

Fig.2 The SS Schenectady

fracture research. Dr Egan was flying back to Europe from a conference in the USA. He awoke as the plane crossed the coast of France and saw a group of people looking out of one of the windows. The captain approached him and said 'Dr Egan, I understand that you are an expert on fatigue and fracture, I would like to ask your advice about a problem we have, could you come and look out of this window'. From the window a crack about 1m long could be seen in the top of the wing. Dr Egan said 'I can give you two pieces of advice: firstly get this plane on to the ground as fast as you can; and secondly slow down!' It is to be hoped that advice from experts in fracture is generally less conflicting. The crack in question in fact grew another 75mm during the flight.

The study of fracture has developed over the last thirty or forty years, and detailed research really started with the investigation of brittle fractures in Liberty ships and tankers which occurred during and after the war (Fig.1). The problem of brittle fracture in welded construction first became prominent in the 1930s with the failure of a number of bridges in Belgium, but investigation of these was overtaken by the war. Of the two types of failure investigated most widely at The Welding Institute, namely fatigue and fracture, fatigue is by far the more costly to industry, although brittle fracture is the more spectacular.

Figure 2 shows the SS Schenectady, which failed in port by brittle fracture under virtually zero load because of a drop in the ambient temperature. Such failures of welded ships gave a strong stimulus to fracture research, particularly in Britain and the USA. It was noted that riveted ships were susceptible to this type of failure. The ship classification societies were heavily involved in the early investigations, and they collected data from many ships that did fail. This made it possible to correlate the Charpy impact properties of plates in which failures initiated and in plates in which failures arrested.

It was found that initiation occurred generally in plates having a Charpy energy of less than 10 ft lb at the service temperature, while arrests were found generally in plates giving more than 20 ft lb. This finding led to the adoption of Charpy testing in the rules of the ship classification societies. Unfortunately there are ways of designing and fabricating ships which make it possible to avoid using Charpy impact tested material. This may save expense but it means that failures are still occurring, although they are not widely publicised. At least four major brittle fractures in ships have come to the author's knowledge within a year.

In one case a ship off the Canadian coast broke in two, and the bow section was floating free. The Canadian Coast Guard considered it a hazard to shipping and accordingly sank it, for which subsequently they were sued by the owners of the ship. The stern section was eventually towed back to Rotterdam to have a new bow welded. Another failure occurred recently in South African waters in a ship that was being driven hard aroung the Cape in heavy seas and poor visibility. During the night the crew in the engine room felt some disturbance to the ship's motion, but gave it no particular attention. The captain rang down to request more revolutions as the speed had dropped, but the engine room replied that the engine speed had not changed. When dawn broke it was found that the front half of the ship had broken off and they had been steaming with only the rear half. These examples lead to consideration of matters of more fundamental importance.

CONDITIONS FOR FRACTURE

In early work it was realised that there were three conditions required for brittle fracture - a defect, a stress at right angles to that defect, and a microstructure that was susceptible to fracture under the environmental conditions ruling at the time.

The importance of welding is then immediately apparent, because almost by definition all welds are defective. Furthermore, in welded structures which have not been post weld heat treated it must be assumed that there will be residual stresses present up to the yield level. Thus welding introduces defects and a high stress at the weld, and it can also lead to a reduction in fracture toughness of the parent material.

Some of the factors affecting fracture behaviour are illustrated in Fig.3. Structural steels are considered in the main, because they are the most widely used tonnage materials and also it is in these types that the most significant brittle fracture problems are encountered.

Firstly there is an effect of specimen size, the small specimen showing a higher fracture toughness, i.e. an increase in toughness at the lower temperature. Secondly, in structural steels there is a marked effect of strain rate, so that the toughness for static tests is considerably higher than it is for dynamic. This is particularly important in terms of the overall philosophy that has to be adopted with regard to fracture control. Thirdly there is the well known transition temperature effect, and incidentally the shift in transition temperature caused by an increase in strain rate may be as much as $80^\circ C$, which is most significant in lower strength steels.

FRACTURE CONTROL

In considering the general philosophy to be adopted regarding fracture control there are two broad approaches available. One is to prevent fracture initiation and the second is to ensure that if a crack does initiate it will subsequently arrest.

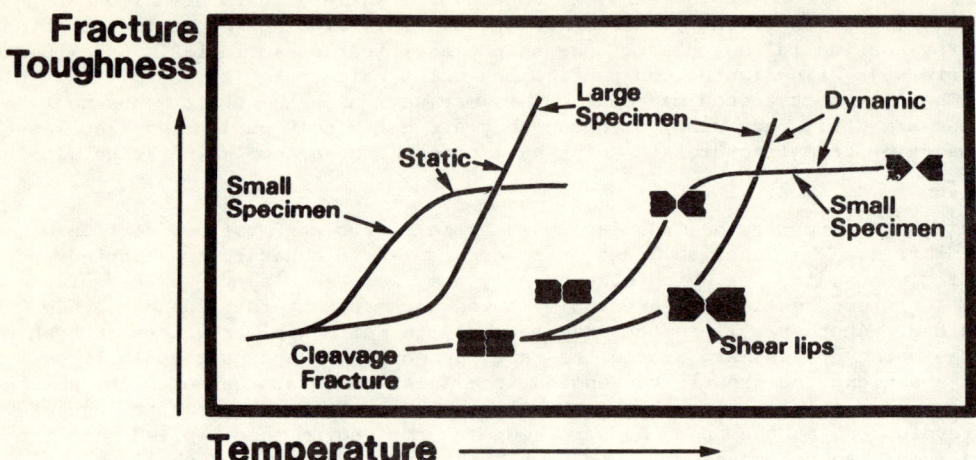

Fig.3 Factors affecting brittle fracture

<u>Crack arrest</u>

The arrest approach assumes that there may be some local area of embrittlement or brittle enclave in the surrounding tough matrix. The toughness of this local brittle region is not known, and it must be ensured that, if fracture does initiate in this region, it will be arrested when it propagates into the surrounding tougher matrix.

Figure 3 indicates that in homogeneous material arrest will be obtained only if the stress driving the crack is reduced by propagation of the crack itself. This is because at the tip of the running crack there will be a high strain rate, so that a shift from high or relatively high toughness at initiation under the slowly applied load to low toughness at the tip of the crack will have already occurred. Thus the arrest philosophy is applicable only when there may be local regions of embrittlement, or to other circumstances where the very propagation of a crack will reduce the stresses that are driving it.

Material of initially good properties is required also if the arrest approach is to be adopted. It is an expensive approach in terms of the material used, and in fact many conventional steel structures - bridges, buildings, offshore platforms etc. - are operating at temperatures which are too low to give arrest if a running brittle crack did initiate in them. For this reason, at The Welding Institute the initiation approach to fracture has been adopted to a large degree. That is to say that the various regions in the welded joint are examined to ensure that the initiation toughness of all these regions is sufficiently high to prevent fracture from starting.

The deficiencies of the Charpy test are also apparent from Fig.3. The test is of course carried out on a specimen of a fixed size, namely 10mm square. Already this has introduced an arbitrary factor, as specimen thickness affects toughness, but only one thickness is sampled. Furthermore, the Charpy test is carried out at a quite arbitrary strain rate, which bears no direct relation to the loading rate likely to be experienced in service. It is not necessarily relevant even to impact loading, because in the actual structure the impact could be at a higher or a lower rate than that used in a test.

A final factor which affects Charpy test results is the use of a blunt notch. It is known that notch acuity is particularly important in determining fracture resistance and thus the Charpy test is again arbitrary. This is not to decry the test, which has been very successful in the past, although only through correlations with service performance. We therefore consider some other types of test available.

The initiation approach

As emphasised above, the implication of an initiation approach must be that there is no region of the structure in which a fracture will initiate, because it is not constructed in expensive material which will bring about crack arrest if propagation does begin. This has most important implications in construction. Figure 4 illustrates schematically the type of variation that may be experienced within one welded joint. This means that the toughness of the whole weld cannot be defined and only the toughness of regions within the weld can be considered. These regions represent the parent steel, the different types of microstructure of heat affected zone, and the weld metal. In Fig.4 the weld metal is shown to

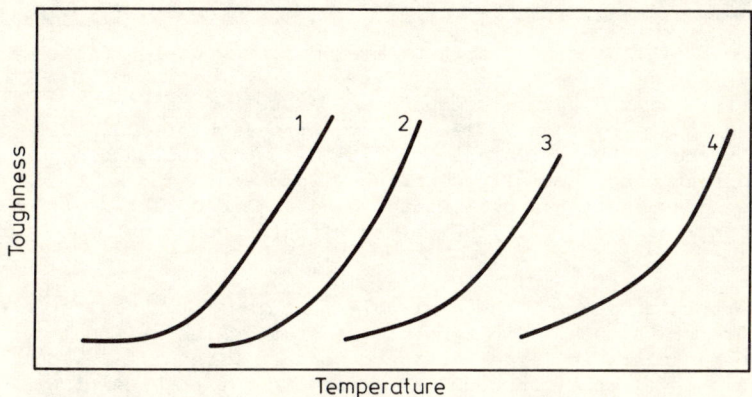

Fig.4 The variation in toughness in a welded joint: 1 - parent plate; 2 - fine grained HAZ; 3 - coarse grained HAZ; 4 - weld metal.

be the most brittle region, and although this is commonly found it is not universally true. Indeed cases have been known where the parent steel itself was the most brittle element, and anything done to it via the weld thermal cycle in fact improved its properties. Naturally steelmakers try to make better steel than that.

Wide plate testing

The importance of welding in brittle fracture gave great impetus to the development of testing techniques to assess fracture resistance. In the late 50s and early 60s Wells at The Welding Institute and workers at Illinois University made a great step forward with the development of the wide plate test.

Figure 5 shows The Welding Institute's wide plate test facility. The testpiece

Fig.5 The Welding Institute's wide plate testing facility

is a large plate a little under 1m square welded to two massive end beams. The beams are pushed apart by four hydraulic actuators each of 1000t capacity, so that in essence the set-up forms a compact tensile testing machine of 4000t capacity. The test plate normally contains a weld and has some kind of defect inserted. Specimens that have been tested have included plates with a longitudinal weld containing a defect, sometimes with the addition of a transverse weld.

This test incorporates many of the desirable features for a realistic test of brittle fracture resistance. One important factor is the residual stress, and the wide plate testpiece is large enough to contain within it residual stresses up to full yield level that can occur as a result of the deposition of the weld. The test is carried out on full thickness plate and as mentioned above this is also an important consideration. It is possible to add to the residual stress by applying a longitudinal stress using the actuators.

Effects of welding can be taken into account in the test but in the form described here it is not entirely satisfactory because the only form of embrittlement that is really considered is the straining and ageing which occurs at the tip of the crack as the weld is deposited. Nonetheless, this is an important form of embrittlement in some types of steel and the wide plate test was the first in which fractures which were known to occur at stresses below yield in mild steels in service were demonstrated in the laboratory. The test method was particularly useful in investigation of the failure illustrated in Fig.6, which shows the Fawley oil storage tank. A weld defect had been inadequately repaired and the tip of the original defect was strained and aged by the repairing process, which produced embrittlement that was particularly amenable to assessment by the test.

In the application of the wide plate test to pressure vessel steels as in the rules laid down in the relevant British and ISO standard, the minimum reliable design temperature is plotted against the Charpy test temperature. These rules were based on correlation between performance in the Charpy and in the wide plate tests. As noted previously, the only really satisfactory way of using the

Fig.6 The Fawley oil storage tank

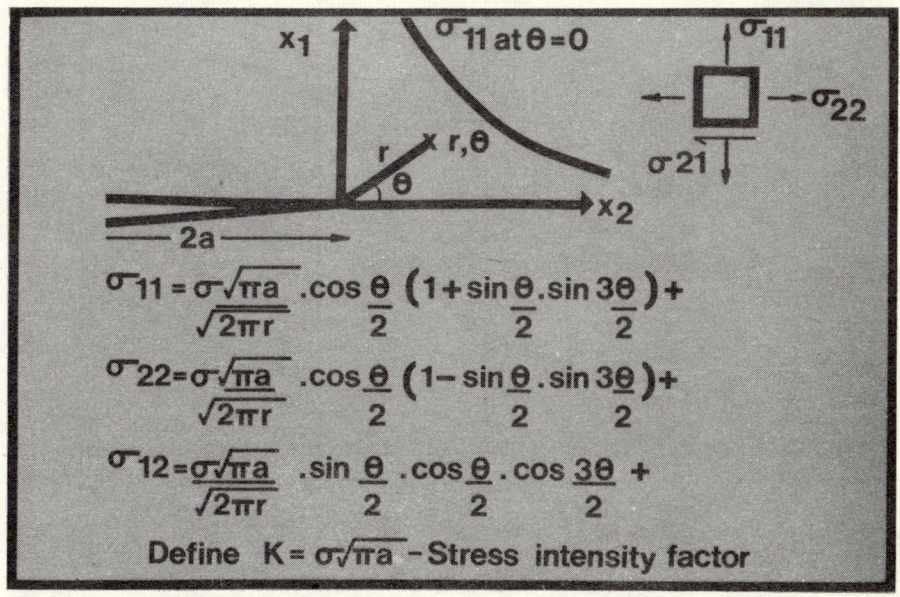

Fig.7 Stresses at a crack tip expressed in polar coordinates

Charpy test is through correlation, because it does not itself represent true service conditions. Account is taken of the effect of thickness in that in thinner sections material of such good properties is not required because toughness rises rapidly with decreased thickness. The beneficial effect of the removal of the residual stress by post weld heat treatment and the tempering that the post weld heat treatment gives to heat affected zones and weld metals is also included. Stress relieved materials can be used at lower design temperatures for the same Charpy properties than can materials in the as-welded condition.

The wide plate test in its cross welded form also offers the possibility of assessing the coarse grained heat affected zone of the transverse weld and the weld metal itself. However, a major disadvantage of the test is its cost, which can range from aroung R2500 - 4000 per test. Because it is extremely expensive, fracture mechanics based tests have considerable attractions.

FRACTURE MECHANICS

The basis for fracture mechanics was laid down by Griffith in the 1920s, his work using an energy balance approach. However, engineers found such concepts difficult to apply and a major step forward came with Irwin's studies of the stress analysis of cracks. He considered a stress analysis of the zone surrounding the tip of a crack. If the position at which it is wished to determine the stress is expressed by means of polar co-ordinates, r and θ, centered on the crack tip, the tensile stresses in the x_1 and x_2 directions and the shear stress can be expressed as indicated in Fig.7. Readers unfamiliar with fracture mechanics should not be daunted by these expressions. The important point is that all that is represented here is simply a function of r and θ, that is it is a function of the position at which we are trying to determine the stress, and there is a single

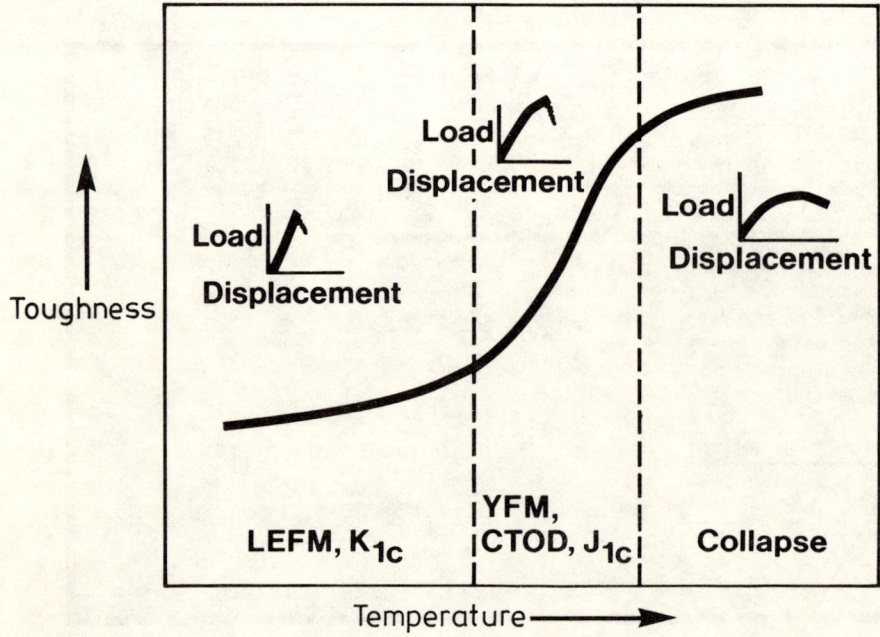

Fig.8 Transition curve

common factor K, the stress intensity factor, which for the given geometry is given by the expression:

$K = \sigma\sqrt{\pi a}$

Thus what Irwin showed is that the whole stress field aroung a crack tip can be expressed in terms of the single parameter K.

This may seem a highly satisfactory solution but strictly it applies only to materials which are ideally elastic and it assumes that the concept of a stress at the crack tip is meaningful. The expressions in Fig.7 indicate that as r tends to zero in fact all the stresses will tend to infinity, and no real material can tolerate these conditions, so that it is clear that this is an idealised approach based on linear elasticity. However, Irwin showed that provided that the plasticity of the crack tip was limited in terms of the overall geometry of the specimen or of the structure the relationships could be applied successfully. Unfortunately, it is found that for welded structures materials are needed which do not obey this linear elastic relationship and display some plasticity prior to fracture.

Because the linear elastic fracture mechanics approach is applicable only to rather brittle material, or to materials which are so thick that the plastic zone size can still be considered as being small compared with the overall geometry, elastic plastic fracture mechanics has been a major research area over the last 15 years or so in all parts of the world. Transitional behaviour is shown in Fig.8, which indicates that linear elastic fracture mechanics is applicable to relatively brittle materials on the lower shelf, while yielding fracture mechanics, which is the area of most widespread current research, is applicable to the transitional regime. For operation at the upper shelf the safety of the structure is really considered in terms of resistance to plastic instability or plastic collapse. The paper by the author, Dawes, Archer and Kamath later in this volume deals with methods used for plastic collapse.

Applications of fracture mechanics

The great advantage of fracture mechanics to the practising engineer should be emphasised. It provides the essential relationship between the three conditions for fracture - the defect, the stress and the material. Figure 9 represents these aspects and the beauty of fracture mechanics is that we can go around the triangle shown in any way we like. For example, we can fix the applied stress or strain in a structure, i.e. the design is fixed. The design can be put in front of non-destructive testing experts and they can quote size of defect that can confidently be expected to be found in the structure. Having determined this the fracture toughness requirement for the material can be specified. Alternatively, the structure may be built and a subsequent non-destructive examination

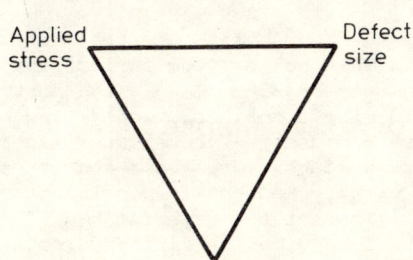

Fig.9 The interrelated variables of fracture.

may reveal defects larger than those normally permitted under the fabrication code. It is then necessary to determine whether the defects can be allowed to remain in the structure, and this is a question on which extremely large sums of money may depend. This is particularly true when structures used in the offshore oil industry are considered.

When British Petroleum were about to commence work in the Forties field in the Northern North Sea, which is a particularly arduous area of offshore development, it was decided that material properties of their rigs would be specified in terms of a fracture mechanics approach. The welding had to be qualified according to fracture mechanics requirements and the specification was fixed in terms of the known design stresses and the assumed worst defect of a buried crack about 25mm long. Because of the great thickness at the weld areas in the structures this specification caused a great deal of difficulty to the fabricators in attempting to meet it. Some development work was required to produce consumables which would give the required fracture toughness properties, and in fact the only way that the specification could be met was by post weld heat treatment of critical intersections, i.e. the regions of maximum stress concentration. The cost of that operation was aroung R2 million.

Nevertheless, the structures were built, and towards the end of the construction period ultrasonic examination began to reveal defects in the welds. For various reasons these were particularly difficult to repair and it became necessary to apply some further engineering judgement. The defects were assessed on the basis of fracture mechanics, which means that having first gone around the triangle of Fig.9 in one direction the engineers then went around it in the other. It was decided that the defects could be accepted and allowed to remain in the structure and this ensured that the platform in question was delivered on time in the year that it was required. This is particularly significant in North Sea operations because of the small weather window, meaning that structures can be launched only in the summer. Missing a summer launch causes a delay of about one year, and BP's investment in the Forties field was about R1.2 billion. Interest charges alone on capital employed could be about R120 million.

SERVICE FAILURES

Finally, consideration of some service failures illustrates the significance of brittle fracture. Figure 10 shows an ammonia converter which failed during preliminary hydrostatic testing. The failure initiated in a weld and the defect which initiated the fracture was about 10mm long in a region where the structure was 178mm thick. The lesson to be learnt here is that in this case it was no good trying to avoid a brittle fracture problem by increasing the severity of inspection. Non-destructive testing cannot reveal with any certainty the sizes of defect that can initiate fracture under these conditions if the material has been wrongly chosen. The solution to the brittle fracture problem lies in the correct selection of material.

It should be possible to rely on non-destructive testing to find the types of defects that caused the failure shown in Fig.11. This boiler drum failed from a nozzle and the initial defect was in fact 88mm deep by 338mm long. The most modest exponents of non-destructive testing would expect that they would find a defect that size. In this case the problem arose because the vessel was inspected at the wrong time. It was examined prior to post weld heat treatment and it is believed that the crack was some form of reheat crack. Had the inspection been timed correctly the crack would have been found. In fact the material had extremely high ductility, and this failure occurred only on the fifth hydrostatic test. Had the fifth test had not been carried out and produced a failure there

Fig.10 Failure of an ammonia converter with detail of the initiation point below

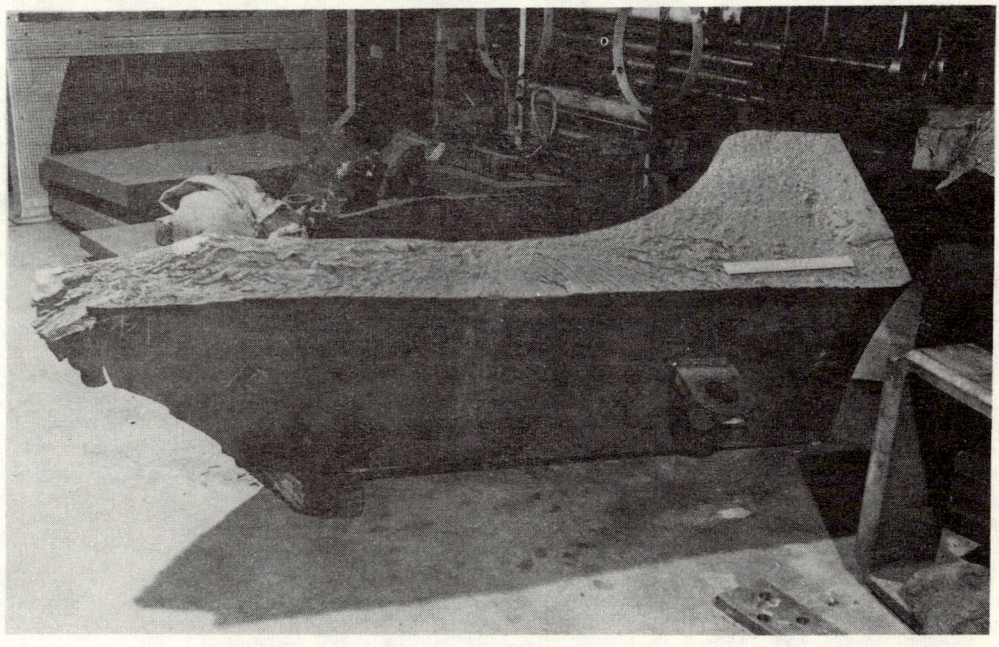

Fig.11 Boiler drum nozzle failure.

is every likelihood that the drum would have given perfectly satisfactory service with this large defect contained in it.

A final example is not concerned with welding at all but relates to a winch shaft from a pipe lay barge operating in the North Sea. The shaft contained a keyway, and the failure illustrated the importance of materials properties regarding brittle fracture. The steel that formed the shaft was some of the most brittle ever tested at The Welding Institute. In the first Charpy test carried out on a specimen extracted from the shaft the testpiece broke into three pieces and did not fail at the notch at all. Design was also a significant aspect of this failure, as the machinist who made the keyway seemed to have taken the greatest pleasure in the accuracy with which it had been machined. He had used the sharpest possible tool and the radius in the corner of the keyway was less than 0.025mm. Naturally this was a long way outside specification but it is possible that had the keyway been correctly machined even this particularly poor material would have given adequate service performance.

SECTION 1

Some Problems of Fracture

FRACTURES IN SPRINGS OF STEEL STRIP

G. Persson

Development-Strip Products, Sandvik AB, Sandviken, Sweden

ABSTRACT

Fractures occurring in the fabrication and use of springs of high carbon and stainless steel strip are discussed. Reasons for failure during spring forming are e.g. coarse microstructure, overrolling and hydrogen embrittlement. Fatigue, wear and corrosion may lead to failure of springs in use. Studies in a compressor valve simulator, SIFT, show that stainless chromium steel is superior to carbon steel regarding impact fatigue properties. Influence of static strength, microstructure, surface and edge condition on the bending and tensile fatigue strength is described. In a 13% Cr steel oxide inclusions of about 15 µm size close to the surface initiate early ($<2 \times 10^6$ cycles) fatigue cracks and interior, larger oxides start cracks at a later stage.

INTRODUCTION

Springs may break not only in use but also already during manufacture. Why do they break even if material and stress were selected in such a manner that they should not? There are several possible reasons, and some of them will be discussed in this paper with reference to broken springs of steel strip.

A visual inspection or a study at low magnification may sometimes give an acceptable explanation of the fracture. In other cases a more sophisticated investigation, e.g. a fractographic examination with a scanning electron microscope (SEM), is required to reveal the cause.

However, it is not always possible to definitely establish the reason; severe corrosion, wear or secondary fractures may have obliterated the original crack.

Strip steels used for springs are mainly 0.75-1% carbon steels type AISI 1074 and 1095, and cold rolled austenitic stainless steel type AISI 301. The carbon steels are either continuously hardened and tempered in strip form, or piece hardened after blanking and forming. Other strip steels used for springs are hardened and tempered silicon alloyed steel and stainless chromium steel, type AISI 420, among others.

Microstructures of AISI 1095 and AISI 301 are shown in Fig. 1 and 2. The tensile strength of the steels depends primarily on the carbon content and heat treatment or, for the austenitic steel, on the degree of cold reduction by rolling. It may be as high as 2000 N/mm^2, or even more.

Fig. 1. Microstructure of a hardened and tempered AISI 1095 spring steel. 500x

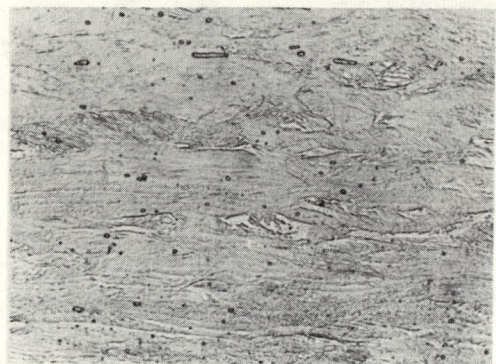

Fig. 2. Microstructure of a cold rolled AISI 301 spring steel.   750x

FRACTURING DURING SPRING FABRICATION

Cracks may appear during forming of high strength steel strip (Ref. No. 1). Fig. 3 is an example where too small a bending radius was used in relation to tensile strength, ductility and strip thickness. It should be noted that the bendability of cold rolled spring steels is less good along than across the rolling direction (2). Bending blanked items along the rolling direction may lead to fracture, especially if the burr side is stretched (1). Hardened and tempered grades have much less directional dependence in their bending properties (2).

If severe shaping is required, a comparatively high ductility, i.e. low tensile strength, should be selected. Alternatively, the spring should be formed from annealed stock and subsequently piece hardened.

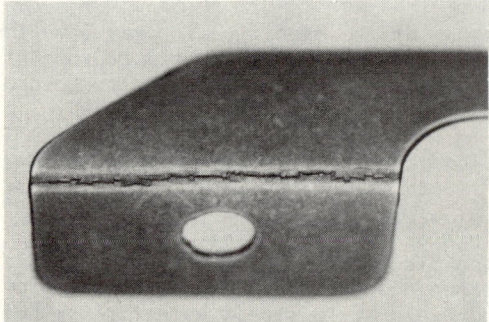

Fig. 3. Breakage due to small bending radius in an AISI 1095 steel spring.   5.5x

Even if the tensile strength is suitable for a given forming operation, fractures may occur. This may be the case, for instance in a carbon steel hardened from an excessively high temperature. It has a coarse microstructure, Fig. 4, and a poor ductility.

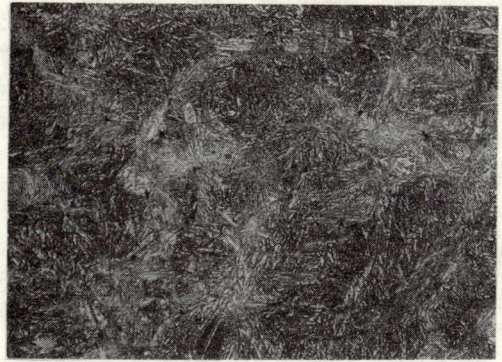

Fig. 4. Coarse structure in an AISI 1095 spring steel due to too high hardening temperature. 500x

Fig. 5. Longitudinal crack due to pipe in the ingot of an AISI 301 spring steel. 0.4x

An overrolled austenitic spring steel is characterised by pronounced shear bands in the microstructure at an angle of 45° to the strip surface. Traces of these bands may be visible at right angles to the rolling direction (1). Such steel may have a brittle behaviour when formed, and cracks will follow the shear bands.

A different type of fracture that may sometimes occur during spring fabrication is illustrated in Fig. 5 The splitting of the strip was due to insufficient cropping of the ingot top before strip rolling. This type of defect is limited to a short part at one end of the total strip length.

Material with the defects just mentioned is, of course, to be sorted out at the inspection before delivery.

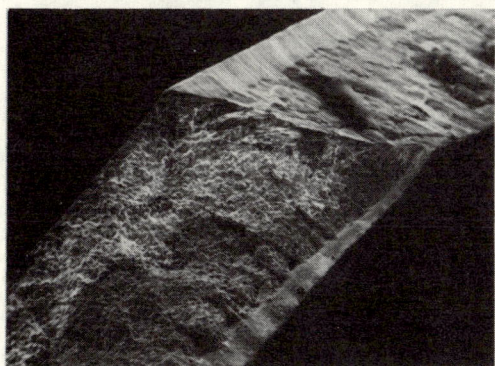

Fig 6a. Fracture due to hydrogen embrittlement in a zinc-coated AISI 1074 steel spring. 35x, SEM

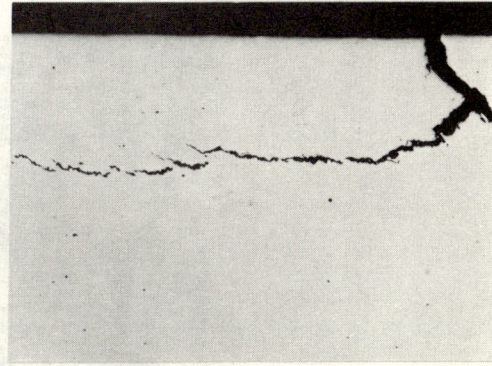

Fig. 6b. Cross section of fracture in Fig. 6a. 35x

Fig. 6a and b show a fracture in a spring made from hardened and tempered steel AISI 1074, electrolytically zinc coated after blanking. Hydrogen diffused into the steel as a result of the electrolytic process and caused the embrittlement. This can be avoided by "baking" the parts at 100-150°C for , say 5 h, immediately after the coating.

Fracturing in the use of springs

The service life of a spring is limited by fatigue, wear, corrosion and permanent deformation, sometimes in combination. Since a spring is in general exposed to varying load, fatigue is probably the most common reason for premature failure.

Fatigue fractures are characterised by three stages, viz. crack initiation, stable crack growth and rapid final failure. Fig. 7 shows a spring of steel strip, which failed owing to fatigue. The crack started near the strip centre and grew slowly perpendicularly to the stress direction. The main part of the fracture surface is rough and comprises the final failure. In Fig. 8 part of the crack growth region is shown at high magnification. Sometimes striations due to repeated plastic deformation at the crack tip are visible in this area (4). The final failure can have brittle or, as in Fig. 9, ductile character.

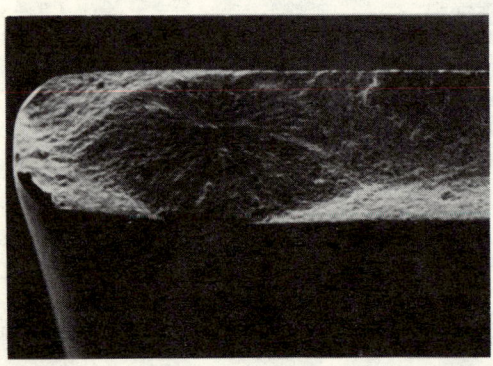
Fig. 7. Fatigue fracture in a steel spring initiated near strip centre.   60x, SEM

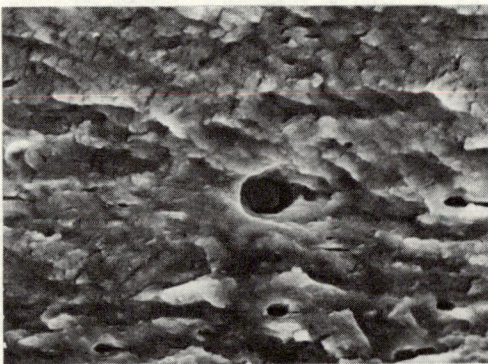
Fig. 8. Fracture surface in the stable crack growth region.   2500x, SEM

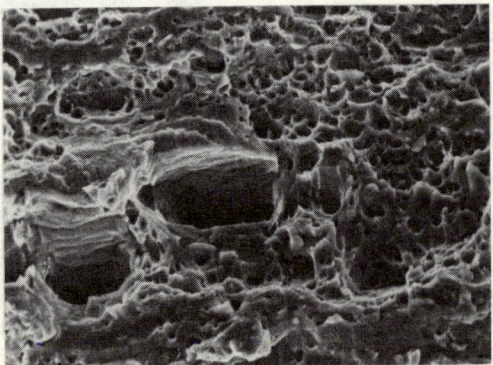
Fig. 9. Final fracture with dimples. 800x, SEM

Influence of load conditions

Fatigue data of spring steels are commonly presented in S-N diagrams, e.g. Fig. 10. In the sloping portion the crack growth stage is a more dominating part the less the number of cycles to fracture, i.e. the higher the stress. Typical crack propagation rates are 10^{-3} to 10^{-5} mm/cycle in this region.

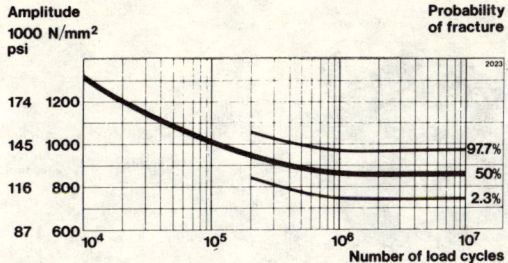

Fig. 10. S-N curve for a modified AISI 420 steel in reversed bending. Tensile strength 1810 N/mm^2.

The initiation stage is predominant at long fatigue life, and the horizontal part of the S-N curve, the fatigue limit, can be said to represent the risk of crack initiation.

In recent years, fatigue crack growth has received attention at high and also at very low propagation rates, less than 10^{-6} mm/cycle. The growth rate da/dN can generally be expressed as

da/dN = AΔK^m

where A and m are constants. The connection to the stress range $\Delta\sigma$ is given by

$\Delta\sigma = \Delta K\sqrt{\pi}/2\sqrt{a}$

for a circular crack with radius a.

Below a threshold value ΔK_{th} no growth will take place. For spring steels ΔK_{th} is 3-5 MNm$^{-3/2}$; (3, 5). This means, for instance, that there will be no growth of cracks of less than 25µm diameter in a spring exposed to a stress range of 1000 N/mm^2, if ΔK_{th} is 4 MNm$^{-3/2}$.

The position of the S-N curve is dependent on the type of loading. The higher the prestress, the lower is the permissible cyclic stress amplitude. Similarly, ΔK is dependent on the prestress, particularly near the threshold value (3, 6). It is therefore difficult to judge the risk of fatigue failure when stress conditions differ from a known case.

At low fatigue loads just one crack is generally observed while at high loads often several cracks may appear. In the sample shown in Fig. 11 two cracks, initiated on opposite sides during reversed bending, joined to form the final fracture. A straight, stable growth part is evident, perpendicular to the rolling and stress direction. Another crack parallel to the first one was initiated at approximately the same distance from the spring edge.

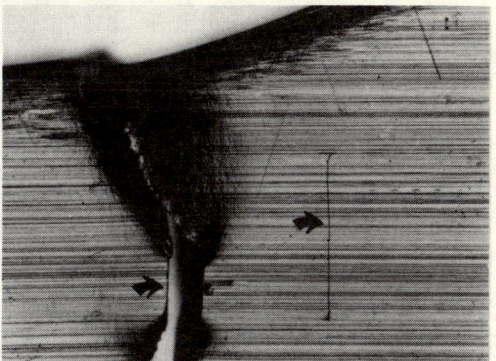

Fig. 11. Fatigue fracture and fatigue crack in an AISI 1095 steel spring. 40x

If the spring is exposed to repeated impacts, the stress can hardly be measured or calculated with any accuracy. This situation prevails in flapper valves in compressors - a special, highly stressed kind of flat spring. The impact fatigue fracture has a very jagged shape, Fig. 12, in contrast to the comparatively straight bending or tensile fatigue cracks.

Fig. 12 Two identical valves failed owing to bending (left) and impact (right) fatigue. 1x

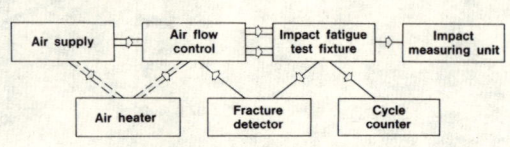

Fig. 13 Sandvik impact fatigue tester (SIFT). Block diagram.

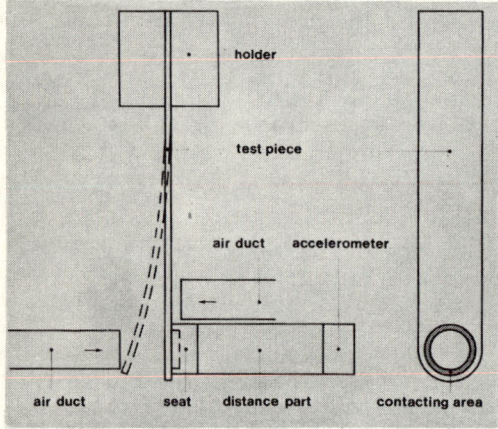

Fig. 14 Impact fatigue test fixture, principle

Fig. 15 Radial impact fatigue crack with two small fractures outside the seat contacting area. 60x, SEM

Tests in a 250 Hz simulator, SIFT, (7) driven by compressed air, Fig. 13 and 14, reveal that impact ratigue cracks can be initiated outside the ring shaped contact area between test piece and seat, Fig. 15. The cracks were radial at an early stage, Fig.16. They were also initiated on the rear side of the valve, not at all in contact with any object. The fractures occurred close to the part of the seat where the last contact took place - not necessarily the extreme specimen tip. This was proved by turning a slightly oblique seat in different directions, Fig. 17, (8). The explanation is probably that the impact velocity was highest in the direction of inclination, similar to a whip-lash effect.

It has seldom been possible to relate the impact fracture to surface defects or metallurgical faults. It is also quite difficult to observe microstructural changes due to impact stresses in the hardened and tempered valve steels. In soft model materials very heavy local deformations were observed (9), possibly responsible for the impact fatigue crack initiation.

Fig. 18 shows that a stainless chromium valve steel resists impact fatigue a good deal better than a carbon steel grade at the same tensile strength, 1850 N/mm^2.

Fig. 16. Radial crack (dark area) initiated outside the contact area. 200x, SEM

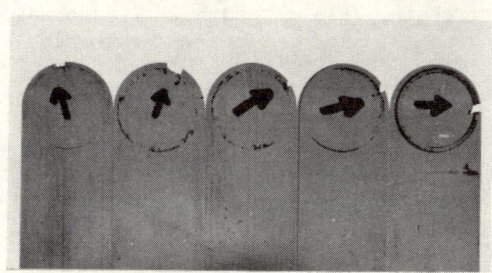

Fig. 17. Impact fractures in different directions of seat inclination. 0.75x

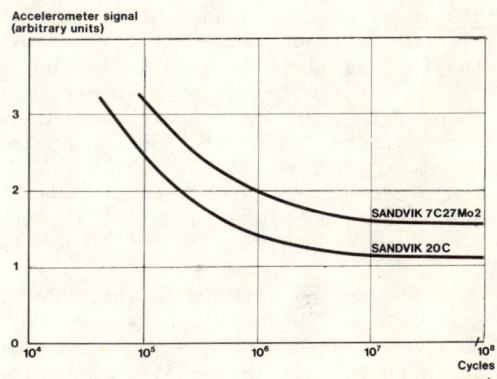

Fig. 18. Impact fatigue test of an AISI 1095 (SANDVIK 20C) and a modified AISI 420 (SANDVIK 7C27Mo2) steel.

Influence of environment
─────────────────────────

The risk of fatigue failure in springs is influenced by the atmosphere. An elevated temperature will reduce the fatigue strength, although rather slightly up to 200°C (2, 10). The negative influence of corrosion is more serious. Already humid air may reduce the spring life considerably without any visible corrosive attack (11, 12). The mechanism of crack growth in high-strength steel when water vapour is present, seems to be hydrogen embrittlement (13). Under corrosive conditions there is no longer a fatigue limit. The drop in fatigue strength is most pronounced for carbon and low alloy steels, less for stainless chromium steels. Austenitic stainless steels are best in this respect.

However, stress corrosion cracking may appear mainly in stainless steels. Fig. 19 shows a spring of an austenitic steel strip which failed owing to stress corrosion after only a few days in a marine atmosphere. Salt crystals can be seen close to the crack.

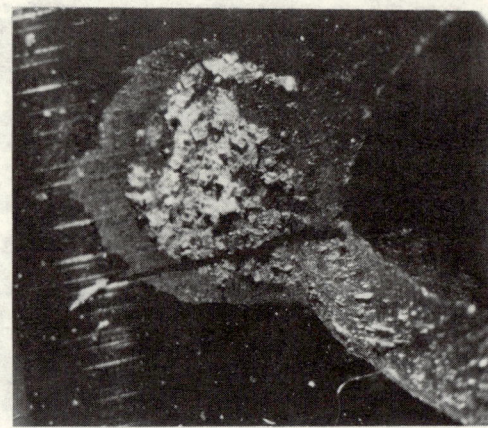

Fig. 19. An AISI 301 steel spring, failed owing to stress corrosion. 20x

Influence of material parameters

Static strength. High tensile strength implies high fatigue limit. i.e. reduced risk of fatigue crack initiation (10). In the case of reversed bending stresses, the fatigue limit is about half the tensile strength. However, the notch sensitivity increases with the tensile strength. Since notches cannot be avoided, there is, in practice, an upper limit for the tensile strength. This strength limit depends, for instance, on steel type and spring design.

The crack growth rate increases with the tensile strength at stress intensities near the threshold value ΔK_{th}. The threshold value itself drops with increasing tensile strength.

When the spring life is in the order of 10^3 to 10^5 cycles, i.e. when the sloping part of the S-N curve applies, the ductility seems to be of importance in addition to the strength. The crack propagation rate is less influenced by the tensile strength in this spring-life range than at high-cycle fatigue (3).

Microstructure. The fatigue strength of spring steel increases to some extent with diminishing grain size. This can be explained by the fact that a fine-grained material is more ductile.

Structural inhomogenities in the form of segregations or decarburisation, Fig. 20, are material defects that may lead to fatigue failure. The fatigue limit can be reduced by up to 50%, if ferrite appears in a surface layer (14). The origin of decarburisation may be an error in the strip manufacture or in the piece hardening.

Non-metallic inclusions. Steel inevitably contains non-metallic inclusions, i.e. oxides and sulphides. Fatigue cracks hardly ever start at sulphides (15). Very thin elongated sulphides in steel strips are in general harmless. However, oxide inclusions may initiate fatigue failures, Fig. 21, provided they are sufficiently large in relation to the fatigue stress. At a constant ratio between fatigue stress and tensile strength the critical size is smaller the higher the tensile strength.

The influence of oxide inclusions on fatigue crack initiation was studied under fluctuating tensile loading ($R \approx 0$) for a modified AISI 420 steel (0.4%C, 13%Cr, 1%Mo). The 0.5 mm thick steel strip was hardened and tempered to 1835 N/mm² tensile strength. The material was manufactured with an unusually high oxide inclusion content, which, in the finished strip, was measured by optical image analysis for particles up to 16 μm length. Larger particles were counted manually in a microscope, Table 1. The particles were in general about twice as long as they were wide. Their composition type was $(MnO)_x SiO_2$. The fatigue testing was performed at four different stress levels. More than 90 per cent of all fractures occurring before 2×10^6 load cycles were initiated at oxides, Table 2, most of them at a distance less than one inclusion radius from the surface.

Fig. 20. Surface decarburisation of an AISI 1095 spring steel. Two samples are put together. 400x

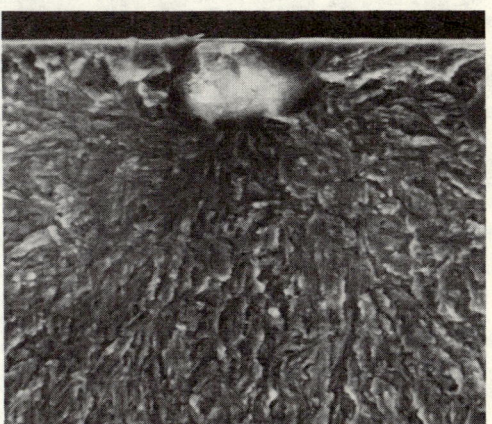

Fig. 21. Fatigue fracture in an AISI 1095 steel spring, initiated by a large oxide inclusion. 1000x, SEM

TABLE 1 Distribution of oxide inclusions. Number of inclusions/mm² with a length larger than indicated

Length (μm)	6	8	10	12	16	40	50	60	70	80
$\bar{x}$	17.4	9.2	5.1	3.4	1.9	0.13	0.059	0.022	0.011	0.004
s	3.5	2.4	1.6	0.8	0.5	-	-	-	-	-

TABLE 2 Result of fluctuating tensile fatigue test (R≈0) at 2×10^6 load cycles

| Stress amplitude N/mm² | Number of test pieces | No. of test pieces broken due to |||| Average incl. size μm | Calculated crack size μm |
		surf. incl.	inter. incl.	mech. def.	other reasons		
570	160	37	5	5	0	17.3	15.6
585	100	31	5	3	2	19.2	16.8
600	75	33	5	1	0	20.1	18.2
625	60	39	1	1	1	20.1	18.8

The fatigue test was continued beyond 2×10^6 cycles for a number of the samples at the two medium load levels. At this stage the fractures were initiated primarily from interior oxides, Table 3.

TABLE 3 Size distribution of inclusions initiating fracture, all test pieces

Position	Stress N/mm²	Number of inclusions per size class upper class limit, µm									
		10	12.5	16	20	25	32	40	50	64	Total
Surface	570±570	1	3	8	10	12	2	0	1	0	37
	585±585	0	6	7	2	13	1	3	1	0	33
	600±600	1	3	16	4	12	3	2	0	0	41
	625±625	1	6	16	2	11	3	0	0	0	39
	Sum	3	18	47	18	48	9	5	2	0	150
Interior	570±570	0	1	0	1	0	1	1	0	1	5
	585±585	1	0	3	3	5	13	2	0	0	27
	600±600	0	0	1	1	3	7	1	0	0	13
	625±625	0	0	0	0	0	1	0	0	0	1
	Sum	1	1	4	5	8	22	4	0	1	46

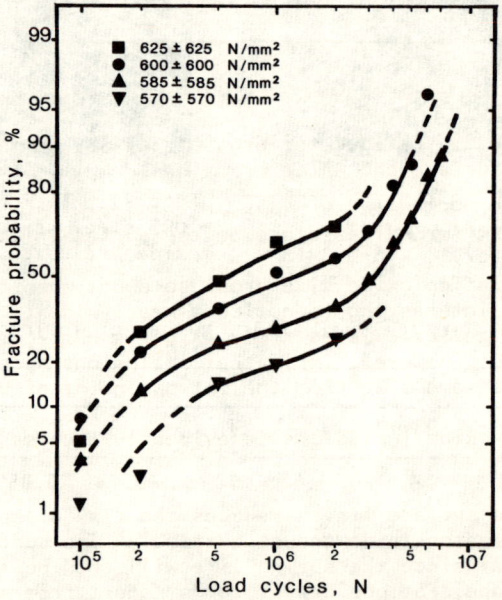

Fig. 22. Fatigue failures due to oxide inclusions in a modified AISI 420 steel.

The probability of oxide-initiated fatigue failure at the four stress levels is shown in Fig. 22 as a function of the number of load cycles. The curves, which have a fairly complicated shape, give the risk of fracture at the end of each regarded load cycle interval. In Fig. 23 one curve is split into separate parts for fractures initiated at the surface and in the interior of the strip. The latter has the same slope as previously found in the fatigue limit region (17).

The risk of crack initiation will increase with the size of the inclusions. At the same time, of course, the relative number of inclusions will decrease as their size grows. The product of these factors leads to a distribution function for the observed crack initiating oxides giving a median size of about 25 µm for internal in-

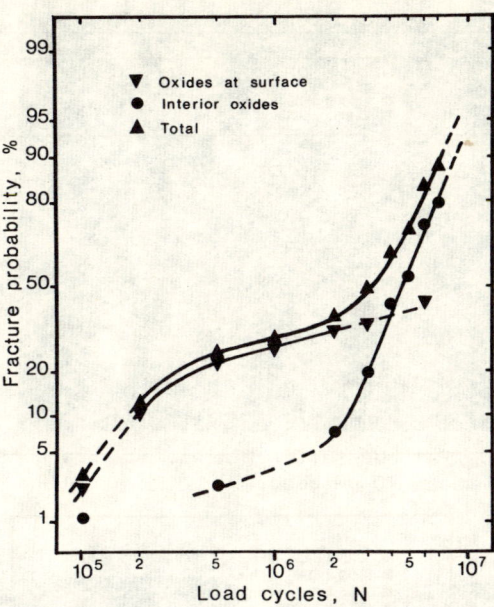

Fig. 23. Fatigue failures at 585±585 N/mm² due to surface and interior oxides.

clusions and about 15 μm for oxides close to the surface, Table 3.

The average crack initiating oxide size drops somewhat when the applied stress is raised, Table 2. It is interesting to notice a nearly identical influence on a circular critical crack size, calculated from $\Delta K_{th} = 4$ MNm$^{-3/2}$. The similarity between the values does not, however, imply that inclusions necessarily should be regarded as cracks of the same size from the fatigue point of view.

A low content of large oxide inclusions is evidently necessary in a high-quality spring steel.

<u>Surface finish.</u> A good surface finish reduces the risk of spring failure (10). There is, however, no reason to demand an extremely fine surface finish (R_a less than, say, 0.10-0.15 μm) since the spring life will then be limited by other factors, e.g. the edge shape. Fatigue cracks seldom start from surface defects less than 2-3 μm deep.

Mechanical defects in the surface may give fatigue failure in the same manner as already mentioned for oxide inclusions. Fig. 24 shows that a defect, about 10 μm deep has started a fatigue fracture. One observation worth mentioning is that surface defects near the edge seem to initiate fatigue cracks easier than surface defects of the same size in the middle of a spring exposed to bending stresses, cf. Fig. 11. The lateral contraction is likely to be responsible for this behaviour.

A different type of surface defect, so-called gouges, Fig. 25a, are more serious the deeper they are. On one occasion the fatigue life was about 10^6 cycles for 10 μm depth and about 10^5 cycles for 30 μm depth, Fig. 25b.

A similar effect of other defects like roll marks has been described elsewhere (1). Of course, such large defects should not be present in a good spring steel.

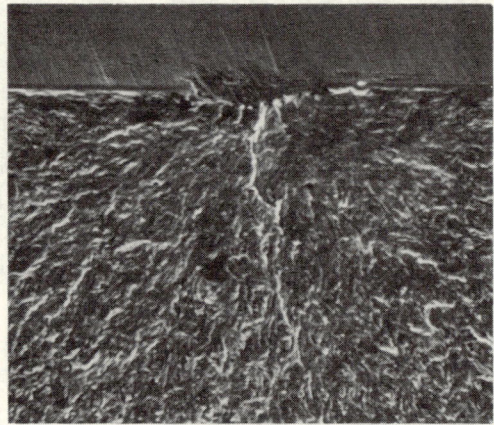

Fig. 24. Fatigue failure initiated at a surface defect, about 10 μm deep. 420x, SEM

Fig. 25a. Gouges in the direction of rolling in the surface of an AISI 420 spring steel, leading to fatigue failure. 20x

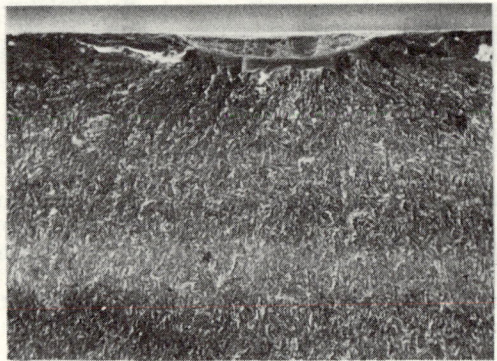

Fig. 25b. Do. Fatigue crack surface. 300x, SEM

Sometimes springs are worn to the extent that fatigue failures occur. Fig. 26 illustrates this fracture mechanism in a flat spring. The remedy is to design the spring in such a way that wear is avoided in highly stressed regions.

In high-carbon steels the temperature may be so high during the wear process that rehardening occurs. The new martensite is very brittle and cracks easily. Fig.27. Fig. 28 a and b illustrate a coiled spring where this phenomenon led to fatigue fracture. The wear was caused by inadequate lubrication.

Still another reason for fatigue cracks is a corroded steel surface, Fig. 29a. The corrosion attack, up to 5 μm deep, had taken place during the spring manufacture and the spring broke under regular, non-corrosive conditions, Fig. 29b.

Edge quality. Fatigue cracks often start from edge defects. A careful edge treatment is therefore essential in highly stressed springs. Fig. 30 exemplifies bad edge geometry of coil springs and Fig. 31 shows a fatigue crack initiated at an edge irregularity. Even if the edge has a good shape, problems may arise in carbon spring steel, if friction hardening occurs in the edge.

Fig. 26. Fatigue crack initiated at a wear mark in an AISI 1095 steel spring. 3x

Fig. 27. Brittle martensite with cracks formed by friction hardening in the surface of an AISI 1095 steel. 100x

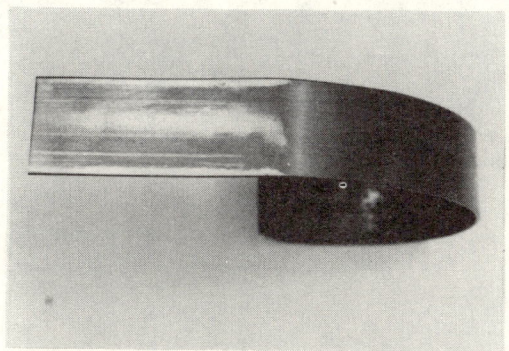

Fig 28a. Coiled spring of an AISI 1074 steel broken owing to heavy wear and friction hardening. 0.7x

Fig. 28b. Do. Fatigue crack initiation site. 100x, SEM

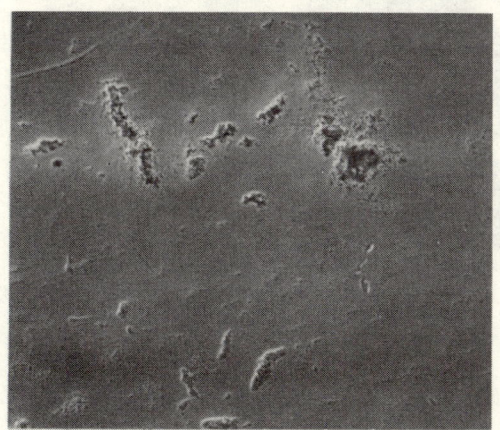

Fig. 29a. Corroded surface of an AISI 1074 steel spring. 300x, SEM

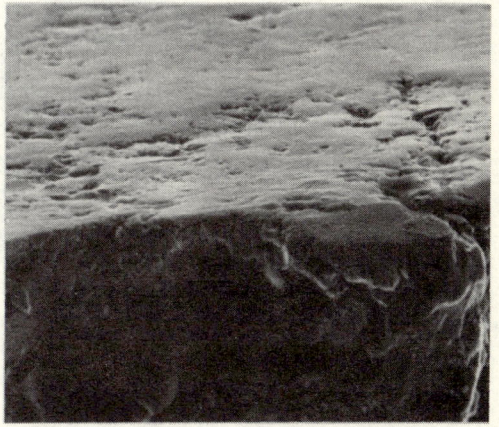

Fig. 29b. Do. Fatigue fracture initiated at a corrosion pit. 300x, SEM

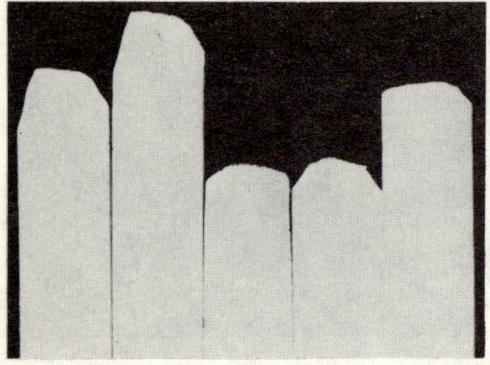

Fig. 30. Edges inadequately prepared for highly stressed springs. 60x

Fig. 31. Fatigue fracture starting from an irregularly rounded edge. 50x

Fatigue fractures may also start at the edge of blanked springs. In the case illustrated by Fig. 32 the clearance between punch and die was too small, which caused a wide shear zone with grooves giving stress concentrations. Remaining blanking burrs (1) or burrs due to wear (4) are other possible sites for fatigue crack initiation, particularly if the burr side is exposed to tensile stress.

Grinding, tumbling or shot peening will improve the edge shape and set up compressive stresses, both favouring a long fatigue life. It should be observed, however, that an initially high compressive stress may be relaxed to some extent by the fatigue stress (16).

Fig. 32. Blanked edge with grooves, initiating fatigue failure 40x, SEM

CONCLUSIONS

The risk of failures in springs of steel strip is minimised if
- the spring design is such that stresses and material properties are balanced against each other
- a material with homogeneous microstructure, metallurgical cleanness, good surface and good edge finish is selected
- notches and hydrogen pick-up are avoided in the spring manufacture
- the springs are not exposed to unnecessary wear or corrosion attack in their use.

REFERENCES

1. G. Persson. Springs Mag., 17 (1978) No. 2, p. 9-27
2. Sandvik Steel Catalogue No. 3,41, Sandviken 1974
3. R.O. Ritchie. Metal Sci., 11 (1977), p. 368-381
4. W.R. Wagner. Springs Mag., 17 (1979), No. 1, p. 54-66
5. R. Dusil, B. Appell. Proceedings, 1976 Purdue Compressor Technology Conf., p. 82-90
6. C.J. Beevers. Metal Sci., 11 (1977), p. 362-367
7. M. Svenzon. Dissertation, Uppsala Univ. 1976
8. M. Svenzon. Proceedings, 1976 Purdue Compressor Technology Conf., p. 65-73
9. I. Smith. Proceedings, 1978 Purdue Compressor Technology Conf., p. 111-115
10. P.G. Forrest. Fatigue of metals, Pergamon Press, Oxford etc. 1972
11. H.H. Lee, H.H. Uhlig, Met. Trans., 3 (1972), p. 2949-2957
12. W.C. Clinton, O.R. Iobst. IBM Systems Development Div., Endicott Lab. TD 01.368, 1965
13. R.P. Wei, J.D. Landes. Mat. Res. Standards, 2 (1969) p. 25
14. P. Funke, W. Heye. Stahl u. Eisen, 96 (1976) p. 28-32
15. J.J. Hauser, M.G.H. Wells. Mech. Working and Steel Processing (1976) p. 314-319
16. R. Johansson, G. Person. Proceedings, 1976 Purdue Compressor Technology Conf., p. 74-81
17. D. Dengel. Dissertation, Techn. Univ., Berlin (1967)

METALLURGICAL FAILURES IN THE MINING INDUSTRY

R. N. Taylor* and W. J. van den Berg**

*Engineering Department, **Mining Department, Technical Development Services, A.A.C. of S.A. Ltd., Welkom, R.S.A.

ABSTRACT

The paper deals with two case histories of failures as practically experienced in the R.S.A. mining industry by an in-house Research & Development Organisation of a major mining house. The first case history deals with premature failure problems experienced with a high energy output hydraulic rockdrill from a major mining equipment supplier. Attempts to overcome the failure problem (which is essentially one of fretting corrosion with possibly some cavitation) are described in full and the results obtained to date analysed. The second case history deals with the failure of raise borer drill rods. Although reamer stems and stabilisers fail occasionally, drill pipe failures are rarely experienced. During 1976 a number of drill rod failures were experienced both in R.S.A. and elsewhere, which led to a series of investigations. The results of these investigations in so far as they are relevant and determination of the probable cause of failure are described.

INTRODUCTION

The paper deals with two case histories of failures as practically experienced in the South African gold mining industry by an in-house Research & Development Organisation of a major mining house. The two failures, one in high energy output hydraulic rockdrills and the other in raise borer drill rods/pipe, are related to equipment which is regarded as essential to a mechanisation programme being introduced into the Anglo American Corporation of South Africa Ltd's South African gold mines. Both failures occurred in equipment which it was considered was well proven in its application, within the South African mining environment, and not in equipment essentially of a developmental nature. The failures concerned in each case involve items from major international mining equipment manufacturers or suppliers and related to equipment not merely in use by the Research & Development Organisation applications section but also owned and operated independently by gold mines of the A.A.C. of S.A. Ltd. group. In each case, in efforts to define the cause, and determine a solution to the failure, over and above the facilities existing within the mining house and those available to the equipment manufacturer and supplier, extensive assistance was sought by all parties from leading educational and research institutions, and to this approach is attributed such success as has been achieved in dealing with the failures. In the case of the failure associated with the hydraulic rock drill, it would appear that this failure is one which is peculiar to the South African gold mining application and conditions: the failure, however, in the raise borer drill rods would appear to be of a more international nature.

CASE HISTORY 1: FAILURE IN A HIGH ENERGY OUTPUT HYDRAULIC ROCKDRILL

The first case history deals with failure in a high energy output rotary percussive hydraulic rockdrill. Hydraulic rockdrills have been under investigation by the A.A.C. of S.A. Ltd. virtually since their appearance on the hard rock mining scene: initial trials were carried out in 1974 with rockdrills mostly mounted on 2-boom rail-mounted development "jumbos". These trials demonstrated clearly the superiority of hydraulic rockdrills for high-speed tunnelling work in the South African gold mining environment. In particular, the advantages of hydraulic rockdrills were seen as: lower energy consumption levels, high performance output, longer rockdrill life, ability to reliably achieve long planned maintenance intervals, improved drill steel and accessory economy, and improved operator environment with regard to noise, visibility etc. In consequence, of the approximate 30 "jumbos" in use in A.A.C. of S.A. Ltd. gold mines for development or roofbolting purposes, better than 80% are electro-hydraulic utilising rotary percussive rockdrills from a variety of major international manufacturers/suppliers. In general, hydraulic rockdrills are favoured with a percussive energy output of about 15 kW (achieved with a blow frequency of between 3000-4000 blows per minute and a blow energy of 300-200 joules), and rotational output energy of up to about 4 kW (100-200 r.p.m. with a torque of up to 200/250 Nm): these require about 6,5-8,5 kN of static thrust under South African gold mining conditions to "liberate" their performance. The most commonly drilled hole is a single pass of about 3,7m and 43mm diameter.

During the last quarter 1977, two hydraulic rockdrills of the same type were returned from an operating A.A.C. of S.A. Ltd. gold mine between their 5000 drilled metres planned maintenance interval to the Technical Development Services Engineering Department (who undertook their servicing) with the complaint that they were leaking oil from their component interfaces. Investigation revealed that these drifters which were amongst the oldest of a "fleet" of about 65 (and known to have completed at least 30 000-35 000 drilled metres) were suffering from extreme pitting of certain component housing interfaces which was resulting in hydraulic oil leaking from internal pressure porting through the rockdrill. Set out in Fig. 1 are diagrammatic representations of the mechanical forces acting upon the rockdrill during the drilling operation, construction and operation of the rockdrill, and mode of mounting of the rockdrill in so far as they were subsequently determined to have bearing on the failure. Shown in Fig. 2 are examples of typical failures experienced by the principally affected components of the rockdrill. By the first quarter 1978 it was apparent that this failure was not isolated and would affect the entire rockdrill "fleet" with increased meterage drilled, necessitating repair, rockdrill or major rockdrill component replacement at or before 40 000 drilled metres.

It is perhaps relevant at this point to consider the financial implication and import of this failure. Operating within the A.A.C. of S.A. Ltd. South African gold mines were a "fleet" of 65 rockdrills at a purchase price relevant to the first quarter 1978 of approximately R18 000 each; this represents an investment of approximately R1,17 million. Justifications for the adoption of hydraulic rockdrills had been based on a rockdrill (and/or all major rockdrill components) life of 200 000 drilled metres giving a predicted cost of 18,75c/metre drilled, inclusive of amortisation and maintenance cost. Against this can be compared a "worst case" equivalent cost determined at this time of approximately 70c/metre

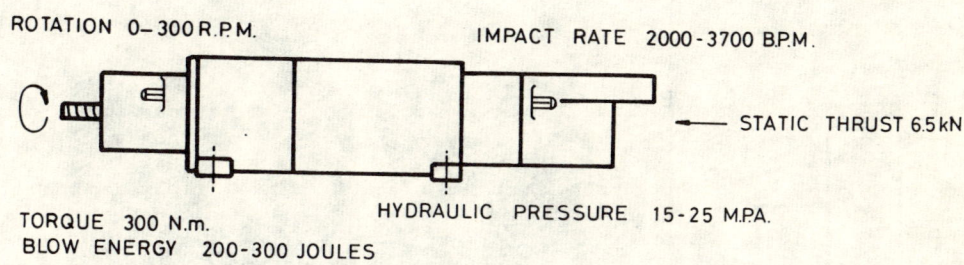
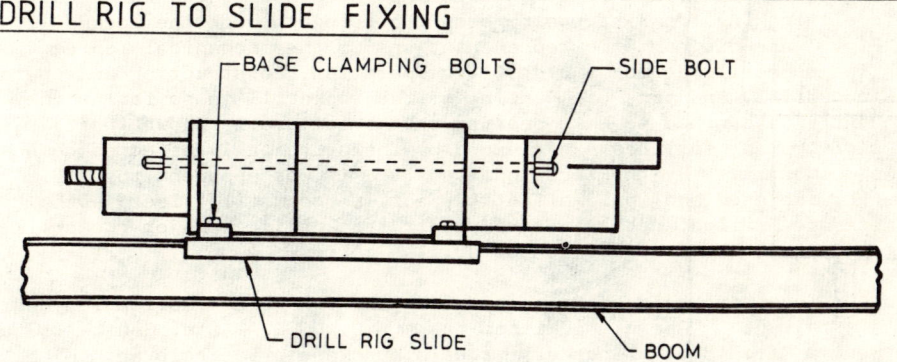

Fig 1 Diagrammatic representation of general hydraulic drifter arrangement

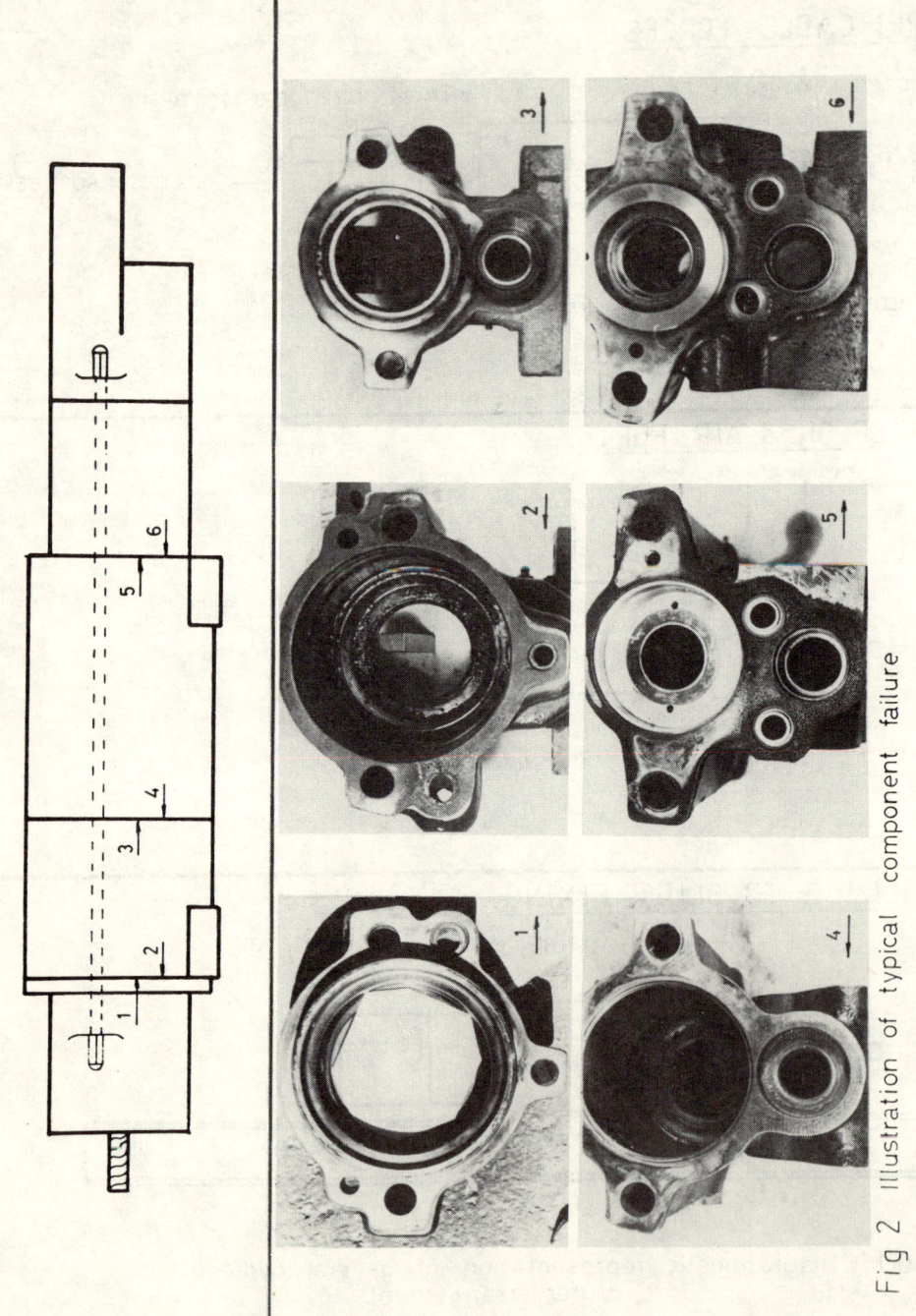

Fig 2 Illustration of typical component failure

drilled for this hydraulic rockdrill; it was considered that costs for an equivalent pneumatic rockdrill were approximately 12c/metre drilled.

The failure experienced most severely and seriously affected the rockdrill cylinder, intermediate housing, gear housing and cover: the cost of replacement of these parts being approximately R14 000 or 80% of a new rockdrill price. Replacement/repair of these rockdrill components became necessary at approximately the same time, normally before 40 000 drilled metres. All housing components of the rockdrill are alloy steel castings of some form which, post machining, have received various heat treatment. This is most importantly reflected in the mating faces where the surface hardness has been raised to approximately 60 Rockwell C compared to approximately 25 Rockwell C for the matrix. It would appear that post heat treatment the mating face receives a final polishing grind. Through discussions with the manufacturers of the rockdrill it became apparent that although identical rockdrills operated under various conditions throughout the world, and even under other conditions within South Africa, its gold mining conditions were the only ones to produce this failure phenomenon. At this stage, A.A.C. of S.A. Ltd. South African gold mines operated a little less than 10% of the total world population of this hydraulic rockdrill.

From the initiation of investigation to define the cause and to determine a cure for the failure, close co-operation has been maintained between the rockdrill manufacturer and the Technical Development Services Engineering Department: various possible solutions conceived by either party have been field-tested jointly on rockdrills operating in A.A.C. of S.A. Ltd. gold mines.

Immediate investigation explored the possible effect of airborne water in the compressed air being delivered to ventilate the chuck and lack or loss of torque of the side iron bolts. These considerations were soon shown not to be relevant to this failure. From an early stage technical assistance had been sought by the Technical Development Services Engineering Department from educational and research institutions in South Africa and by the end of the first quarter 1978 it was satisfied that the cause of the failure had been defined. It was felt that the failure was one caused by lack of rigidity within the rockdrill housing components caused by the construction method, most especially the use of only two side iron bolt constraints operating high above the base of the rockdrill: this resulted in fretting corrosion, possibly combined with some cavitation to cause ultimate failure by high pressure oil leakage across the mating interfaces of the housing from internal porting. As can be seen in Fig. 3 the appearance of the pitting suggests a combination of mechanical damage and chemical attack. The pitting of the interface decreases from the bottom of the housing towards the location of the sidebolts, indicating flexing of the assembly with the rockdrill in operation. This cyclic flexing causes localised mechanical damage to the protective oxide film on the interface and also allows low pH flushing water, or more likely environmental moisture, to enter between the mating surfaces. Localised corrosion attack of the metal mechanically exposed then occurs: the "river pattern" evident suggests that liquid subsequently trapped in pits or between the interface is squeezed out at high pressure resulting in erosion and/or with some cavitation effect.

Essentially, attempts to determine a solution to the failure have been directed in three areas: firstly, in various surface treatments of the interfaces, secondly, in affording additional restraint to the base of the rockdrill, and thirdly, in the renovation or repair of affected surfaces. During the first quarter 1978 sealant of a rubber-based type was provided by the manufacturer and since this date, anaerobic sealants and lead oxide have been tried: various bitumen tapes and polyurethane sealing around housing joints have also been investigated. These treatments have resulted in no marked improvement in rate

Fig 3 Illustration of typical rockdrill component failure

of failure. In the second quarter 1978 more substantial efforts were made with the manufacturer providing housing components for 2 rockdrills, one N1 hardened with the other zinc-plated; the South African supplier made available a specially adapted rockdrill mount offering additional restraint to the rockdrill base; the Technical Development Services Engineering Department provided a rockdrill mount surface ground to which was dowelled a rockdrill. Although the testing of these variants became protracted (and to some extent is still continuing), the evidence is that adoption of any of the variants does not significantly affect the rate of failure of component housings.

With regard to reconditioning of affected component housings, machining and grinding of the affected surfaces to remove the damaged areas has been the accepted method of repair, and alternative to other costly replacement, since first occurrence of the failure. This has normally been carried out without subsequent heat treatment and is not satisfactory as (as previously noted), machining through the heat treated surface zone results in a decrease in hardness from 60 to 25 Rockwell C, and this results in accelerated rate of interface deterioration and failure. Also machining instructions obtained from the rockdrill manufacturers during first quarter 1978 allow so little permissible machining of housing components as to be impractical. In practice, permissible machining practice has been developed and is at the discretion of the technician personnel servicing the rockdrills: many rockdrills presently in operation have been machined far beyond that permissible without apparently suffering any ill effect. Such renovation at best extends the housing component life before failure by $\pm$ 15 000 drilled metres.

Rebuilding of the affected interfaces by various methods had until recently not been successful as the inevitable heat involved had resulted in distortion of the component housing. During the third quarter 1978, technicians employed by the Technical Development Services Engineering Department developed a rebuilding process for the rockdrill intermediate housing using a normal flame sprayed corrosion resistant metal surfacing process. The process involves the conventional spraying technique applied to the rockdrill intermediate housing immersed in a water bath, and is applicable only to the intermediate housing, as of the affected component housings, this is least critical with regard to distortion. After trials of various spray powders, the powder composition preferred is one in which nickel and cobalt are predominants.

Field testing has shown that the treatment of the intermediate housing faces results in significantly reduced deterioration of the mating faces of the cylinder and rotation housing. Although not definitively confirmed by field trials as yet, it is estimated that this will result in a life before failure of these housing components of at least three times that previously experienced i.e. $\pm$ 100 000 drilled metres. Experimentation is continuing with a view to the development of this technique so that it can be applied to other affected component housings.

It is of interest to note that this failure appears to be peculiar to the South African gold mining application of the rockdrill: no other application (and presently more than 1000 rockdrills are in operation worldwide) in any part of the world has reported similar failure. It is not immediately clear why this should be the case: consideration has been given that the mode of rockdrill operation used (high frequency mode rather than the medium frequency mode generally preferred elsewhere) could have some influence. Although some investigation is being carried out in this area, general drilling economics are thought to mitigate against the use of other than the rockdrill's high frequency mode of operation in South African gold mining applications.

CASE HISTORY 2: FAILURE IN RAISE BORER DRILL ROD

The second case history deals with failure in 10" (255mm) diameter high strength raise borer drill rods, in the double pin subs. The Technical Development Services Mining Department has been engaged in a wide range of "boring" activities since 1971: its present "boring" fleet consists of two hard rock full face tunnel borers capable of boring 3,5m diameter tunnels, six raise borers of various boring capabilities in hard rock of between 1,8m and 3,66m diameter, and three blindhole borers capable of boring 1,5m and 2,13m diameter holes. This fleet in present-day cash terms can be represented by a replacement value in excess of R13 million. The Technical Development Services Mining Department offers a boring service where required to the A.A.C. of S.A. Ltd. gold mines: certain gold mines of the group, however, themselves own, operate, and maintain their own raise borers to meet their requirements.

In South African hard rock boring applications 10" drill pipe with 11" (280mm) stabilisers is typically used on the smaller type raise borers capable of up to 2,44m diameter holes. During the reaming sequence of the boring operation, typically, loading of the drill string is high, often approaching the drill string manufacturer's cautionary operating zone, with depending on a variety of circumstances and conditions, maximum operating thrust up to 2900 kN and maximum operating torque up to 204 000 Nm. Set out in Fig. 4 is a diagrammatic representation of typical South African hard rock raise boring practice, showing the various operating parameters applicable. As can be seen, drill string and reamer head stabilisation is achieved by the use of 11" stabilisers adjacent in the string to the head, with reamer head stinger protection being afforded by a saver sub: additional stabilisation is afforded when necessary and is normally a hole length dependent. In view of drill string and reamer head stabilisation, failure in the drill string components during operations is rare although occasional failures in stingers, saver sub and stabilisers do occur. It is generally agreed that in the South African hard rock raise boring conditions and elsewhere, these failures are a result of corrosion fatigue; elsewhere sometimes, these are aided by, and attributable to, operator abuse especially during the "collaring" process at the commencement of reaming. As any failures of the drill string during boring operations are highly undesirable for obvious reasons, all drill string components are subjected to rigorous non-destructive testing methods during all stages of their life i.e. during manufacture, upon receipt as new, after each hole, etc., using generally both magnetic and fluorescent particle examination techniques and ultrasonic testing.

Negligible legitimate failure occurred in 10" diameter high strength drill pipe (other than attributable to malpractice, negligence or abuse) until approximately mid-1975. From this time to mid-1977 approximately 18 legitimate failures occurred throughout the world, of these, two in operations in South Africa of the Technical Development Services Mining Department: all failures occurred in the epoxied pin end of the double pin sub. At this time there were estimated to be approximately 7500 10" diameter high strength drill pipes in the field, thus giving a failure rate of rather less than 1/4%; however, in view of the importance of a zero failure rate and the common mode of failure of the drill pipe double pin sub, some urgent investigations into the cause were initiated independently by the manufacturers and raise borer supplier and the Technical Development Services Mining Department. Shown in Fig. 5 is a typical example of failure experienced in the drill pipe double pin sub.

In the manufacture of drill string components, the material used is A.I.S.I. 4340, 4330 (vanadium modified-vacuum degassed) and Timken designation 4333 M6. In the case of the 10" drill string components, saver subs and stabilisers are made from A.I.S.I. 4330 (vanadium modified-vacuum degassed): 10" drill pipe is manufactured

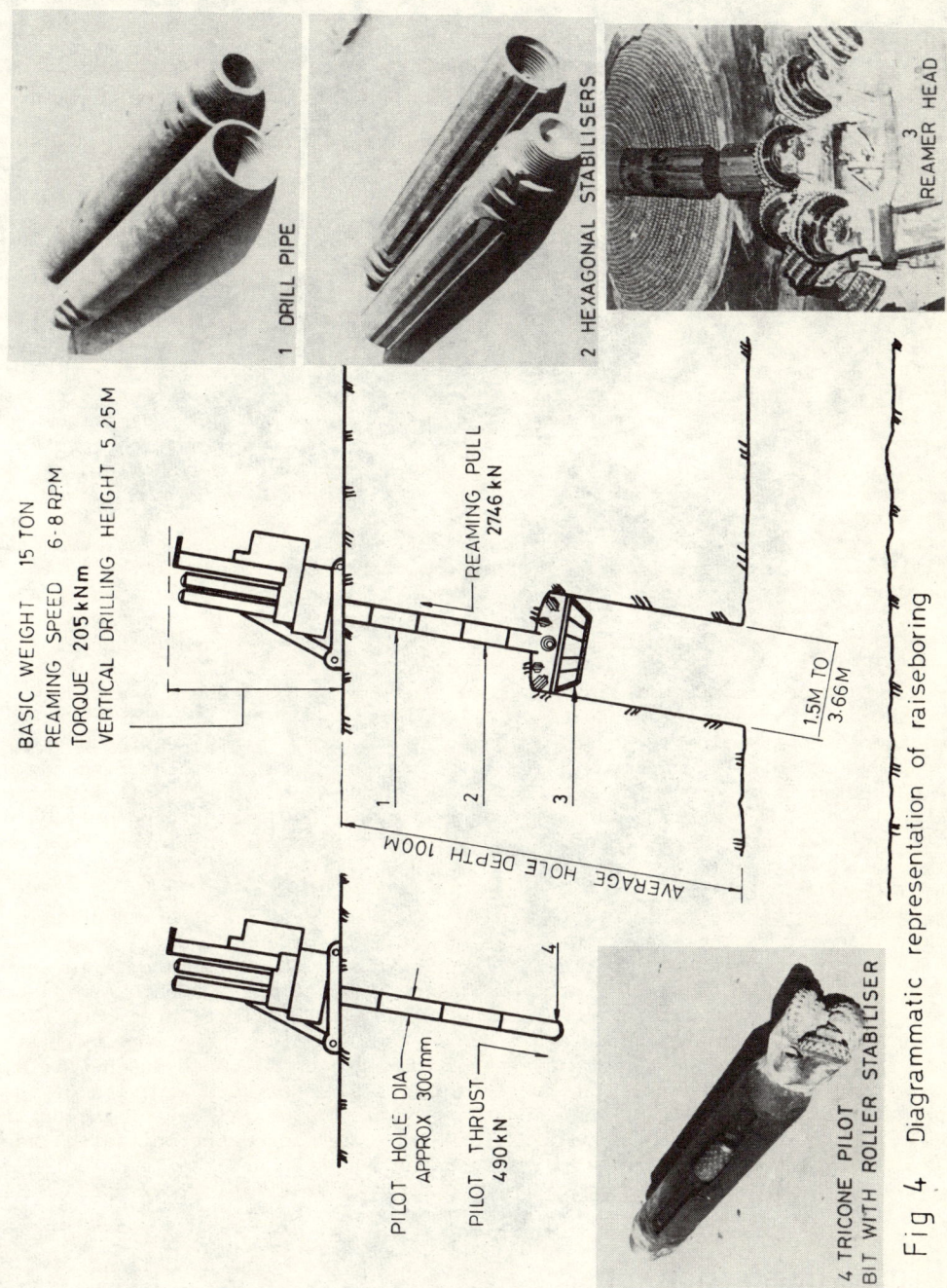

Fig 4. Diagrammatic representation of raiseboring

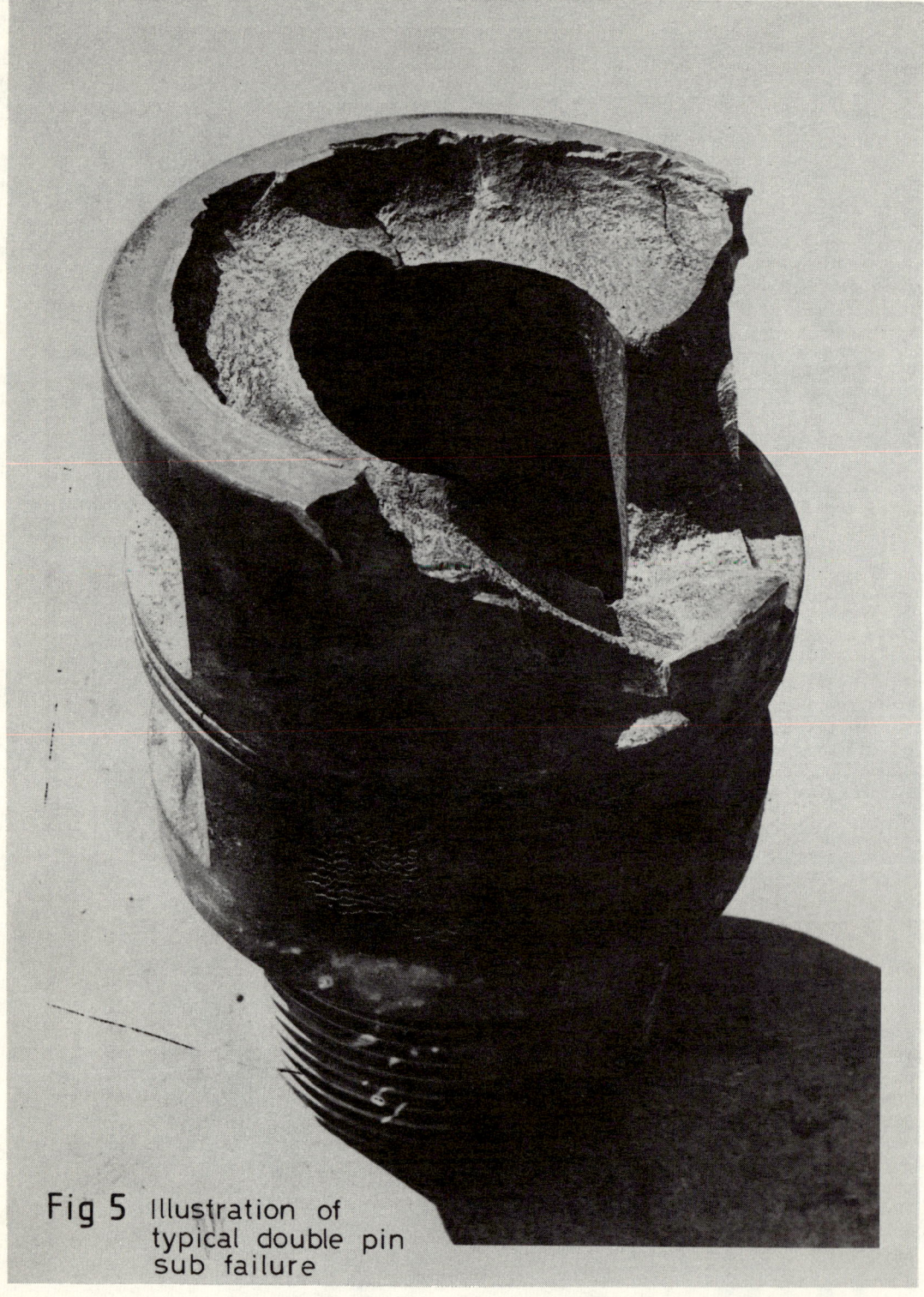

Fig 5 Illustration of typical double pin sub failure

from A.I.S.I. 4340 and this is obtained in the form of "hot rolled tube". The threaded connection called the double pin sub is made from Timken designate 4333 M6, or alternatively A.I.S.I. 4330 (vanadium modified-vacuum degassed) solid stock, and this is torqued into one end of the drill pipe with an epoxy resin adhesive to 244 000 Nm, thus forming the 10" drill pipe with a male-female end. In the manufacture of 10" drill pipe components the material used receives its heat treatment prior to machining. Heat treatment is carried out either by the steel suppliers or by the manufacturer of the drill pipe and produces a drill pipe component with hardness of 40 Rockwell C. Where hard facing on drill pipe components is required (as in the case of the wear strips on hexagonal stabilisers), care is taken that any necessary pre-heating etc. does not affect the heat treatment that the component has already received. All threaded connections are subjected to an etching or pickling process with a phosphoric acid solution (known commercially as "Kem-plating"), and this subsequent surface receives a final sand blast with a fine grade sand. The purpose of this process is to "condition" the thread and shoulder surfaces of the drill pipe components to accept better the thread compound (zinc based grease) necessary under this extreme pressure application. Because of the necessity of a high quality product, the manufacturers exercise throughout strict quality control standards; steels are supplied from the mills with written certification and are subjected to physical and microscopic examination upon receipt and, when applicable, prior and post heat treatment; non-destructive testing methods using eddy current and ultrasonic testing techniques are used prior to machining; a high standard of quality control during machining is maintained using conventional control techniques.

Investigations carried out into the failure of the double pin subs by the manufacturer, a raise borer supplier and Technical Development Services Mining Department, made use of three American and one South African educational and/or research institutions. Throughout the investigation free exchange of information was practised by all parties and it is felt this, the early commencement of investigation, and the use of independent educational and research institutions, contributed significantly in the investigation to define the cause and thus determine a solution to the failure. Although most of these investigations did not agree about a sole cause of the failure, most identified a number of common factors relevant to the failure.

It was agreed that the chemistry of the steel used in the manufacture of the double pin sub was within the specification range and that its hardening was satisfactory. It was determined also that its mechanical properties i.e. tensile, impact and fatigue characteristics, were representative of the class of steel and its heat treatment. The fracturing of the double pin sub initiated in the fillet areas of the machined surface with the fracture initiation being intergranular in nature. Fig. 6 shows in detail a typical fracture surface and origin of fracture.

It was the consensus of the investigations that the hydrogen content of the steel in the near fracture zone was high whilst in the bulk of the steel it was low; also that untempered martinsite was present in the fillet zone and extended 0,05mm into the machined surface zone. Also identified were an abnormally large number of inclusions for a vacuum melted product, variously ascribed as non-metallic, manganese sulphide and tramp elements. A majority of the investigations reported that crack propagation in the initial stages was by stress corrosion cracking and that subsequent macro cracking developed via fatigue or ductile fracturing. Examples of microscopic examination of a section displaying light etching zones characteristic of the untempered martinsite found are shown in Fig. 7. Examples of branching networks of fine cracks characteristic of corrosion cracking, are similarly shown in Fig. 8.

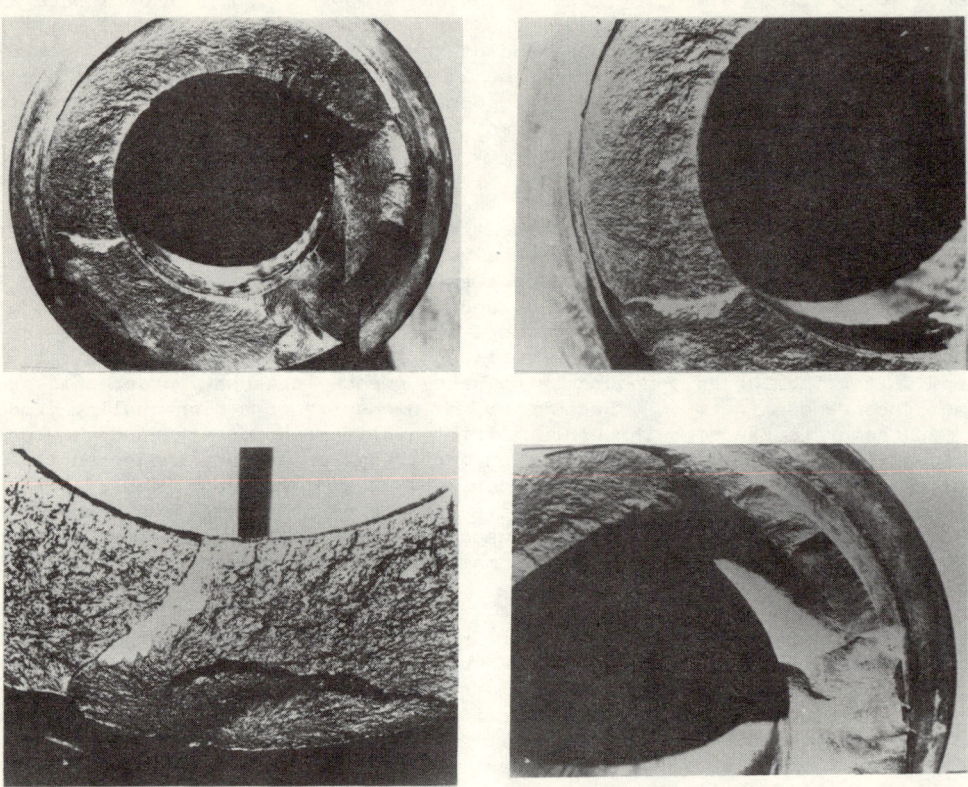

Fig. 6. Typical fracture surface and origin of fracture

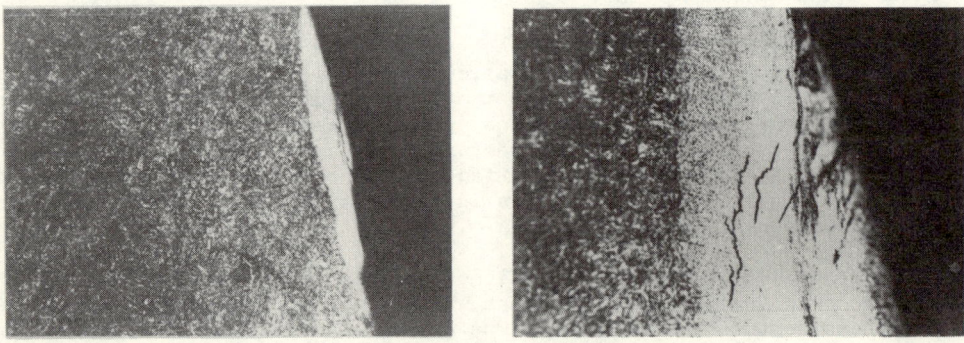

Fig. 7. Light etching zones characteristic of untempered martinsite

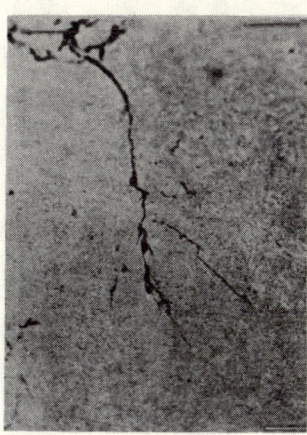 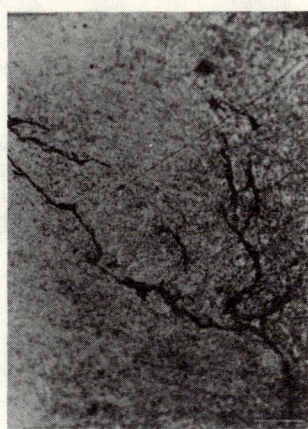 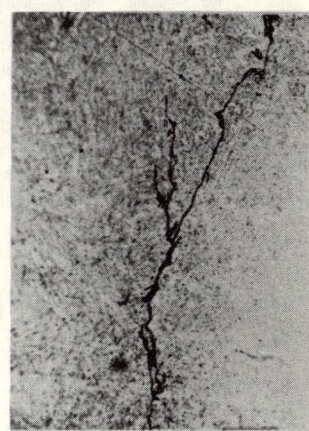

Fig. 8. Branching networks of fine cracks characteristic of corrosion cracking.

Consequently, explanations setting out the cause of failure ranged from temper embrittlement as a result of tramp elements causing high susceptibility to hydogen embrittlement and stress corrosion, to hydrogen embrittlement, to untempered martinsite embrittlement. Subsequent investigation by the manufacturers centred on determination of the hydrogen source for hydrogen embrittlement and the process which led to the formation of untempered martinsite.

Although samples submitted for investigation were inevitably heavily rust-infested as a result of the hostile operating environment, it was felt that hydrogen formed as a result of rusting did not solely account for this accumulation. Whilst approximately 60% of the hydrogen was found in easily removed near-surface accumulation, a significant portion was near the surface, ingested into the surface layers. An obvious possible source of hydrogen was from the thread "pickling" or "etching" process ("Kem"-plating), from the phosphoric acid pickling media. Investigation into alternative pickling media, or other alternative treatment processes failed to reveal any suitable for the application which did not use "hydrogen-rich" preparations. In an effort to drive out any hydrogen ingested during the "Kem"-plating process, the manufacturers had some double pin subs baked at a temperature of approximately 200°C for a period of two hours. It is not immediately clear what effect this had and how successful this procedure was.

For untempered martinsite to form at or near the surface of the steel, the steel would have had to have been heated subsequent to its heat treatment to at least the lower critical temperature for the steel of about 720°C. As double pin subs are not directly subjected to any heat treatment during manufacture from solid stock it was considered that this could only occur during the machining operations. An investigation into the machining operations showed that it was the manufacturer's procedure to machine the double pin subs on N.C. machines, although it was possible in time of peak demand that manually-operated machines would be used. It appeared that whilst it was unlikely that high temperatures could be generated during machining on N.C. machines, this might not be the case with manually-operated machines where the quantity of cutting compound, the depth of cuts, cutting speeds, condition of cutting tool, etc. are to some extent dependent on the operator's discretion. It is interesting to note that subsequent to the probable period of manufacture of the double pin subs which failed, natural development by the manufacturer of his facilities had resulted in his acquisition of additional N.C. machines affording less critical N.C. machine capacity.

Although in this particular case no clear cut cause of the failure could be conclusively determined, through investigations by a number of independent educational and research institutions, common factors did emerge which provided a sound basis for a multi-point approach in determination of a solution to the failure. To some extent the investigations were complicated by the extremely low incidence of failure and whilst legitimate failure in double pin subs is not presently being experienced, it is conceivable that the phenomena which resulted in the earlier failures are still present, inherently, in drill pipe being manufactured, or in use, today.

ACKNOWLEDGEMENT

The authors would like to thank the Anglo American Corporation of South Africa Limited, and Mr. E. Schmid, Consulting Engineer, Research & Productivity, for permission to present this paper.

It should be noted that whilst this paper has been compiled on a basis of the experiences obtained by the authors with the Technical Development Services, A.A.C. of S.A. Ltd., the views and opinions expressed in it, are those of the authors, and do not necessarily represent those of the Anglo American Corporation of South Africa Limited.

The authors acknowledge the assistance of L.K. Moger, G.A. Kermis and J.W. Phillips of T.D.S. Engineering Department, A.A.C. of S.A. Ltd., in the preparation and compilation of this paper.

PROGRESS IN EXTENDING THE LIFE OF STEEL COMPONENTS EXPOSED TO HIGH TEMPERATURE ENVIRONMENTS

D. Davidson

Highveld Steel and Vanadium Corporation, Witbank, R.S.A.

ABSTRACT

The paper outlines experience in the Steel-making industry where readily available materials such as carbon steel and Cr/Ni steels are successfully employed in high temperature environments for lengthy periods in such a way that process plant campaign life is optimised. Optimisation is achieved by attention to design detail and appreciation of the properties of the materials involved, as well as an intimate knowledge of mode of failure and the influence of the environment on the performance of materials. It has been found more economical to either combine heat-resisting steel with refractories, to extend the life of the composite or to arrange rapid change of components, than to resort to exotic costly solutions to component failure problems!

INTRODUCTION

This paper discusses experience gained and improvements made to process plant components exposed to environments where temperatures range between $800°C$ and $1300°C$ and where the atmosphere could be of a reducing nature.

THE INCENTIVE FOR ECONOMY AND IMPROVEMENT

A problem that the engineer in the continuous process industry is confronted with is to extend the useful life of components by whatever means possible, within a certain cost framework, so as to allow the process to continue uninterrupted for as long as possible.

Usually the loss of revenue due to loss of production from continuous processes far outweighs the cost of even exotic solutions to fracture problems. However, most continuous processes have to be interrupted from time to time, to deal with the effects of wear, so the most economical solution to the fracture problem occurs when the failure of components due to fracture coincides with the campaign life of the process plant unit. In the industry in which I am employed the life of refractory components, which are the elements exposed to wear, is continuously being extended, resulting in the need for other elements of the structures, which are usually ferrous materials, having to keep pace in order to obtain an optimum campaign life for the process plant unit.

COMPONENT DESIGN

Designers frequently visualise the component they are designing as an item which exists at room temperature and then select permissable working stresses that would be acceptable at the elevated temperature at which the component will be expected to work.

This approach sometimes leads to other factors such as change in shape due to expansion, while the component is being heated up, being ignored. As in any design situation all factors defining the environment in which the component will be required to work must be adequately defined and duly considered at the design stage.

In this context the following examples of failures are given. Hot ducting which collapses as the result of fracture at its supports, due to lack of allowance for expansion. The fracture of a water cooled component allowing water to enter a furnace resulting from too high a temperature gradient through its walls. These are catastrophic situations that have occurred in practice.

MATERIALS

When considering the available range of high strength materials for use in the pyrometallurgical/heavy engineering industry at elevated temperatures, the range is prescribed by cost, availability and properties, in that order. Low carbon steels can be used successfully at elevated temperatures but the components have to be water or sometimes air cooled, and in the event of cooling system failure, catastrophic failure of the component would be expected. The so-called "heat resisting" Cr/Ni steels also have a place, but they too have limited life at high temperature and cooling or shielding has to be resorted too. Cobalt and titanium are sometimes encountered in heat resisting alloys, but they are costly and not readily available.

Besides strength at elevated temperature, other properties such as coefficient of expansion, thermal conductivity and resistance to oxydising or reducing atmospheres have to be taken into consideration when selecting a material. These properties are compared in Table 1.

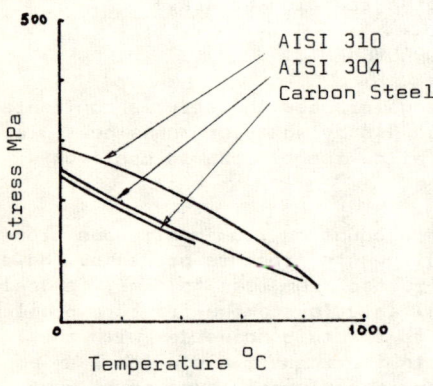

Fig. 1
Yield Stress

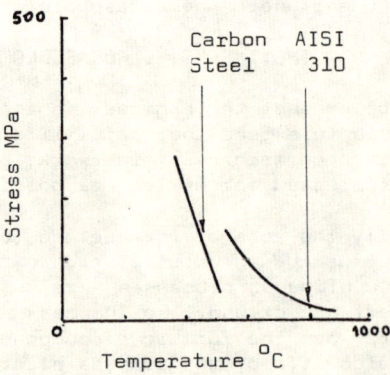

Fig. 2
10 000 Hours
Rupture Stress

TABLE 1   Some Properties Of Materials Commonly Used At Elevated Temperature

Type	Composition				
	C	Mn	Si	Cr	Ni
Carbon Steel	0,25	0,65	0,10	0,25	0,30
Cr/Ni AISI 304	0,08	2,00	1,00	19,0	10,0
Cr/Ni AISI 310	0,25	2,00	1,50	25,0	20,0
	Thermal Conductivity W/mK				
	$0°C$	$200°C$	$400°C$	$600°C$	
Carbon Steel	49	46	42	37	
Cr/Ni AISI 304	14	18	21	23	
Cr/Ni AISI 310	10	14	17	19	
	Coefficient of Expansion $\times 10^{-6}/°C$				
	$0°C$	$200°C$	$400°C$	$600°C$	
Carbon Steel	11,0	13,0	15,0	-	
Cr/Ni AISI 304	15,5	17,0	17,5	18,5	
Cr/Ni AISI 310	14,5	15,5	16,5	17,0	

The advantages of using carbon steel material for components subject to elevated temperature environments are lower cost, high thermal conductivity and hence less massive components. While the disadvantages include surface temperature being limited to a maximum of about $450°C$, poor creep resistance and low tensile strength. See Figs. 1 and 2 for temperature dependence of yield stress and 10 000 hour rupture stress.

Austenitic chrome nickel steels offer better creep resistance and higher tensile strength at elevated temperatures. However, the poor thermal conductivity of these steels can result in surface overheating in thick sections and proneness to thermal fatigue failure.

The different coefficients of expansion and different rates of change of this factor account for the inadvisability of submitting large composite carbon steel/chrome nickel steel welded components to high temperature.

CASE STUDIES

Kiln Combustion Air Pipe

The first case study involves the development of a combustion air pipe design suitable for a campaign life approaching 36 weeks in a rotary kiln at gas temperatures of $1150°C$ where the gas is reducing in nature.

The rotary kiln is 4 metres in diameter and 60 metres long, it is refractory lined and fired by a pulverised coal burner. A mixture of iron ore, coal and fluxes is fed into the rotary kiln so as to partly reduce the ore. During this process, the coal is charred and the volatiles given off by the coal are burned to provide secondary heating. The burning of these volatiles requires the addition of air, which is blown into the kiln through combustion air pipes, which protrude through the kiln shell. The amount of this air is regulated to control the temperature at 1150°C.

In 1968, when the rotary kiln plant first commenced production the refractory lining was a high alumina brick and the combustion air pipes were each fabricated from three castings. The castings were a sand cast nozzle section, sand cast base section and spun cast barrel, all welded together to form a single 'pipe'. The castings were heavy walled, nominally 25 mm thick and the material was a heat resisting steel containing 0,32%C, 0,9%Mn, 1,4%Si, 12%Ni, 25%Cr, while the weld metal contained 0,34%C, 21%Ni, 24%Cr.

It was found that some of this design combustion air pipe failed by cracking after only eight weeks of continuous unprotected exposure to the hot, reducing atmosphere in the rotary kiln.

Metallurgical examination of the failed pipes suggested crack propogation from porosity at or near the surface of the centrifugally cast barrel section through precipitated grain boundary carbides, due to the thermal stresses encountered in service. The welding was considered to have played a part in the failure, while sigma phase precipitation was not thought to be present.

The short life of these combustion air pipes represented a catastrophic situation which had to be rectified as soon as possible, since the loss of production due to frequent rotary kiln outages was considerable.

Further design and metallurgical investigations were carried out and an extensive programme of tests was undertaken. A single piece, sand cast design was employed and the materials listed in Table 2 were used in successive trials:-

TABLE 2 Composition of Heat Resisting Steels used in Trials

No	C	Si	Mn	Cr	Ni	Co
1	0,28	0,8	1,3	25,0	12,5	-
2	0,30	1,5	0,8	25,5	19,5	-
3	0,40	1,5	0,8	15,5	39,0	-
4	0,08	0,8	0,8	28,0	-	50,0

Of the materials in Table 2 only No. 4 proved to give a significantly longer life than the others in the completely unprotected state.

In addition to experimenting with various materials, variations in the design of the combustion air pipes were made and tested. The design changes included reducing the pipe wall thickness, shortening the pipe, the incorporation of external or internal ribs and an elliptical cross section in place of the circular cross section. Various combinations of these design changes were experimented with but shortening or reducing the length of the pipe was the only change that gave any significant improvement in life. This improvement was due to reduced bending stress values (air pipe is subjected to bending stress reversal with each rotation of the kiln).

During these trials, detailed physical and metallurgical examination of the failures continued, from which it became clear that cracking started in the barrel of the pipe just above the burden bed level in the rotary kiln and this cracking occurred mainly on the side of the pipe where the hot gases passing through the kiln impinged. Inter-granular cracking persisted and carbide as well as sigma phase precipitation were recorded.

Extending the life of Steel Components 37

Fig. 3
New Fabricated
Air Pipes

Fig. 4
Used Fabricated
Air Pipes

It was noted that the kiln atmosphere is both reducing and carburising, containing sulphur gases from the coal (probably as hydrogen sulphide), carbon monoxide, carbon dioxide and nitrogen. Increased carbides were found near the external surface of the pipes confirming the carburising effect. It was concluded that the pipes failed by a combination of the following:-

> Alternating fatigue stresses,
> Sigma phase formation,
> Formation of chromium carbides,
> Reduced strength of material as a result of operating temperatures,
> Stressing due to alternating temperature changes.

Having established that cast steel air pipes in the alloy 25 Cr/20 Ni gave an average life of 13 weeks and the alloy 28 Cr/50 Co, 20 weeks, whereas the refractory lining of the rotary kiln could survive for a longer period, it was clear that further development was required. The next development was a trial with a fabricated steel pipe using 25 Cr/20 Ni steel plate, the contention being that the mechanical properties of the material could be better controlled in a rolled plate than in the sand cast pipes. Also a fabricated pipe could have a much thinner wall than a casting, which was felt to be advantageous with the heat resisting steels, due to their lower thermal conductivity. At the time of this development, trials with steel heat shields were being conducted on the cast steel pipes with some success. It was therefore decided that the fabricated pipe should have a heat shield. In fact, the heat shield was built into the pipe in the form of a double wall construction. This outer pipe (heat shield) was found to be sacrificial, thus lengthening the life of the pipe. (See Figs. 3 and 4). Nevertheless, this new design gave an average life of 15 weeks which was better than the casting in the same alloy and at considerably lower cost than any of the cast pipe designs.

Continuing with the development of the shielding principle, various types of castable refractory were applied to the outside of the air pipes, which again allowed the life to be extended until, at the present time with new developments in castable refractories, a life of 35 weeks has been achieved. With such a long life being achieved, the rotary kiln refractory lining life and combustion air pipe life become equal and the optimum campaign period is achieved.

The events related so far refer mainly to the first 5 year period of the rotary kiln operation when kiln campaign lives were increased from the earlier catastrophic 12 weeks to 22 weeks. When once it had been decided to standardise on the less costly fabricated combustion air pipe, advancement of the campaign period was achieved mainly by experimenting with different refractories, some of which are listed in Table 3. There has been a gradual shift away from refractory bricks to castable linings for the kiln shell. Similarly the air pipes have had bricks bolted to them and many castables and refractory anchoring techniques tested in this time.

The environment inside the rotary kiln is very hot. Surfaces are readily abraded by flyash and the motion of the burden. In the event of any maloperation temperatures can soar, resulting in fusion and slagging taking place. In addition, severe temperature stratification and cycling, axially and radially, has been observed. Consequently the gradual advancement of the campaign life from 22 weeks to the present 35 weeks has been achieved by continuous experimentation, but still requires a short shut-down of 3 to 4 days at some time between the 17th and 25th week for minor repairs to refractories.

Fig. 5 Single Piece Roof Beam

Fig. 6 Crack in Roof Beam

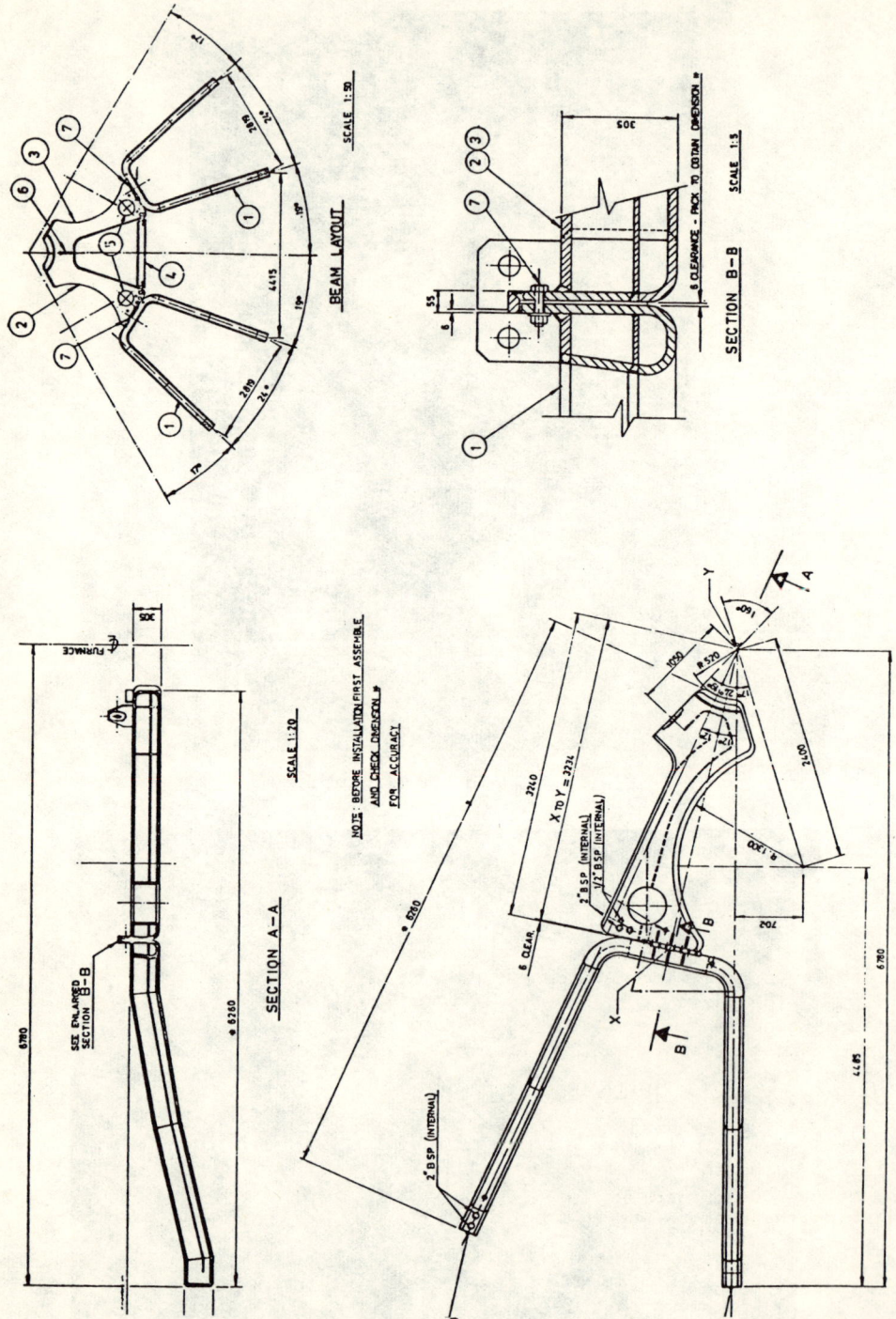

Fig. 7 Two Part Roof Beam

The advancement of locally produced refractories has kept pace with the demands of the industry and the presently available high alumina castable based on corundum offers excellent temperature and wear resistance.

TABLE 3   Comparison of some Castable Refractories

APPLICATION	MAIN CONTENTS %				COLD CRUSHING STRENGTH MPa	MAX WORKING TEMPERATURE °C
	Al_2O_3	SiO_2	ZrO_2	CaO		
Abrasion resistance	37	47	-	13	22	1400
Abrasion resistance (Corundum)	75	13	-	8	61	1500
Mechanical Strength	41	39	-	14	38	1300
High Temperature	55	41	-	4	40	1600
High Strength	46	40	-	13	41	1300
Very High Strength	61	11	24	4	111	1600

Furnace Roof Beam

The submerged arc electric furnace used for the production of pig iron has a cover to contain the evolved gases. The cover is in the form of a water cooled steel beam matrix with intermediate refractory panels. Since the evolved gases reach temperatures of about 1 000 C, the hot faces of the water cooled roof beams are exposed to this temperature. The water cooled beams are manufactured from a low carbon steel of the type BS 1501 - 151 28B, commonly known as boiler plate.

Fracture has featured significantly in the life and development of these components. Fractures have resulted from arcing between the furnace electrodes and the beams, and between the beams and the furnace charge. In addition fractures have resulted from shape stress raisers and manufacturer's welding technique. The reason for choosing a water cooled beam internal to the furnace for the original design was due to the need for a stable shape component of high mechanical strength suitable for supporting the substantial loads imposed from above and the severe thermal loading imposed from below. Due to the number and size of openings in the furnace cover, the beams have an intricate shape which results in complex fabrication problems.

In addition, the cooling water is circulated through the beam in such a way that it passes along the hot lower surface first and then returns through the upper portion of the beam. (See Fig. 5)

To date, the average life of a beam in service has been three years, at which stage the frequency of outage for repairs to water leaks is so great that complete replacement is required. The interesting point here is that every time a water leak develops, this must be repaired by welding in situ. Weld repair in situ on a hot furnace, probably in an awkward position, is seldom successful and further cracking and leakage soon occurs. The weld repair itself often results in cracks necessitating further repair. (See Fig. 6)

Changing roof beams could mean anything up to 80 hours furnace outage and a lengthy furnace recovery period back to full production after the outage. This loss of production, together with the danger of explosion associated with water leaks, has been the incentive for striving for longer beam life.

The first approach was to consider repositioning the beams externally to the furnace and supporting the refractory roof bricks from the beams, but space considerations did not allow this. Coating the underside of the beam with refractory was also tried but found to be unsuccessful. The best approach appeared to be to design a beam that would last as long as the existing beams, but which would be easier to replace, thus reducing the furnace outage time. The solution was a two part beam such that the centre section could be removed without seriously affecting the furnace cover refractories, thus reducing the time required. By splitting the beam into two parts, the individual parts became simpler, easier to manufacture and the water passages became more regular and hence more effective. (See Fig. 7)

Long shut-downs for roof beam replacement are no longer required since the centre vulnerable sections of the beams may be replaced individually during separate shut-downs. Also the retention of the low carbon steel water cooled beam concept has kept the component cost down while giving an acceptable life.

Furnace Charging Shaft

An interesting component failure and development also associated with the furnace cover, discussed above, is the charging shaft piece or port through which the furnace charge is fed into the furnace.

Originally a heat resisting iron casting with loose spade plates inserted around it's periphery, to control the shape of the charge pile inside the furnace, was used. This design experienced a short life due to distortion and cracking as a result of direct exposure to the hot furnace gases. (See Fig. 8)

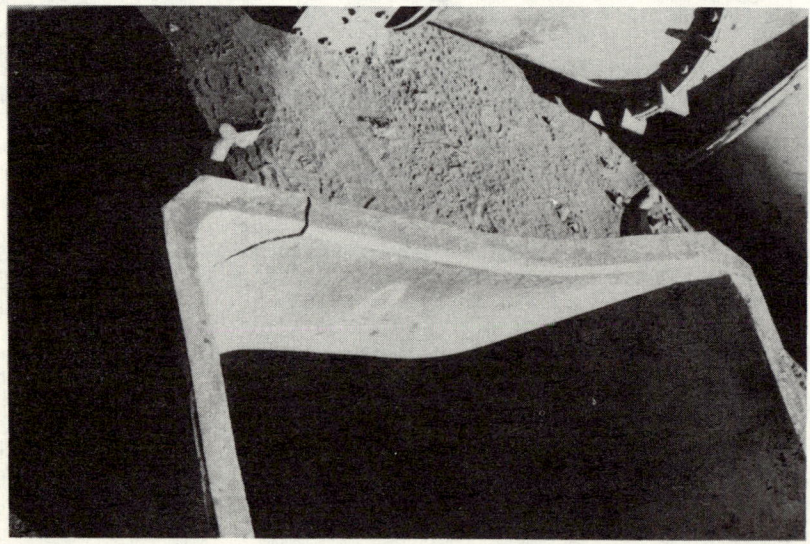

Fig. 8   Cracked Charging Shaft Casting

Fig. 9 Failed Reclaimed Charging Shaft

The life of the unit was extended by resorting to chrome/nickel steel castings with refractory linings but the failure rate has remained high, mainly due to the difficulty of retaining the refractory and subsequent overheating of the surface of the thick walled casting. The overheating of the surface results in distortion and the formation of cracks in restrained portions of the casting, e.g. corners.

The component has been re-designed to better accommodate the refractory and to allow freedom of expansion. During the development of this redesign some charging shaft pieces were fabricated from reclaimed parts (See Fig. 9), but a low carbon steel plate was welded to the chrome nickel steel centrepiece. Inadequate precautions were taken during welding. Also the refractory soon fell away inside the unit, resulting in the materials being exposed continuously to gases at about 1 000°C. Differential expansion was restrained and cracking resulted.

CONCLUSION

Engineers can learn to live with fracture by experimentation with different materials and attention to design detail. However, at elevated temperatures fracture cannot be eliminated altogether, particularly when working with commonly available materials, and it is under these circumstances that the best compromise, probably the combination of design and materials that gives the most economic life, will remain the solution to fracture problems for some time to come.

ACKNOWLEDGEMENT

I wish to thank the Management of Highveld Steel and Vanadium Corporation for permission to publish details of component life development programmes.

UNUSUAL FRACTURES IN THE MINING INDUSTRY

D. D. Howat

National Institute for Metallurgy, Randburg 2125, R.S.A.

ABSTRACT

In the mining industry safety is a predominant consideration in the design, inspection, and maintenance of equipment used for the hoisting of men. Fractures of such equipment are comparatively rare, mainly because high safety factors are incorporated in the design. Two unusual examples of such fractures, both of the fatigue type, are discussed. In one, a 60-ton shaft fractured at the junction of a hexagonal splined section with a cylindrical section. This was due to a failure during machining operations to blend the sides and the root radius of a circumferential groove separating the two sections of the shaft. It is possible that the fatigue crack developed over a period longer than 20 years.

In the second example, fracture occurred on the side of a bridle fabricated from 12in x 3½in steel channel. Two main factors were involved. Firstly, severe impact forces were transmitted to the side of the channel by metal-to-metal contact on the return of the skip to the vertical position after discharging ore. Secondly, there was severe corrosion, as a result of contact with acidic mine water, of the plastically deformed steel in the area of impact on the channel.

In addition, a description is given of an unusual type of fatigue fracture involving the formation of zig-zag cracks in the mantle of a crusher.

KEYWORDS

Fracture; fatigue; fatigue crack propagation; 60-ton 1000mm shaft; corrosion fatigue; mining equipment.

INTRODUCTION

Fractures in the mining industry, particularly in deep-level hard-rock mining, are probably more frequent and varied than in any other industry. The most critical areas for fracture are in the hoisting of men, ore, and materials. Up to 80 men may be crammed into a one-man cage to ascend or descend distances of up to 3000m. A catastrophic fracture will endanger the lives of all those involved, and create immense damage in the shaft necessitating weeks of repair-work and resulting in a very serious loss of production. Consequently, great care, attention, and close inspection are imperative in those critical areas. In addition, any accident involving the hoisting of men and materials must be the subject of an official

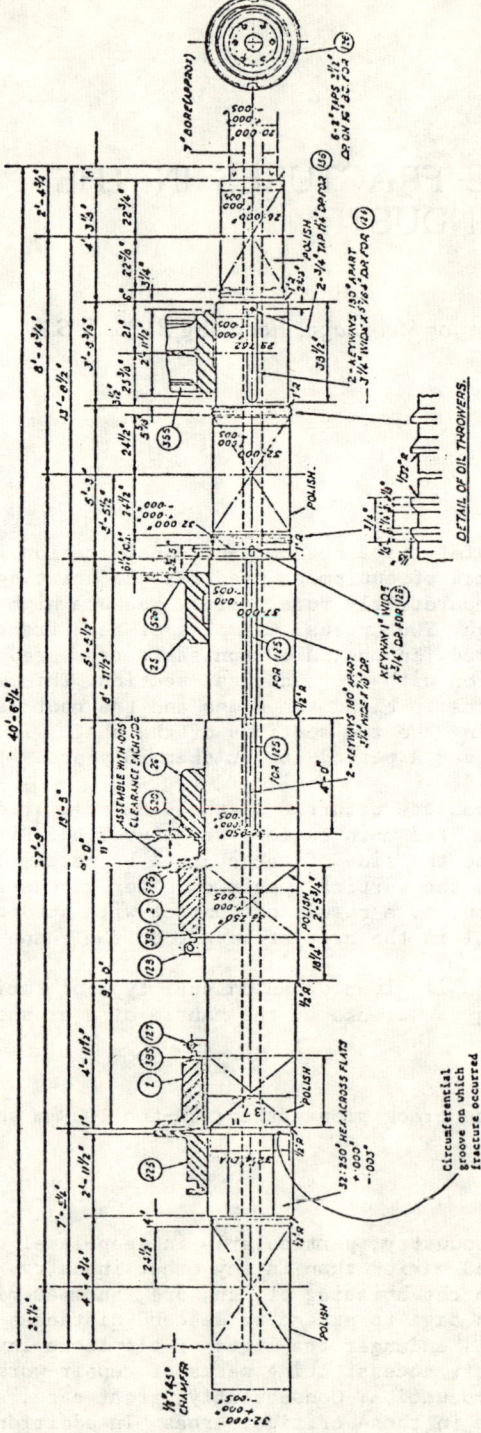

Fig. 1. General arrangement of double drum shaft showing the location of the fracture

enquiry by the Office of the Government Mining Engineer.

Special care must be taken with the design of plant and equipment to ensure that risks of catastrophic fracture are reduced to the absolute minimum. Inevitably, since engineers are extremely conservative, very high safety factors are built into the design.

Nevertheless fractures do occur (normally of the fatigue type, as with the majority of failures of engineering components), and three that were examined recently presented rather interesting features.

FRACTURE OF A DOUBLE DRUM SHAFT FOR KOPPERS HOIST

The first of these concerned a double drum shaft for a Koppers hoist, a line drawing of the 40-ton shaft being shown in Fig. 1. The shaft is 40ft $6\frac{3}{4}$in (12 169mm) in length overall and $38\frac{1}{4}$in (1146mm) in diameter over a length of 9ft (2700mm) at the widest zone. The material specification on the drawing calls for open-hearth, forged steel with the following mechanical properties:

U.T.S. 70 000 lb/in^2 (minimum) 482,65 MN/M^2
Yield stress 42 000 lb/in^2 (minimum) 293,04 MN/M^2
Elongation 16 per cent (minimum)
Hardness 170 HB (approximately).

The shaft was installed at the mine in December, 1950 and failed suddenly on 7th July, 1977, after $27\frac{1}{2}$ years' service.

The plane of the fracture coincided with the junction of a circular portion 37in (940mm) in diameter by 4ft $11\frac{1}{2}$in (1488mm) in length with a length of shaft of 2ft $11\frac{1}{2}$in (888mm) on which hexagonal flats $32\frac{1}{4}$in (820mm) in width had been milled. The location of the fracture is shown in Fig. 1. These hexagonal flats form splines on which the clutch slides to engage the winding drum. As shown in Fig. 1, between the circular length of shaft, which is 37in (940mm) in diameter, and the portion on which the hexagonal flats had been milled, there is a circumferential groove that the drawing indicates shall be $3^3/_8$in (60mm) in depth, 1in (25mm) in width, and with a root radius of $\frac{1}{2}$in (12,7mm).

Machining of such a groove was apparently considered the best method for separation of the two adjacent sections of the shaft, since it eliminated the machining problem of blending the hexagonal flats into the circumferential section.

Only visual examination of the fractured shaft was possible. General views of the component pieces and of the two mating faces of the fracture are shown in Figs 2, 3, and 4. It is clear from Figs 3, 5, and 6 that the crack initiated at the side of the circumferential groove adjacent to the cylindrical portion of the shaft, but not at the bottom of the groove. As will be obvious from Figs 2 and 4, the fracture is of the fatigue type with a large smooth cylindrical area and a rough portion immediately adjacent to the bore of 7in (172mm) diameter where the final sudden fracture occurred. The rough area of sudden fracture is about 12 per cent of the total cross-sectional area of the shaft. The smooth area shows virtually no ripple or beach markings, which frequently charaterize fatigue fractures. The most feasible explanation is that the fatigue crack propagated very slowly and in very small steps. Figs 2 and 4 show three distinct radial 'steps' marked A, B, and C in Fig. 4, A being the deepest step. This indicates that, initially, there were at least three separate fatigue cracks that later merged into one. Fig. 7 is a close-up of steps A and C, from which it is clear that the edge of step A was worn fairly markedly before final fracture occurred.

Reference was made earlier to the fact that the crack appeared to have initiated

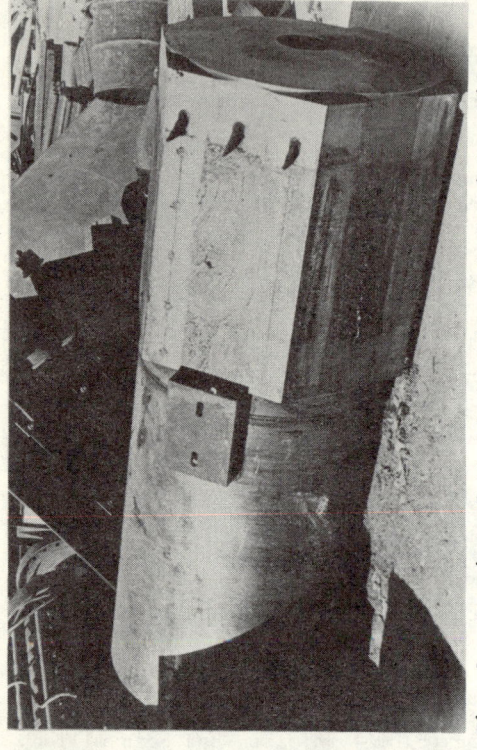

Fig. 2. Photograph of the long piece of double drum shaft (portion on right-hand side of the fracture in Fig. 1) showing the surface of the fracture

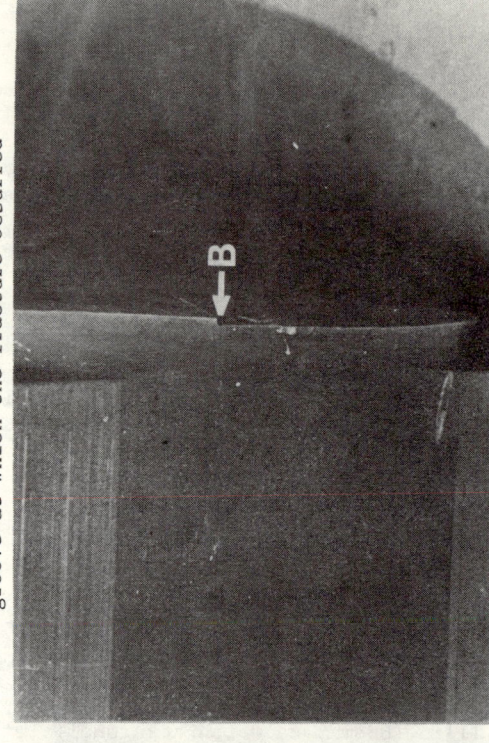

Fig. 3. Portion of the double drum shaft with splines on which the clutch travels, and showing the annular groove at which the fracture occurred

Fig. 4. A view of the surface of the fracture at the annular groove on the piece of shaft shown in Fig. 3

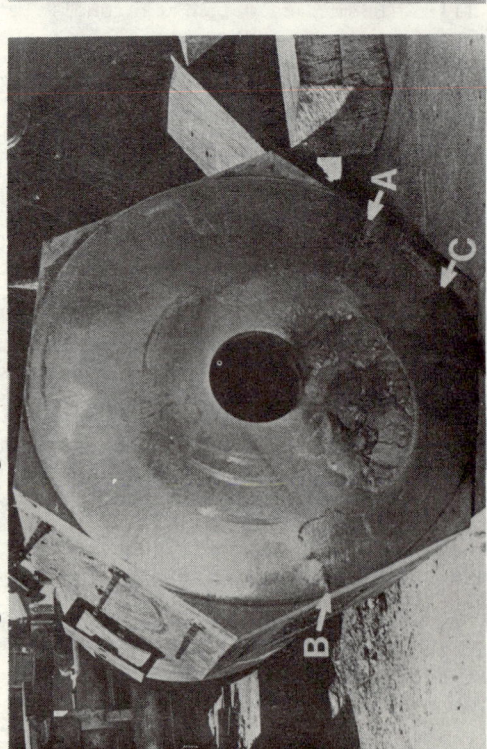

Fig. 5. A view of a portion of the annular groove on which the fracture occurred (area round step B)

Unusual Fractures in the Mining Industry 49

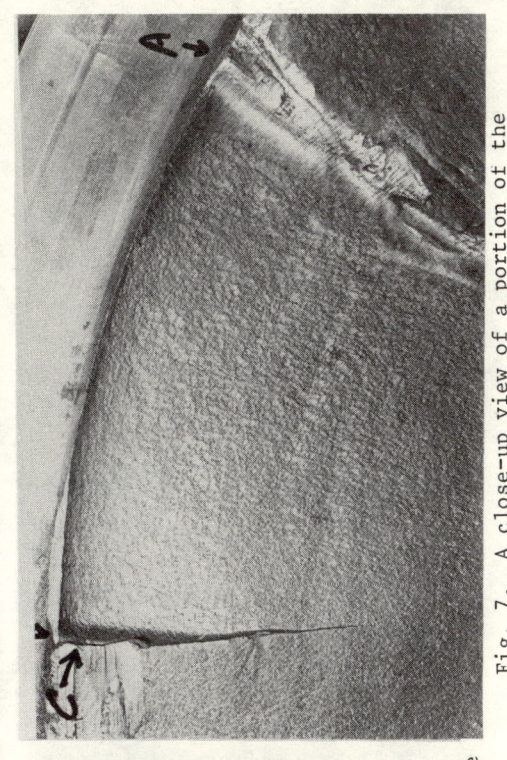

Fig. 6. A close-up view of the portion of the annular groove at the central portion in Fig. 5. This shows the extremely sharp edge at the periphery of the fracture surface

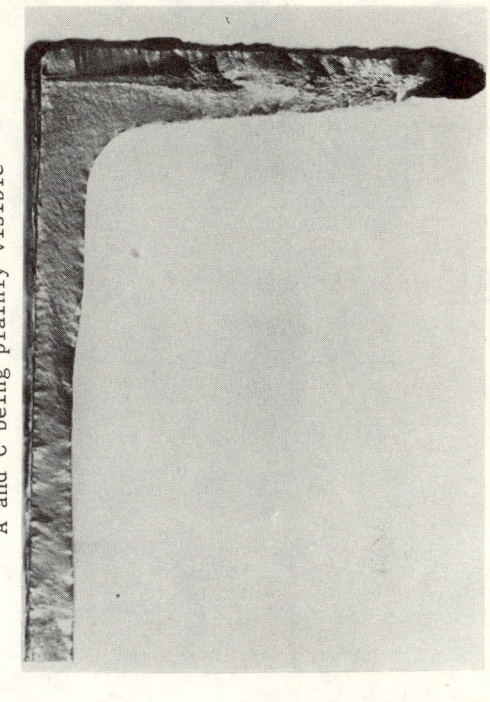

Fig. 7. A close-up view of a portion of the fracture surface shown in Fig. 2, steps A and C being plainly visible

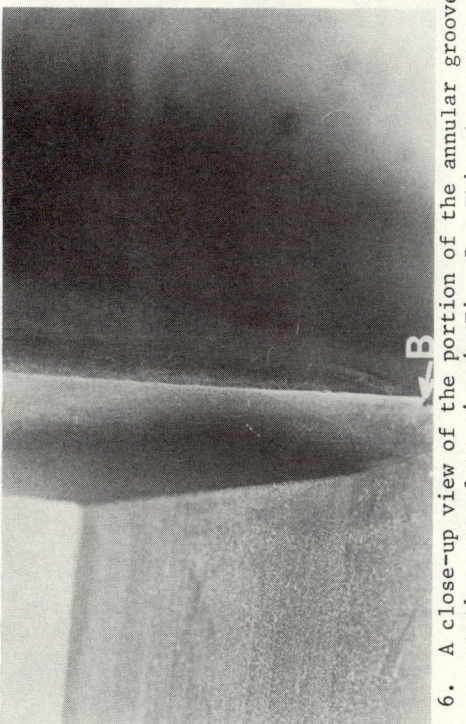

Fig. 11. A view of part of the fracture surface in the steel channel. The fracture appears to have been initiated at the top left-hand corner. Reduction in the thickness of the side of the channel by impact of the buffer plates is also evident

Fig. 12. A view of the remainder of the fracture surface in the steel channel. The area of sudden final fracture is at the lower right-hand corner

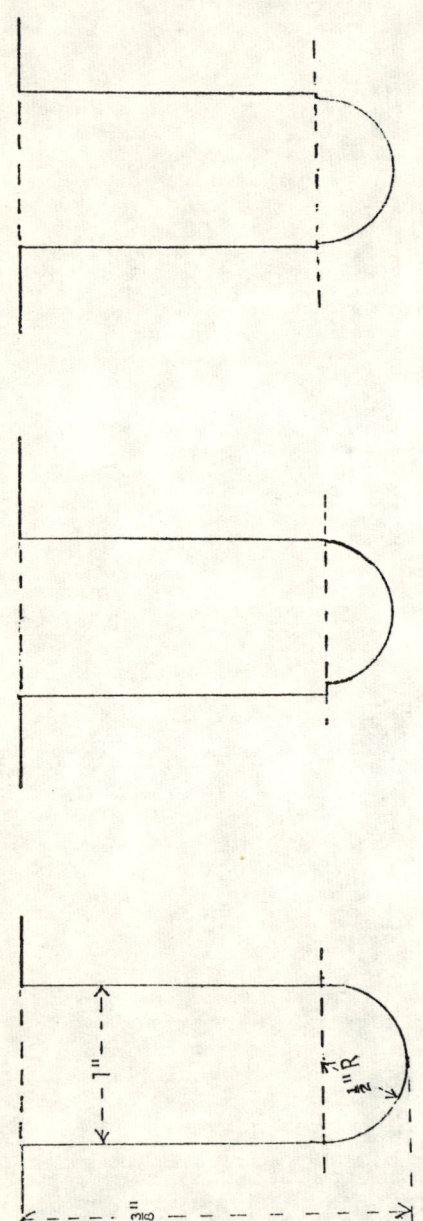

(a) As designed

(b) Suggested mode of formation of a notch on the side (sides) of the circumferential groove by use of a slightly undersize radius tool to form the bottom of the groove

Fig. 8. Annular groove on winding drum shaft

on the side of the circumferential groove and near to the bottom of the groove. This is clearly established by Fig. 5, step B being clearly visible. From these photographs, it is very clear that the circumferential edge of the fracture was very keen and sharp. Fig. 6, a close-up photograph of the edge shown in Fig. 5 just above step B, illustrates that the circumferential edge of the fractor is extremely sharp and well defined.

All the evidence points to the existence on the side of the groove, and near the bottom, of a clearly defined continuous circumferential line that acted as a 'stress raiser' while the shaft was in operation on the winder. When the circumferential groove was machined, it can be assumed that the two sides were formed to the appropriate depth by a side-cutting tool. The root of the groove was then formed by a round-nose tool with a radius of $\frac{1}{2}$in (12,7mm). By careful operation, the straight sides of the groove could be blended accurately with the radius at the root of the groove. However, it would appear that this accurate blending of the sides with the root radius of the groove was not effected smoothly, and a step was produced on one side, or possibly both sides, of the groove near the bottom. In Fig. 8 an attempt has been made to indicate how such a step might have been formed on the side or sides of the groove, by the use, say, of a radius tool that was slightly undersize.

If it is assumed that the fatigue crack was initiated as a result of the high concentration of stresses around such a step or notch near the bottom of the groove, a number of intriguing questions arise. The first of these is obviously how soon the crack was initiated, and how much time elapsed before final fracture. There is no record of any modification in the power of the driving motor, of increase in the speed of hoisting, or of increased pay loads in the skip. It is therefore possible that the crack initiated at a very early stage in the history of the hoist and propagated at a very slow speed over a period possibly longer than 20 years. The absence of any ripple or beach marks on the fracture surfaces suggests a very slow speed of crack propagation.

An ultrasonic test was carried out on the drum shaft 19 months before the fracture occurred, and the inspector's report stated that no defects were found. The most feasible explanation for the inspector's failure to detect the crack was that the ultrasonic test could be carried out from only one end of the shaft. The distance of the detector from the crack was 7ft 6in (2250mm).

The spare shaft available was immediately fitted into the winding gear, but this shaft and several others in use had been manufactured to the same drawing by the same manufacturer at about the same time. As was obvious from the experience with the shaft that fractured, ultrasonic detection could not be relied upon to give conclusive evidence of the existence of a crack. Without dismantling of the whole winding-gear assembly, it was not feasible for adequate visual examination to be carried out of the circumferential groove, and detection of a crack by dye-penetrant was very uncertain. Magnetic crack detection was virtually impossible, so there was no alternative for the mine except to continue with the existing shafts.

When additional shafts were ordered, particular attention was directed to a change in design that would eliminate the need for the circumferential groove. This presented considerable practical problems, but it was finally decided to eliminate the groove and accept a step between the portion featuring the hexagonal flats and the adjoining circular portion of the shaft - of 37in (940mm) diameter - the step to have a polished radius of $1\frac{1}{2}$in (38mm). These features of the new design are shown in Fig. 9. When the shaft is installed, a concomitant requirement will be to increase the radius on the entry bore of the winding drum.

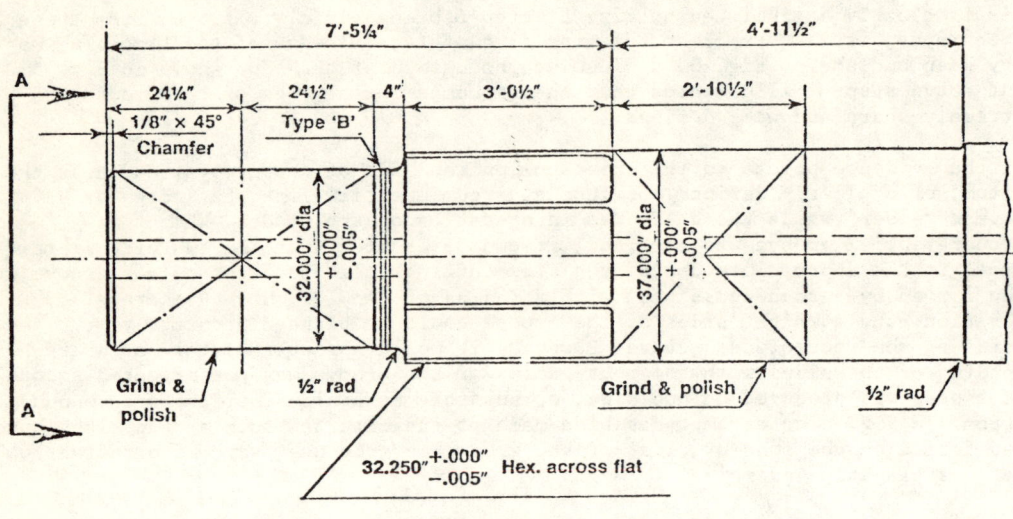
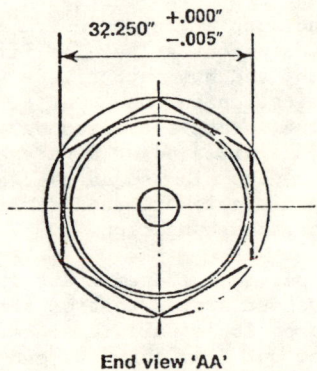
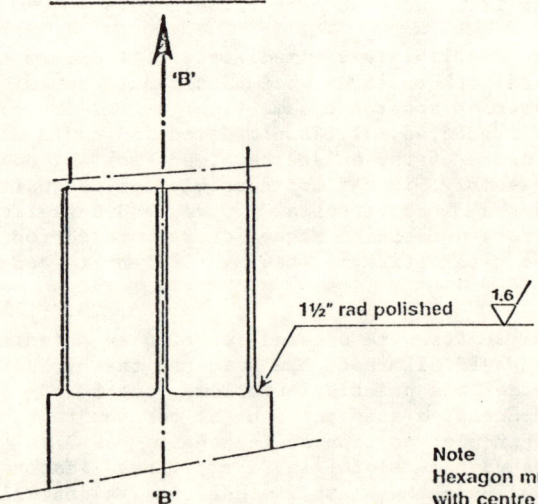

Fig. 9. Drawing showing the change in design of the double drum shaft to eliminate the annular groove

FRACTURE OF A JOB'S BRIDLE

A Job's bridle is the component used to carry the ore-skip from the loading zones underground to the discharge at the top of the headgear. Usually fabricated from steel channel, the two sides of the bridle are about 26ft (7800mm) in length and are rigidly fixed together by the top and bottom cross heads of the bridle, the maximum width being 4ft 9¾in (1448mm). In this particular bridle, the main frame was fabricated from 12in x 3½in (300mm x 89mm) channel, which runs in the vertical timber guides in the shaft compartment. When the skip, which is loaded with about 5 tons of ore, reaches the top of the headgear, the top of the skip follows the guides on the tipping path and the payload discharges into the shaft bin. When the bridle begins to descend, the empty skip returns to the vertical, coming to rest against the buffer plates on the side of the channel of the bridle. The skip, weighing about 2½ tons returns to the vertical almost under gravity so that the striker plates on the skip impact with great force on the buffer plates on the channel. Figure 10 illustrates a portion of the bridle and indicates the location of the fracture on the steel channel in relation to the buffer plates. The position of the striker plates on the side of the skip is also shown.

Fracture of one side of the bridle occurred when a fully loaded skip was being hoisted, causing the broken bridle and the ore-skip to become jammed in the shaft compartment. Before the portion of the bridle attached to the rope could be hoisted up the shaft, the unbroken side had to be flame-cut near the bottom. The remaining portion of the bridle and the ore-skip were finally hoisted after about 3 weeks.

Visual examination of the part of the bridle that was hoisted first showed that the fracture had occurred on the steel channel of the bridle along a line 9ft (2700mm) from the top and just beneath the bottom edge of the shaft bracket of the man-cage. Photographs of the surface of the fracture are shown in Figs 11 and 12, which clearly indicate a fatigue fracture, the origin of which was at the top left-hand corner as shown in Fig. 11. The fatigue crack had propagated along the side and along the base of the channel. Final sudden fracture occurred after the fatigue crack had propagated as far as part of the other side of the channel (see Fig. 12) over an area estimated at 10 to 15 per cent of the original cross-section of the channel.

A noticeable feature of the channel at the fracture surface was that the thickness of the 'struck area' on the side of the channel had been markedly reduced.

Before an explanation is attempted of the mechanics associated with the fracture, reference can be made to the arrangement of the steel components involved. Buffer plates of 12in x 5in (300mm x 127mm) with four bolt holes are fixed to the side of the channel so as to coincide with the slightly curved striker plates bolted to the sides of the skip. As shown in Fig. 13, the buffer plates project about 35mm above the base of the channel. A modification was made to the design as shown in the original drawing of the bridle, in that a flanged plate 155mm x 8mm with a 37mm flange is bolted to the 12in (300mm) portion of the channel, so that the 37mm flange acts as a reinforcement for the buffer plate. Another view of these assembled components is given in Fig. 14. The top plate is the buffer plate, and there is clear evidence of the indentation on this plate produced by the repeated impacts of the striker plate on the skip. This indentation is most severe on the outer edge, but is distributed over a length of 217mm and over a width of 50mm on the channel and about 35mm on the flanged backing plate bolted to the channel. When the buffer plate and flanged backing plate were removed, it was evident that the transmission of the force of impact through the buffer plate had produced an indentation of 305mm x 50mm on the 3½in (89mm) side of the channel. This had occurred in spite of the additional reinforcement provided by the flange of the

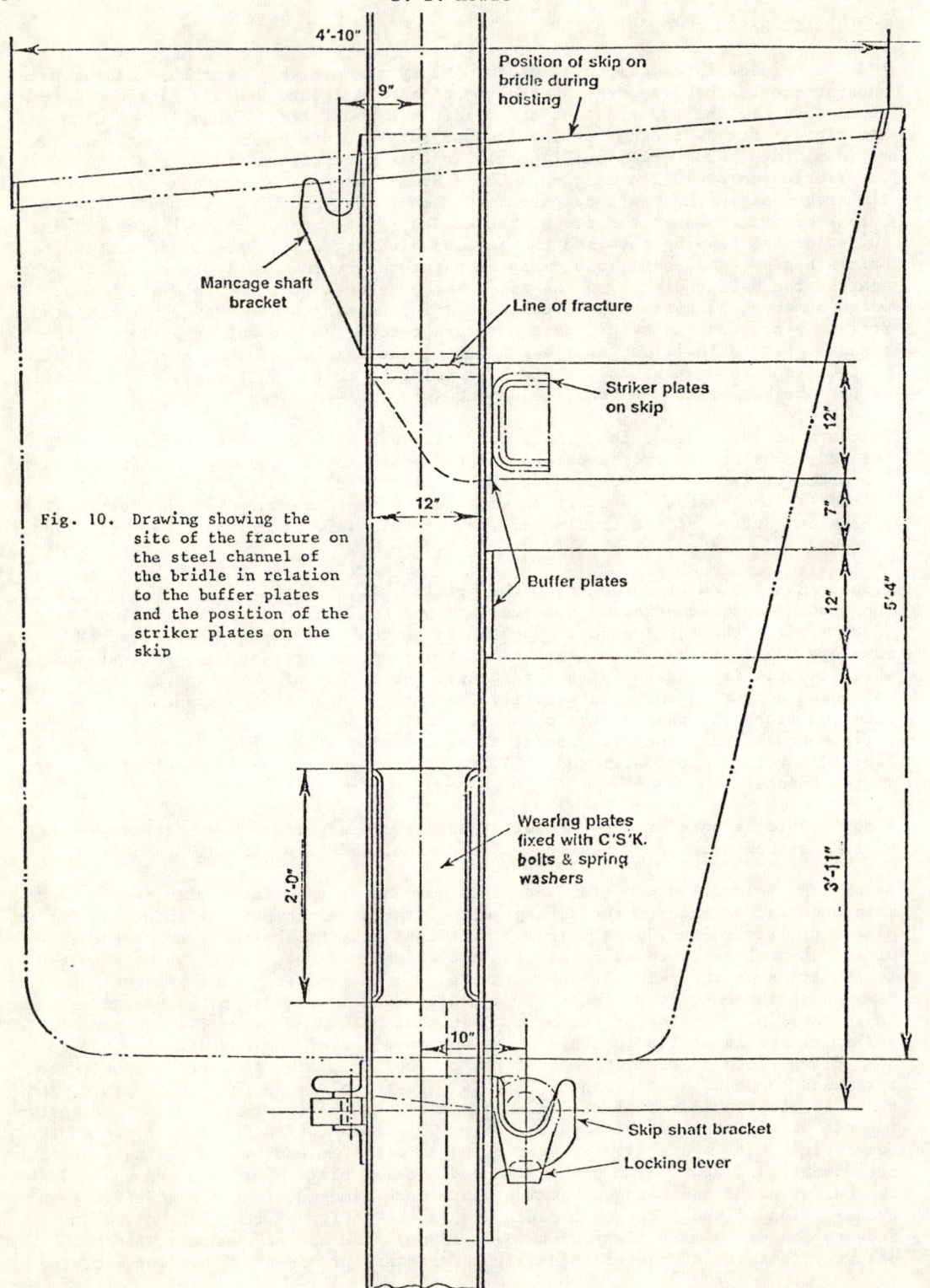

Fig. 10. Drawing showing the site of the fracture on the steel channel of the bridle in relation to the buffer plates and the position of the striker plates on the skip

Unusual Fractures in the Mining Industry

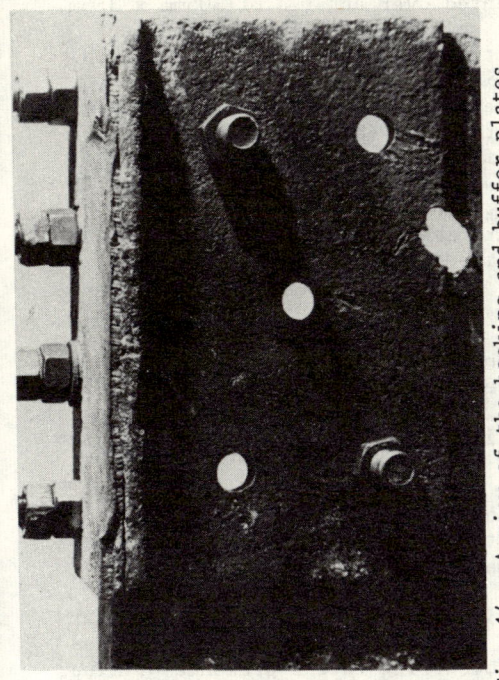

Fig. 14. A view of the backing and buffer plates showing severe indentation of the buffer plate caused by the impact of the striker plate on the skip

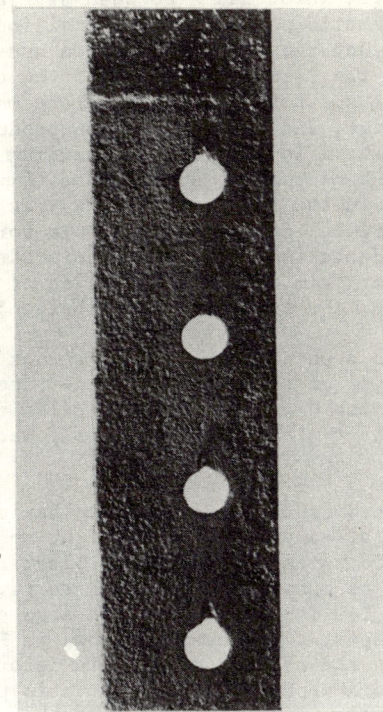

Fig. 16. A photograph of the 'struck side' of the channel. The fracture occurred at the left-hand side

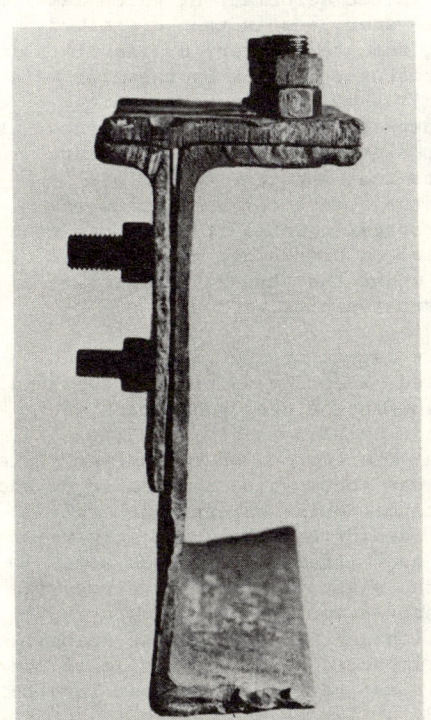

Fig. 13. The arrangement of the buffer plate and backing plate on the channel. Note the gouge mark and deformation of the buffer plate at the corner of the channel

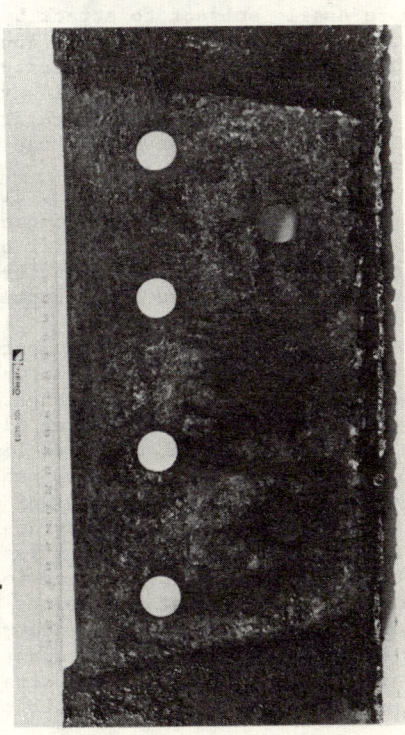

Fig. 15. The indentation on the side of the channel caused by the impact of the buffer plate. The photograph is of an unbroken bridle taken out of service

backing plate, which had also been severely indented over an area of 305mm x 35mm. As might be anticipated from the position of the row of bolt holes, the buffer plate could undergo a rocking motion under impact. Consequently, the indentation produced on the 3½in (89mm) side of the channel is deepest at the corner. A buffer plate and flanged backing plate were dismantled from another bridle taken out of service, and the indentation produced on the 3½in (89mm) side of the channel is shown in Fig. 15. The maximum depth of the indentation is to be found at the left-hand corner, and measures 6mm, corresponding to a reduction of over 30 per cent in the thickness of the side of the channel in that area. On the fractured bridle, it was not possible for accurate measurements to be made of the depth of indentation, since this coincided with the line of fracture. However, it will be clear from the left-hand side of Fig. 11 that a very decided indentation was produced, the estimated depth being 5 to 6mm.

Figure 16 is a photograph of the 'struck area' on the side of the channel, the fracture being on the left-hand side. The difference in the surface appearance of the side of the channel above and below the line of bolt holes clearly indicates that the buffer plate rocked slightly each time it was hit by the striker plate on the skip.

These repeated impacts must cause plastic deformation of the steel, and microscopic examination of samples taken at right angles to the 'struck-surface' area of the channel would be expected to show plastic deformation of the steel. However, examination of many samples failed to reveal any signs of plastic deformation except in a few isolated spots. No sign of plastic deformation was revealed right up to the impacted surface, as shown in Fig. 17. This anomaly can most readily be explained by the effects of plastic deformation occurring in the presence of mine water with a low pH value that, at this particular shaft, is sometimes as low as 2,5. The sequence of events must therefore be that, under impact, the surface layer of the steel in the 'struck area' suffers plastic deformation, which makes it much more easily subject to attack by the mine water, and is then dissolved. This cyclic process, repeated over a long period, resulted in very decided thinning of the side of the channel. The severity of the corrosive attack by the mine water was demonstrated in a rather unusual way. Figure 11 shows that numerous pits formed on the surface of the fracture at the corner at which the fatigue crack was initiated. A section was cut across the corner at right angles to the fracture surface. This section intersected a number of pits and, when the sample was examined microscopically, it was very clear that these were corrosion pits, as shown in Fig. 18. Obviously, the fatigue crack propagated until it was large enough to permit a slight opening and closing of the steel on both sides of the crack. The acidic mine water seeped into the crack and, under the strongly anodic conditions obtaining in that area, formed the corrosion pits.

As with most fatigue fractures, the most difficult aspect of the investigation is the determination of how the crack was initiated. When the section of the bridle containing the buffer plates was brought to the surface, a deep gouge mark was evident on the top edge of the buffer plate, as can be seen in Fig. 13. The formation of the gouge mark, which is 18mm wide at the top end of the buffer plate and extends in a taper form 21mm down the plate, was accompanied by bending of the buffer plate inwards just at the corner of the channel where the fatigue crack apparently initiated. There is no evidence as to when or how the gouge mark was formed, but the balance of probabilities is that the buffer plate struck some unexpected object when the bridle was moving up the shaft. This may have resulted in an impact blow on the outside corner of the channel sufficiently severe to initiate a crack in the thinned and corroded section of the steel in the channel. The crack started from the deepest corner of the impacted area on the side of the channel. Although the root radius at the corner is not severe, there must inevitably be some concentration of stresses that would facilitate the initiation of a crack.

The conclusion is that this fatigue fracture is due to a combination of severe impact stresses transmitted by metal-to-metal contact to the side of the channel, accompanied by severe thinning of the steel section due to corrosion by the mine water of the plastically deformed surface layers of steel. The initiation of the fatigue crack may have been caused by an impact blow on the buffer plate causing quite severe localized deformation of the buffer plate, and a severe blow on the corner of the channel.

The report of the investigation was circulated to resident engineers in the gold-mining industry with the request for comments and notes on any similar failures encountered. In the past, several bridles fractured in a similar fashion and, at one large gold mine, the bridle structures are now renewed at intervals of two years' operating life. On one mine the problem is regarded as being sufficiently serious to warrant complete redesign of the main frame of the bridle. Instead of rolled-steel channel being used, two-angle iron strips are riveted to a back plate, so that a crack developing in any of the component parts of the frame will not propagate into the adjacent parts and will be less likely to cause complete fracture of the whole side of the bridle. It is believed that stresses on the main frame of the bridle are aggravated by the difficulty with which accurate alignment of the timer guides is maintained in the shaft.

The suggestion has been put forward that the severe bumping on the side of the bridle when the skip returns to the vertical position after tipping could be alleviated by a change in the design of the tipping path. There would be decided advantages in a correctly designed transition curve leading from the vertical into the tipping curve, but the very limited space in the tipping area of the headgear makes this desirable change of design impracticable.

Two other suggestions have been advanced. Pieces of conveyor belting may be inserted between the buffer plates on the sides of the channel or between the striker plates on the side of the skip. (This suggestion has been adopted on the mine on which the fracture of the bridle occurred.) Alternatively, some form of shock absorber - of the conventional road-vehicle type - may be fitted to the side of the skip to absorb a proportion of the sudden force of impact.

FRACTURE OF A MANTLE IN A SHORT-HEAD CRUSHER

The short-head cone type of crusher is very extensively used in the gold-mining industry in this country. In this machine, the crushing head is carried on the main vertical shaft, the lower portion of which is held within the eccentric sleeve. The only positive motion imparted to the main shaft is a gyratory movement, no direct rotary movement being applied. There are two main wearing regions in the crusher: the mantle fixed to the head of the main shaft, which gyrates, and the stationary bowl liner rigidly fixed within the bowl. Both are made of austenitic manganese steel, which is practically the only abrasion-resistant steel that will tolerate the severe impacts occurring during the crushing of lumps of ore. The mantle casting is essentially a hollow truncated cone that fits over the tapered head of the main shaft and is locked in place by heavy nuts. A flat circulat plate is fitted to the top of the central shaft and moves with the shaft, so permitting 'choke feeding' of the crusher. The annular space between the mantle and the bowl liner flares outwards from the feed to the discharge, but the gap between the two castings becomes progressively smaller. Crushing of the ore takes place as the mantle follows a gyratory motion within the bowl. In most cases, the mantle wears away fairly rapidly, probably lasting for a throughput of only 35 000 to 45 000 tons of hard quartzitic rock. Fractures of the mantle are fairly rare, but one unusual fracture merits notice. As shown in Fig. 19, the mantle, on removal from the head, showed very marked zig-zag cracks oriented at about 45° to the centre line of the main shaft. It is known that, in the case of a fatigue failure due

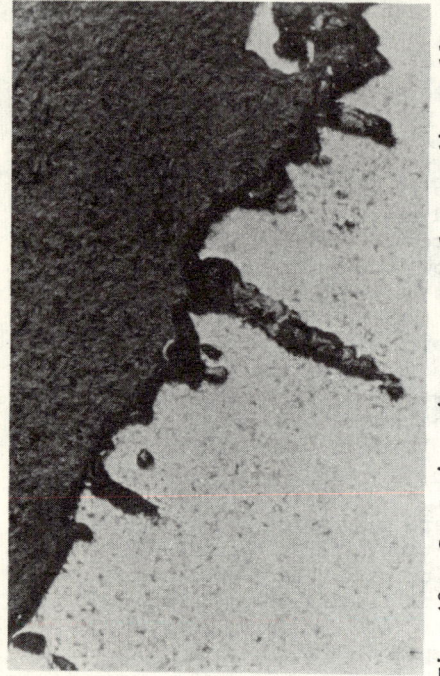

Fig. 17. A photomicrograph of the steel at right angles to the impacted surface of the channel. No evidence of plastic deformation can be noted. (Magnification 150X)

Fig. 18. Corrosion pits on a sample cut diagonally across, and at right angles to, the fracture surface at the top left-hand corner in Fig. 11. (Magnification 50X)

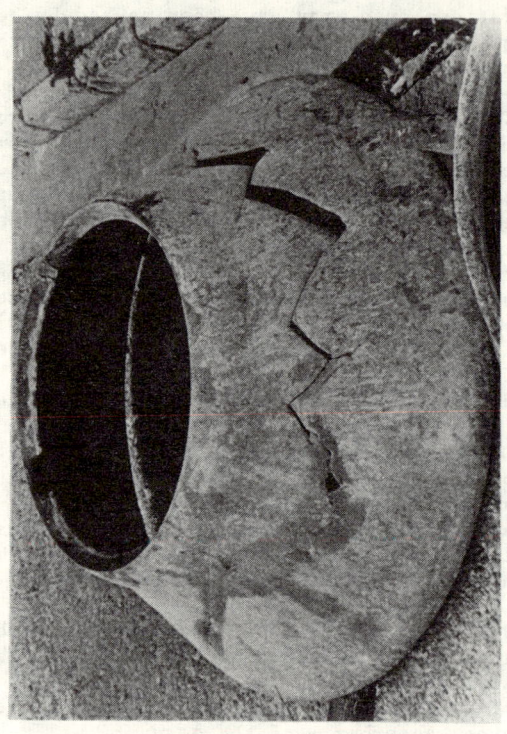

Fig. 19. Zig-zag cracks that developed in the mantle used in the 4 1/4 short-head crusher

to fluctuating or reversing torsion, a crack will develop on a helical place oriented at an angle of about 45° to the axis of rotation. This is due to the fact that a pure shear stress is equivalent to a bi-axial stress system of one compressive and one tensile principal stress, the magnitude of each being equal to the shear stress. A fatigue crack may develop normal to the direction of the maximum tensile stress. Only if the torque is of a reversing character will two cracks develop at right angles to one another and frequently intersect one another. However, the existence of a fatigue crack oriented at 45° to the axis of rotation indicates a unidirectional torque that is fluctuating. In the photograph in Fig. 19 it can be seen that this crack progressed in a zig-zag fashion round the mantle. The crack presumably originated in the area of the mantle shown at the right-hand side of the photograph, and the crack that formed in the early stages increased in width as the crack propagated from right to left. The presence of a fatigue crack oriented at 45° to the axis of rotation is normally taken to indicate a unidirectional torque, i.e. the type of torque that would be expected to prevail in the mantle of a short-head crusher. The zig-zag character of this fracture tends to suggest a decided element of a reversing character in the applied torque. During the normal crushing action, high compressive forces act radially on the mantle. The motion of the eccentric sleeve causes the mantle to follow a small circular path, the radius of which is the radius of gyration. During this time, tensile stresses must be set up over the outer periphery of the mantle. These conditions therefore satisfy the requirement of the simultaneous imposition of compressive and tensile stresses. The compressive stress will remain approximately constant during the whole period of operation, particularly if the crusher is choke fed, and it is difficult for any major reversal of the tensile stess to be visualized. Nevertheless, it appears that some reversal of torque must occur, or it could be expected that the fracture would progress right across the width of the mantle.

The cooperation and assistance of the engineering staff at the mines concerned is greatly appreciated. Thanks are due to Mr D.P. Enright and Miss Helen Paulsen for their work in the metallographic examination of the components.

Grateful acknowledgement is made to the National Institute for Metallurgy for permission to publish this paper.

SECTION 2

Understanding Fracture

THE STRENGTH AND FRACTURE OF TWO-PHASE ALLOYS—A COMPARISON OF TWO ALLOY SYSTEMS

J. Gurland

Engineering Division, Brown University, Providence, R.I., U.S.A.

ABSTRACT

The functional roles of the hard and soft constituents in the deformation and fracture of two-phase alloys are discussed on the basis of two commercially important alloy systems, namely spheroidized carbon steels and cemented carbides, WC-Co. A modified rule of mixtures provides a structural approach to the yield and flow strength. Consideration of the fracture toughness is attempted by means of a phenomenological modelling of the fracture process on the microscale. While there are large differences in properties between the two alloys, the deformation and fracture processes show broad similarities which are associated with the features of the interaction between constituents common to both alloys.

INTRODUCTION

One of the most effective means of strengthening metals is by the introduction of a hard second phase. Unfortunately, in general, the fracture properties (such as fracture toughness) change in opposition to the strength properties (such as yield strength, flow stress and hardness), and the design of alloys which are both strong and tough is a continuing challenge. This article presents several aspects of the problem dealing with the change of properties of two-phase alloys as a function of composition and microstructural variables, with particular emphasis on two commercial alloys at opposite ends of the two-phase composition range, namely spheroidized carbon steels and sintered tungsten carbide-cobalt alloys. The former are classified as dispersion hardened alloys, with particles of iron carbide dispersed in a matrix of ferrite. The latter belong to the class of cemented carbides, with a structure consisting of tungsten carbide embedded in a cobalt-rich binder phase. The natures of the two phases in each alloy are similar, both alloys contain a hard brittle constituent, the carbides, and a soft and ductile constituent, cobalt or iron. However, the relative amounts of the two constituents in each alloy are very different, since the steel contains of the order of 5-16 volume percent of hard particles, and the cemented carbides contain approximately 64-95 volume percent of tungsten carbide. Not surprisingly, the yield strength at room temperature and the effect of temperature on fracture toughness are also different in the two alloys, as presented schematically in Figure 1.

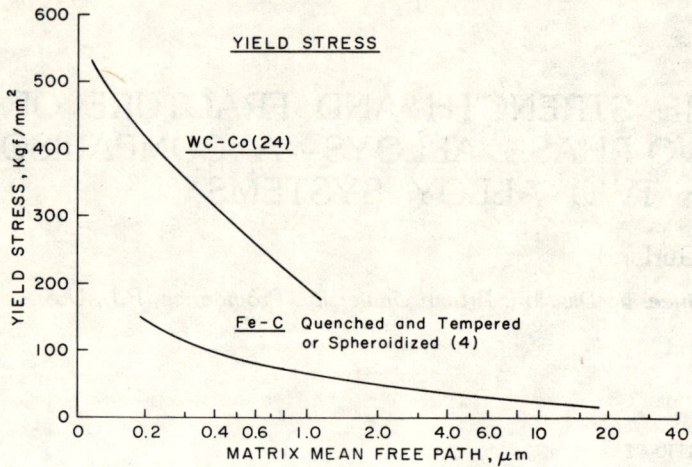

Figure 1a. Schematic comparison of the yield stress σ_y of spheroidized steels (Fe-C) and cemented carbides (WC-Co)

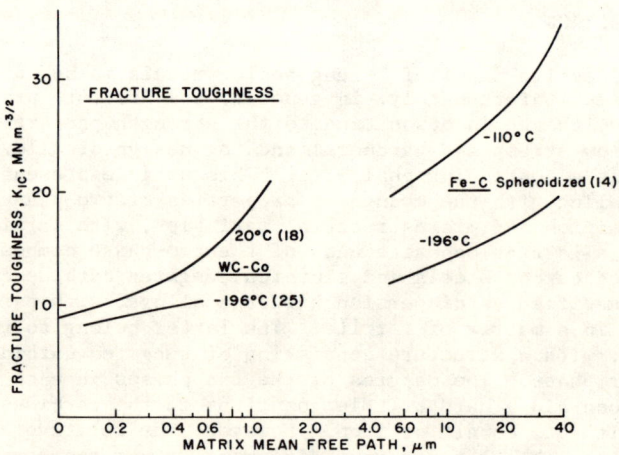

Figure 1b. Schematic comparison of the fracture toughness K_{IC}, of spheroidized steels (Fe-C) and cemented carbides (WC-Co).

It will be shown that in spite of the different functional roles of the constituents in each alloy, there are common features to the interaction between constituents in each alloy during deformation and fracture, and it is hoped that their consideration will lead eventually to generalizations applicable to the group of two-phase alloys as a whole.

1. Yield Strength and Flow Stress of Two-phase Alloys - A Structural Approach

A comprehensive theory of the strength of two-phase alloys, which would predict the yield strength and flow stress of multi-phase alloys from the properties of their constituent phases, has not yet been achieved. The relation between local and average stress and strain values in non-homogeneous microstructures is very complex and depends in detail on the spatial variation of such microstructural factors as the volume fraction, the morphology, the degree of dispersion or continuity and the orientation of the structural elements. However, recently published work of H. Fischmeister and B. Karlsson (1) showed that some useful generalizations may be obtained from the consideration of average deformation characteristics and, in particular, that in certain two-constituent steels and Al-Cu alloys with coarse microstructures the distribution of stress and strain among the constituents obeys the empirical mixture rules first used by Tamura, Tomata and Ozawa (2), namely

$$\bar{\sigma} = f_\alpha \sigma_\alpha + f_\beta \sigma_\beta \qquad (1)$$

$$\bar{\varepsilon} = f_\alpha \varepsilon_\alpha + f_\beta \varepsilon_\beta \qquad (2)$$

where $\bar{\sigma}$ and $\bar{\varepsilon}$ are, respectively, the flow stress and the strain of the aggregate, σ_α and σ_β are the average in-situ stresses, ε_α and ε_β are the average in-situ strains, and f_α and f_β are the volume fractions of the α and β constituents, respectively.

Equation (1) will be used here as a basis for the discussion of the strength properties of two-phase alloys, following, in part, a more complete presentation published elsewhere (3). When restated specifically for the yield strength of a two-phase alloy (with random or isotropic microstructure, i.e. the area fraction of a phase on any plane is statistically equal to the volume fraction of that phase), Equation (1) becomes

$$\bar{\sigma}_y = \sigma_\alpha f_\alpha + \sigma_\beta f_\beta \qquad (3)$$

where $\bar{\sigma}_y$ is the yield strength of the alloy.

The general problem is to calculate $\bar{\sigma}_y$ for any composition from known or estimated values of σ_α and σ_β, at yield of the alloy. In order to do this, certain simplifying assumptions are required which idealize the actual behavior of the constituent phases. The idealized composite alloy consists of a relatively soft and ductile matrix α and embedded coarse particles (1-500 μm) of a hard and ductile phase β. The matrix phase is assumed to be continuous and to deform homogeneously in conformity with the aggregate body. The hard β phase may or may not be continuous depending on the structure of the alloy. The in-situ yield strengths of the two phases are $\sigma_{\alpha y}$, and $\sigma_{\beta y}$, respectively, with $\sigma_{\beta y} \geq \sigma_{\alpha y}$. The magnitude of the in-situ yield strength of each phase is a function of the parameters of the microstructure.

The preceding concepts will now be applied to the two types of two-phase alloys under discussion.

a) Dispersed particle strengthening. The alloy is assumed to consist of equi-axed β particles dispersed in a matrix of α. Since the continuous α matrix is postulated to deform in conformity with the alloy as a whole, it must yield when the alloy yields (or vice-versa), and therefore at yield of the alloy $\sigma_\alpha = \sigma_{\alpha y}$. As in other treatments of dispersion hardening, the excess load-carrying ability of equiaxed hard particles is ignored, so that, as a first approximation the stress in the β particles is equal to the stress in the matrix and $\sigma_\beta = \sigma_{\alpha y}$, also. After substitution in equation (3), the yield strength of the alloy is given as

$$\bar{\sigma}_y = \sigma_{\alpha y} \tag{4}$$

i.e. the yield strength of the alloy is equal to the in-situ yield strength of the ductile matrix phase.

This simple and perhaps trivial statement hides a great deal of complexity, since in general, the yield strength of the matrix is a complicated function of its dislocation structure and internal boundary spacing. For instance, the yield strength of spheroidized steels increases markedly with decreasing iron carbide particle spacings between 10 and 0.2 μm (figure 1). However, a detailed study (4) showed that the yield stress of tempered and annealed spheroidized plain carbon steels is not directly determined by the particles themselves but is controlled by a subgrain boundary network, the mesh size of which is stabilized by the iron carbide particle spacing. The yield strength follows a Hall-Petch type equation,

$$\bar{\sigma}_y = 9.5 + 1.33\, \lambda_{\ell,p}^{-1/2} \quad \text{kgf/mm}^2 \tag{5}$$

where $\lambda_{\ell,p}$ is the spacing of the subgrain boundaries resulting from the heat treatment. In steels in which the subgrain boundaries were eliminated by thermal cycling about the transformation temperature A, the yield stress was found to be predominantly controlled by the ferrite grain size λ_g via the relation

$$\bar{\sigma}_y = 12.4 + 1.87\, \lambda_g^{-1/2} \quad \text{kgf/mm}^2 \tag{6}$$

and the intraboundary particles contributed only a small strain hardening term which increased the value of the friction stress (12.4 kgf/mm^2) over that associated with grain boundary strengthening alone (8.8 kgf/mm^2). The Hall-Petch relations of equations (5) and (6) are compared with that for grain boundary strengthening in figure 2.

The effect of the particles increases rapidly with strain due to the buildup of dislocation density and internal stresses associated with the incompatibility between elastic particles and plastically deforming matrix (5,6,7). As an example, figure 3 shows the relative magnitudes of the components of the hardening stress as a function of strain, according to one theory (6). Components $\Delta\sigma_i$, due to internal stress, and $\Delta\sigma_{sg}$, due to geometrical dislocation density, are associated with the particles. Component $\Delta\sigma_s$ is due to the strain hardening of the matrix. The stress on the particles is limited by the fracture strength of the particles or interfaces.

b) Matrix-reinforced hard-particle skeleton. Increasing the relative amount of the hard phase in a two-phase alloy system leads, in general, to clustering of the particles and eventually to their agglomeration into a continuous skeleton. By analogy with fiber reinforcement, a larger share of the load is transferred

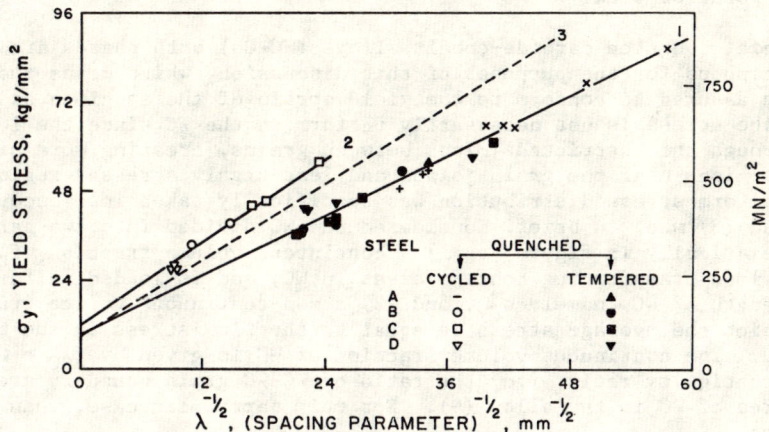

Fig. 2. Yield stress σ_y versus spacing parameter λ. Line refers to equation 5, line 2 refers to equation 6, and line 3 is the Hall-Petch plot for grain-size strengthening in mild steels. (After ref. 4). [Reproduced by kind permission of the Metallurgical Soc. of AIME and Am. Soc. for Metals].

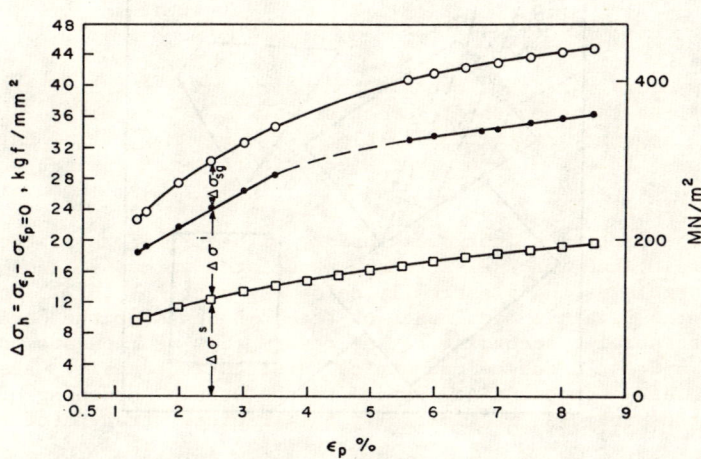

Fig. 3. Magnitudes of the three components of the hardening stress for a spheroidized steel (After ref. 6). [Reproduced by kind permission of Pergamon Press].

to the hard phase as it becomes more continuous, by the action of the matrix shear stress on the particle-matrix interface (3). The stress in the hard phase is limited by the yield strength or fracture strength of that phase.

The yield strength and deformation of the type of alloy with a continuous hard phase can also be considered on the basis of the concepts underlying equation 3, and the strength and hardness of sintered WC-Co alloys will now be discussed from that point of view.

In commercial tungsten carbide-cobalt alloys (WC-Co) both phases are considered to be continuous for the purposes of this discussion, which means that both phases are assumed to conform to the yield strain of the specimen as a whole. However, the stress is not necessarily uniform in the WC since the load is transmitted through the restricted joints between grains, creating more highly stressed regions near the grain joints and less highly stressed regions elsewhere. The non-uniform stress distribution was specifically taken into account by Lee and Gurland (8) who, in brief, considered the WC divided into two parts as indicated schematically in figure 4: 1) a continuous volume fraction, $f_{\beta C}$, which links the load transferring contact areas in WC, and is loaded uniformly to the yield strength of WC, namely $\sigma_{\beta y}$, and 2) a non-continuous volume fraction $f_\beta - f_{\beta C}$, in which the average stress is equal to the flow stress in the binder phase, namely $\sigma_{\alpha y}$. The continuous volume fraction of WC is given by $f_{\beta C} = C f_\beta$, where C is the contiguity ratio, i.e. the ratio of WC-WC grain boundary area and total surface area of WC in the alloy (9). For this particular case, then, equation (3) becomes

$$\sigma_y = \sigma_{\alpha y} f_\alpha + \sigma_\beta f_\beta \qquad (7)$$

$$\text{where} \quad \sigma_\beta f_\beta = \sigma_{\alpha y} (f_\beta - f_{\beta C}) + \sigma_{\beta y} f_{\beta C}$$

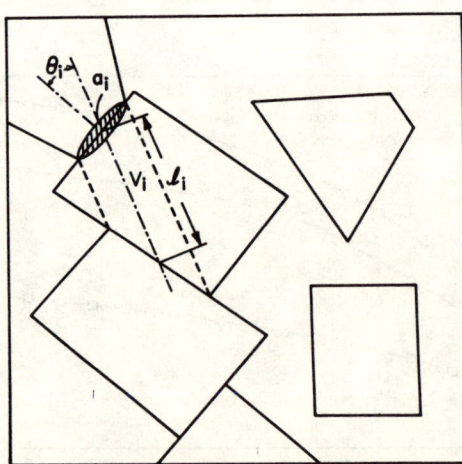

Fig. 4. Structure model for the calculation of the continuous volume in WC-Co alloys (After ref. 3). [Reproduced by kind permission of Elsevier Sequoia SA.].

and, after substitution,

$$\sigma_y = \sigma_{\alpha y}(1 - Cf_\beta) + \sigma_{\beta y} Cf_\beta \qquad (8)$$

Lee and Gurland (8) applied their model to the hardness of WC-Co. The experimentally determined change of hardness with composition is shown in figure 5. For hardness, the equivalent equation is

$$\bar{H} = H_\alpha(1 - Cf_\beta) + H_\beta Cf_\beta \qquad (9)$$

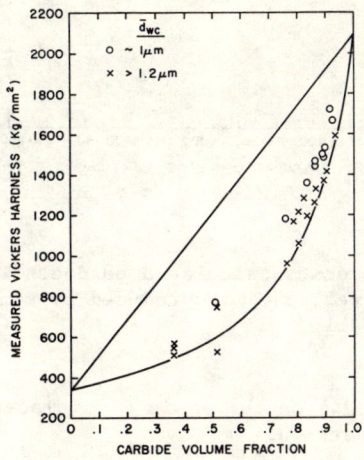

Fig. 5. Hardness versus carbide volume fraction in WC-Co (After ref. 3). [Reproduced by kind permission of Elsevier Sequoia SA.].

where $\bar{H}$, H_α and H_β are the hardnesses, respectively, of the aggregate, the cobalt-rich binder phase and the continuous portion of the WC phase. From limited available data, the in-situ hardness of the matrix follows the equation

$$H_\alpha = 304 + 12.7\, \lambda_\alpha^{-1/2} \qquad \text{kgf/mm}^2 \qquad (10)$$

and for the carbide phase, again from limited data,

$$H_\beta = 1382 + 23.1\, \bar{\ell}_\beta^{-1/2} \qquad \text{kgf/mm}^2 \qquad (11)$$

where λ_α is the mean free path of the binder and $\bar{\ell}_\beta$ is the mean intercept grain size of the carbide. As shown in figure 6, good agreement was found between the measured hardness of the aggregate and the hardness values calculated from Equation 9, 10 and 11, for this alloy system. It is of interest to note that the hardness of the matrix in this cemented carbide follows a Hall-Petch type relation (eq. 10) with respect to mean free path. This may indicate that the structural dependence of the flow stress or hardness of the binder in WC-Co is governed by similar strain hardening mechanisms as the structural hardening of the matrix of the spheroidized steel (eqs. 5 and 6). A conclusion of this kind

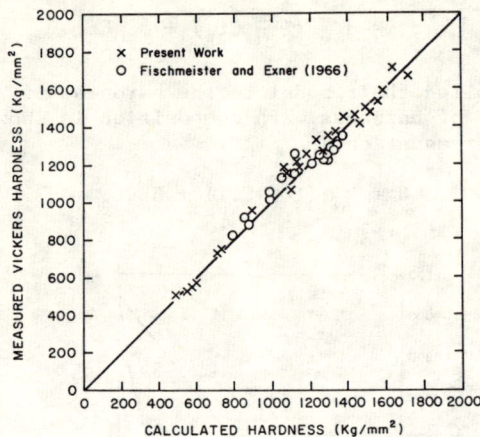

Fig. 6. Measured hardness versus calculated hardness in WC-Co, according to equation 4 (After ref. 3). [Reproduced by kind permission of Elsevier Sequoia SA.].

had been reached by Doi et al.(24) on the basis of dislocation density determination in WC-Co as a function of strain.

2. Fracture and Fracture Toughness of Two-phase Alloys - A Phenomenological Approach.

A rule of mixtures applies only to certain fracture properties of two-phase alloys in which the failure criterion is directly related to the stress distribution among the phases and involves large regions of the specimen. Examples are the ultimate tensile strength of dual phase steels (10) and that of fiber reinforced metals (11). For other fracture properties, such as the fracture toughness, which involve deformation and fracture processes in very localized regions, a comprehensive theoretical framework has not yet been formulated for a general relationship between fracture strength and microstructure. The best one can do, perhaps, at the present stage of development of the science, is to examine the microscopic fracture processes of various alloy systems for similarities, and consider whether these can be developed into predictive relations between microstructural parameters and macroscopic measurements of fracture toughness. In this spirit, the fracture of the two alloy systems, spheroidized carbon steels and sintered tungsten-carbide, will now be considered.

It is generally agreed that the fracture of alloys with hard inclusions starts with the cracking of the inclusions or their interfaces (12,13,14). The failure of inclusions and interfaces is associated with the accumulation of stress and strain in the ductile matrix during strain hardening. The particle or interface stresses increase either because of direct load transfer across the particle-matrix interface or due to the indirect effects of the increasing incompatibility between elastic particle and plastically deforming matrix, until particle fracture or interface decohesion occurs. However, the initiation of cracks or voids at the hard particles is not generally the critical event (although it is often a necessary pre-condition of fracture), but it is the unstable growth of these cracks into the surrounding microstructure which produces the failure of the specimen. The specimen fracture is brittle if the crack growth is by cleavage. The fracture is ductile, if particle or interface cracking is followed by void growth and void coalescence.

The conventional measures of fracture toughness in a thick, deeply-cracked test piece subjected to mode I loading are K_{IC}, the critical stress intensity factor, and G_{IC}, the critical energy release rate. A number of failure criteria have been developed on the basis of models of the fracture events at the crack tip. For instance, for fracture propagation entirely by cleavage, Ritchie, Knott and Rice (15) used the attainment of a critical stress level, σ_f^* over a minimum size scale, X_o, as a local criterion for unstable cleavage fracture. Under brittle conditions, such as low temperature, high strain rate or high concentration of the brittle constituent, when σ_f^* is of magnitude comparable to or smaller than the yield stress, this criterion reduces to

$$K_{IC} = \sigma_f^* \sqrt{2\pi X_o} \tag{12}$$

The authors (RKR) found good agreement between theory and experiment for a high nitrogen mild steel using X_o = 2 grain diameters and a macroscopically measured fracture stress.

With increasing ductility, at higher temperatures for instance, the microscopic mode of fracture initiation changes from cleavage to ductile tearing. An approximate failure criterion was stated by Rice and Johnson (16), namely that crack extension proceeds when the heavily deformed region in front of the crack tip is comparable in dimension to the width of unbroken ductile ligaments in the microstructure, at the onset of ductile rupture. Since the severely deformed region is approximately equal to the critical crack tip opening displacement δ_c, where $\delta_c = 0.5\, K_{IC}^2/\sigma_y E$, and furthermore δ_c is of the same order as the interparticle spacing Δ, one obtains

$$K_{IC} = [2E\, \Delta\, \sigma_y]^{1/2} \tag{13}$$

where σ_y and E are, respectively, the yield stress and Young's modulus of the metal. The experimental data for a number of steels and aluminum alloys have been found to agree with equation 13 (17).

The fracture of spheroidized steels at low temperatures and that of cemented carbides at room temperature will now be examined in the context of the Ritchie, Knott and Rice, and Rice and Johnson criteria. It should be noted here that these two theories are among a number of proposed relations between the fracture toughness parameter and unidirectional tensile properties and microstructural variables which have met with varying degrees of success and are not necessarily generally accepted (13).

a) <u>Spheroidized carbon steel</u>. The observation of the fracture process in notched and fatigue cracked specimens (0.13 - 1.46%C) by Rawal and Gurland (14) revealed two different modes of fracture in the region near the crack tip, at low temperatures. At very low temperatures (-196° to -150°C) the fracture is almost completely brittle, with cleavage crack initiation sites at cementite particles often observed. At somewhat higher temperatures (-150° to -110°C) a fibrous (ductile) fracture region is formed between the fatigue crack tip and the final brittle fracture. Examination of crack tip blunting showed broken particles and voids. At even higher temperatures the specimens fail to meet the plain strain conditions for fracture toughness testing due to the extension of the zone of extensive plastic deformation.

All indications strongly support the conclusion that both the brittle and the ductile fracture processes start with particle cracking or interface decohesion. In effect, the comparison of the stress required for particle fracture or decohesion (1400 - 1670 MN/mm^2) with the estimated maximum stress intensity (~3 σ_y) ahead of the crack shows that the stress conditions for particle or interface cracking exist in these specimens within a few grains or particle spacings from the crack tip.

The critical condition for fracture is the condition for crack growth or propagation from the particles into the ferrite matrix. The RKR model was applied to the brittle initiation case. The critical tensile stress σ_f^* (eq. 12) was equated to the fracture propagation stress of a crack nucleated within a grain-boundary carbide particle as given by Smith (12).

$$\sigma_f^* = \left(\frac{4E\gamma}{\pi(1-\eta^2)\bar{d}} \right)^{1/2} \quad (14)$$

where γ is the effective surface energy of ferrite, η is the Poisson's ratio and $\bar{d}$ is the mean carbide thickness. The size scale X_o (eq. 12), i.e. distance within which the principal stress at the crack tip equals or exceeds σ_f^*, was empirically established to be equal to 1.3 grain diameters. The values of K_{IC} calculated by the RKR procedure are compared with the observed values of K_{IC} in figure 7, they show a fair degree of agreement for all specimens with observed fracture initiation by cleavage.

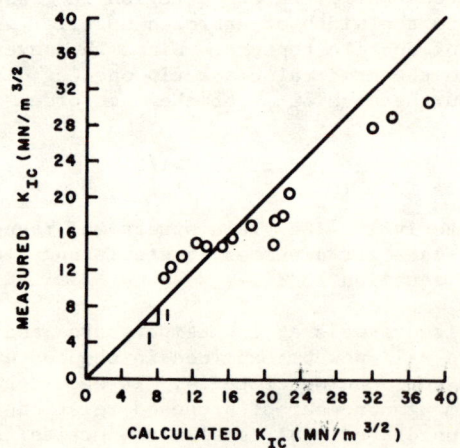

Fig. 7. Measured K_{IC} versus calculated K_{IC} in spheroidized steels, for small scale yielding conditions at X_o = 1.3 λ_y (After ref. 14). [Reproduced by kind permission of The Metallurgical Soc. of AIME and Am. Soc. for Metals].

At higher temperatures or reduced carbon contents, the measured toughness values deviate appreciably from the values calculated by the RKR criterion. The specimens for which the cleavage fracture stress appreciably exceeds the available stress intensification (experimentally, if $\sigma_f^*/\sigma_y > 3.86$) show ductile rupture surface characteristics in the initiation zone. The stress level is sufficient to break the particles or interfaces, but is insufficient to cause cleavage propagation. The fracture here proceeds by void growth and coalescence over a distance more or less equal to the nearest neighbor interparticle distance, Δ, in Equation 13. The nearest neighbor particle distance for randomly distributed uniform spherical particles is $\Delta_2 = 0.36\, df^{-1/2}$ on a plane, and $\Delta_3 = 0.45\, df^{-1/3}$ in a volume (9). Hahn and Rosenfield (17) used the volume spacing Δ_3 in applying Equation 13, but, as shown in figure 8 we found better agreement between experiment and theory when using the planar spacing Δ_2, for the specimens with ductile initiation (solid line). It is surprising to note that the results are insensitive to the average particle size, also that equation 13 with volume spacing Δ_3 agrees with the results of cleavage initiated fracture (broken line). These aspects of the model are further discussed in reference 14.

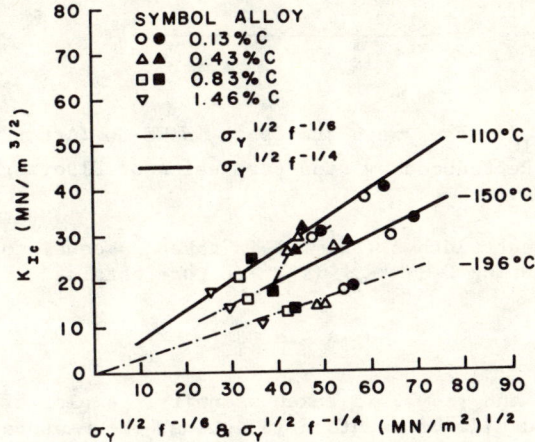

Fig. 8. K_{IC} versus parameters related to nearest neighbor distances of particles in Fe-C, equation 13. (After ref. 14). [Reproduced by kind permission of The Metallurgical Soc. of AIME and Am. Soc. for Metals].

b) <u>Cemented carbide alloy WC-Co</u>. The fracture toughness of sintered WC-Co alloys (5.1 - 36.9 volume percent Co) was measured at room temperature on single-edge notched beam (SENB) specimens precracked by electron discharge machining (18). The fracture toughness increases with both binder content and WC particle size, and it is a function primarily of the matrix mean free path, as shown in figures 1 and 9. In general agreement with other published observations (19,20,21), it was found that the fracture path involves three main modes of rupture: i) interfacial decohesion along WC-WC and/or WC-Co interfaces, ii) cleavage of large WC grains, and iii) ductile rupture of binder ligaments.

Although there are several published fracture theories for cemented carbides based on dislocation mechanisms (19) and fracture mechanics (21,22,23), the present approach attempts to test a particular phenomenological criterion against the measured results. The observed fracture process, namely interfacial decohesion or WC cleavage followed by extensive plastic deformation and rupture of the binder ligaments, is compatible with the ductile fracture criterion of Rice and

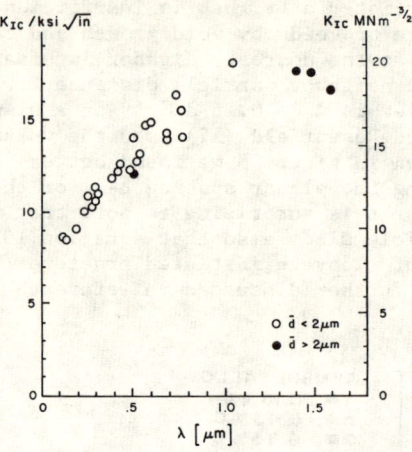

Fig. 9. K_{IC} versus mean free path in WC-Co (After ref. 18). [Reproduced by kind permission of Elsevier Sequoia SA.].

Johnson. If the ligament width Δ (Eq. 13) is taken as equal to the binder phase mean free path λ, and using $G_{IC} = K_{IC}^2 (1-\nu^2)/E$, one obtains

$$\sigma_y \lambda \approx \alpha \, G_{IC} \qquad (15)$$

where α is a constant, and $(1-\nu^2)$ is taken as unity. A plot of the calculated values of the left hand side of equation 15 against the measured critical strain energy release rate G is shown in figure 10. The linearity is fairly good, although the value of the slope (0.24) is appreciably less than the theoretically

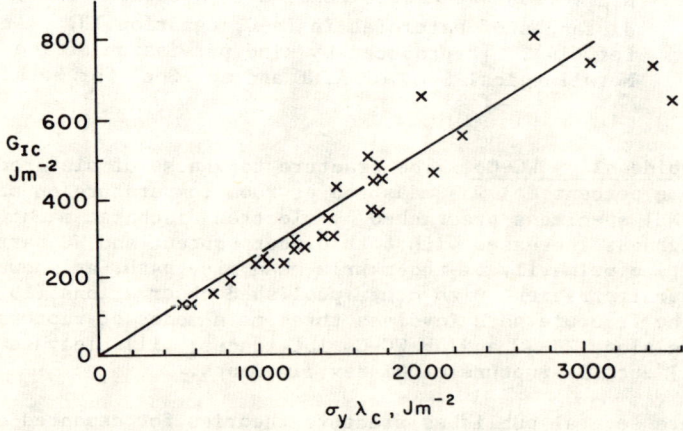

Fig. 10. Critical strain energy release rate G_{IC} versus $\sigma_y \lambda$ (After ref. 18). [Reproduced by kind permission of Elsevier Sequoia SA.].

expected value (between 1 and 2). This discrepancy can be accounted for partially by correcting for a possible overestimate of the ductile ligament thickness and an overestimate of the in-situ yield strength of the binder phase in the calculations of Pickens and Gurland (18).

COMMENTS AND SUMMARY

The resistance to plastic deformation of the two-phase alloys follows a modified rule of mixtures, and, in principle, the yield strength and flow stress of the alloys can be accounted for by the summation of the separate contributions of each phase. In practice, the properties of each phase are subject to the effects of interaction between the phases, and the interaction effects are not always well known. The following general statements apply to the two alloys under discussion.

In dispersion hardened or particle strengthened alloys, the flow strength generally increases as an inverse function of the spacing of structural barriers to the propagation of slip, as, for example, grain boundaries, subgrain boundaries and hard particles. In the dispersion-hardened spheroidized carbon steels, the direct load-carrying contribution of the hard particles is negligible, but they exert a strong indirect effect on the yield and flow strength of the matrix, due to the interaction of the particles with subgrain boundaries and with the dislocation structure of the matrix during plastic deformation.

The strength of a two-phase alloy with an appreciable amount of the hard constituent does not depend only on the strength of the ductile matrix, but is also affected by the load carried by the hard phase. The fraction of the total load carried by the hard phase increases generally with the amount of that phase, as it becomes more continuous. The mechanism is one of load transfer between ductile matrix and hard inclusion. In the case of WC-Co alloys of commercial composition, the carbide phase is continuous, for all practical purposes, and it must therefore deform in compatibility with the alloy as a whole and carry the major portion of the load. At yield, the strength of the alloy is determined by both the in-situ yield strength of the binder phase and that of the continuous portion of the carbide phase. The binder phase is subject to similar indirect strengthening processes as the matrix in the spheroidized steel.

As the plastic deformation proceeds, the increasing plastic incompatibility between the soft and hard constituents causes cracks to form as a relaxation process. The major stages of the fracture processes in spheroidized steels at low temperatures and in WC-Co alloys at room temperature are similar, in spite of their different microstructures. In both alloys, the initial cracking takes place in the carbide or at the carbide-matrix interface, and is followed by crack growth through the matrix. The ductile fracture process in both alloys may correspond to the fracture model underlying the Rice-Johnson criterion, although the plastic deformation during fracture occurs on very different size scales in the two microstructures, and the fracture toughness values are comparable only within a narrow temperature range. At first glance it may seem surprising that the rather brittle cemented carbide and the more ductile steel are compatible with the same ductile fracture model. However, it should be realized that the materials' fracture behavior was compared only under conditions of small scale yielding. In the case of the spheroidized steel, the small scale yielding condition is obtained only at very low temperatures. In cemented carbides, it holds over a wide range of temperatures (including room temperature) because the extensive plastic deformation is restricted to the ductile binder phase which constitutes only 5 to 30 volume percent of the alloy. In the case of the cemented carbides, the localized plastic deformation of the binder phase layers at the crack tip contributes the dissipative work, which accounts for most of the fracture surface energy without greatly affecting the deformation of the bulk of the material which deforms elastically. In both steels and cemented carbides, the linear elastic fracture mechanics are applicable only if the plastic yielding is highly restricted. The conclusions about the fracture process are very tentative because the ductile fracture criteria are not yet sufficiently developed to account quantitatively and with confidence for the fracture behavior of a group of alloys over a range of microstructural parameters.

In summary, the respective role of each constituent is succinctly stated as follows: the hard carbide contributes the strength and hardness; the soft metal provides the toughness. The expected functions of the minor constituents of each type of alloy are aptly described by current terminology, namely "dispersion-hardened" and "cemented" alloys. The hard particles in steels serve to strengthen the matrix in which they are dispersed. The main function of the binder phase in cemented carbides is to cement the aggregate, i.e. to hold together and to transmit load. The matrix or binder phase has the additional important function to maintain the carbide in a desirable state of dispersion such as to maximize the resistance to fracture propagation.

ACKNOWLEDGEMENT

The author's work cited in this article was supported by the U.S. Department of Energy and by the National Science Foundation.

REFERENCES

1. H. Fischmeister and B. Karlsson, Z. Metallkunde, 65 (1977) 311.
2. I. Tamura, Y. Tomota and H. Ozawa, Proc. 3d. Int. Conf. Strength of Metals and Alloys, Cambridge; Inst. of Metals and Iron and Steel Inst., London, Great Britain, (1973) Vol. 1, p. 611.
3. J. Gurland, Materials Science and Engineering, (1979) to be published.
4. L. Anand and J. Gurland, Met. Trans. 7A (1976) 191.
5. L. M. Brown and W. M. Stobbs, Philos. Mag., 23 (1971) 1201.
6. L. Anand and J. Gurland, Acta Met., 24 (1976) 901.
7. Y. W. Chang and R. J. Asaro, Metal Science, 12 (1978) 277.
8. H. C. Lee and J. Gurland, Mat. Sci. and Engnrg., 33 (1978) 125.
9. E. E. Underwood, Quantitative Stereology, Addison-Wesley, Reading, MA (1969).
10. R. G. Davies, Met. Trans. 9 (1978) 671.
11. A. Kelly, Strong Solids, Clarendon Press, Oxford (1966) 189.
12. E. Smith, Int. J. Fract. Mech., 4 (1968) 131.
13. G. G. Garrett and J. F. Knott, Proc. 2nd Int. Conf. on Mechanical Behavior of Materials, Boston 1976, Am. Soc. for Metals, Metals Park, Ohio (1978), Spec. Vol. p. 488.
14. S. P. Rawal and J. Gurland, Met. Trans. 8A (1977) 691.
15. R. O. Ritchie, J. F. Knott and J. R. Rice, J. Mech. Phys. Solids, 21 (1973) 395.
16. J. R. Rice and M. A. Johnson, Inelastic Behavior of Solids, M. F. Kanninen et al. eds. McGraw-Hill, New York, NY (1970) p. 641.
17. G. T. Hahn and A. R. Rosenfield, ASTM-STP 432 (1968) p. 5.
18. J. R. Pickens and J. Gurland, Materials Science and Engnrg., 33 (1978) 135.
19. F. R. N. Nabarro and S. Bartolucci Luyckx, Trans. Jap. Inst. Met. 9 (supplement), (1968) 610.
20. J. L. Chermant and F. Osterstock, J. Mat. Sci., 11 (1976) 1939.
21. M. J. Murray and C. M. Perrott, Proc. Inst. Congr. Hard Materials Tool Technology, R. Komanduri, eds., Carnegie-Mellon Univer., Pittsburgh, PA (1976)

p. 110.
22. R. C. Lueth, <u>Fracture Mechanics of Ceramics</u>, Vol. 2, ed. by R. C. Bradt et al. Plenum Pub., New York, NY (1974) p. 791.
23. T. Johanesson, <u>4th European Symp. for Powder Met</u>., Grenoble (1979).
24. H. Doi, Y. Fujiwara and K. Miyaki, Trans. Metall. Soc. AIME, <u>245</u> (1969) 1457.
25. M. Nakamura and J. Gurland, unpublished data.

FAILURE BY FATIGUE

G. G. Garrett

*Department of Metallurgy, University of the Witwatersrand,
Johannesburg, R.S.A.*

ABSTRACT

The lifetime of many components or structures under service conditions is often limited by time - dependent processes such as corrosion, wear or fatigue. Since failure by these processes can result in significant economic loss and also perhaps a safety hazard to the consumer, much effort has been aimed towards understanding these phenomena, as well as towards developing improved design procedures to guard against their occurrence. In the field of fatigue in particular, considerable advances have been made in recent years in both the understanding of the mechanisms of the process, and in its quantitative treatment, especially in the case of fatigue crack growth. This paper reviews the current status of fatigue, from the mechanistic through to the design viewpoints, with particular regard to the analytical approaches currently available for describing and predicting fatigue behaviour, and to the various factors, both operational and environmental, which affect fatigue performance.

INTRODUCTION

In this paper we are concerned with the problem of fatigue failure and those steps which may be taken to minimise its occurrence. Fatigue failure, of course, results from the repeated application of stress and whilst the engineer may regard fatigue as a particularly insidious form of mechanical fracture, in reality it is a cumulative damage phenomenon: structural deterioration may initiate during the first few stress cycles, but visible signs of crack extension leading to failure may be evident only towards the end of the fatigue life.

This type of failure process is observed in both crystalline and non-crystalline materials, with metals and polymers being of principal concern. Fatigue involves both the initiation and propagation of cracks, and irreversible plastic deformation plays a key role in each stage; it is hardly surprising, therefore, that a completely elastic, i.e. brittle, material is not susceptible to fatigue failure. Regrettably, however, such materials are rarely useful in engineering applications. Thus, although ductility is typically regarded as one of the more important characteristics of metals and alloys, under cyclic loading it leads to failure: yet another example of a foundation principle of materials engineering - The Law of Maximum Cussedness (also fondly known as Sod's Law)!

In considering the various components which are subject to fatigue, a division of these into two categories is sometimes made: firstly, those which comprise the primary load bearing members of structures such as aircraft, cars and trucks, bridges, pressure vessels, etc.; and secondly the remaining bulk of engineering products ranging from can openers to turbine blades. (This latter category can be broadened to include certain types of wear failures which in fact result from 'fretting'.) Of these two categories, generally more attention has been given to structural failures, where it has been estimated that 95% of all failures result from fatigue (Freudenthal, 1970). This high incidence is a result of the fact that very few structures are subject to the static loading which is assumed in their design. Although statistics dealing with the frequency of occurrence of both structural and non-structural fatigue failures, as well as with estimates of the economic factors associated with such failures are very hard to come by, the integrated economic loss and drain on natural resources is indeed very significant.

Fatigue has now been the subject of research investigations for well over a hundred years and despite the very significant progress which has been made, failures continue to occur. This situation results from a number of factors. Firstly, experimental fatigue data are inherently susceptible to wide statistical scatter; consequently, laboratory behaviour cannot be reliably translated into service predictions. Furthermore, it is often difficult to model accurately the precise operational conditions to which a system is exposed over its entire design life. As a result, many designs must be subjected to expensive, time-consuming fatigue-simulation tests (Fig. 1) the results of which are usually adjusted by conservative safety factors to allow for undefined statistical fluctuations in fatigue behaviour and thereby minimise any possibility of failure by fatigue.

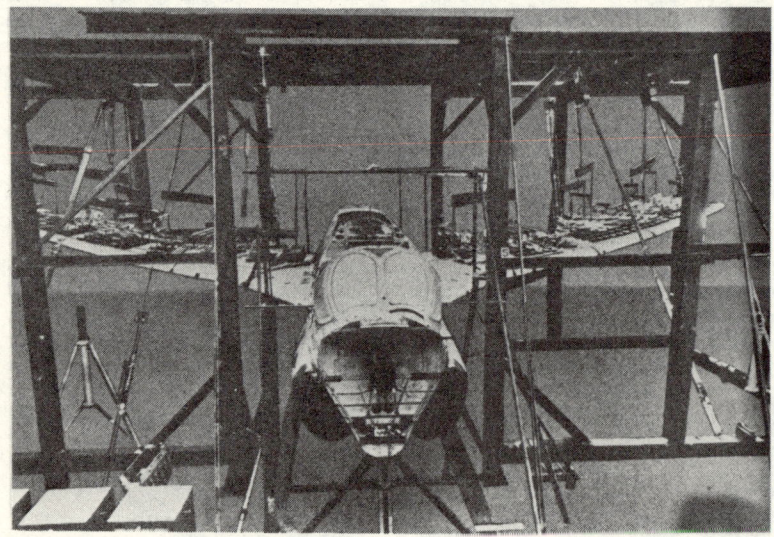

Fig.1: Full scale test of U S navy A-6 air frame (after Sorkin et al, 1973)

Much of the fatigue problem, however, is simply due to poor communication. Many designers do not know how to deal properly with fatigue and do not treat it as a matter of concern until product failures start to occur. Thus, as an educational challenge there is still much to be done and, as will be seen later in the paper, having first gained an appreciation of the factors which affect fatigue life, there are several approaches to design that go a long way to minimising fatigue problems in engineering service.

Fig.2: Typical fatigue fracture surface. Fracture originated at the point on the edge of the sample from which the clamshell markings emanate (courtesy F P A Robinson)

Fig.3: Formation of surface 'extrusions' and 'intrusions' in a sample of Cu-Al alloy tested in fatigue (courtesy D H Avery)

UNDERSTANDING FATIGUE

Fatigue fractures, of the sort shown in Fig. 2, often have very similar characteristics. Generally, fatigue is a surface sensitive phenomenon. The initiation of a fatigue fracture in service is frequently associated with design/manufacturing detail such as abrupt changes in section, high stress interfaces such as keyways or interference fits, or surface defects such as inclusions, scoring or coarse machining marks. In plain, polished specimens cracks initiate through micro-plastic deformation processes producing surface 'intrusions' and 'extrusions', Fig. 3, by means of 'slip' on planes of atoms corresponding with the maximum shear stress - 45° to the applied stress for uniaxial loading - in the specimen surface. For 'high-cycle fatigue', where peak stresses are in the elastic range and the number of cycles required to cause failure is in excess of 10^5, the nucleation of a microcrack in plain specimens may constitute ~80-90% of the total fatigue life; in 'low-cycle fatigue', where the stresses are high enough to cause macroscopic plastic deformation, fatigue life is correspondingly reduced (usually below 10^5 cycles), and the initiation stage may represent only ~30-40% of the total life.

Once initiated, cracks grow in three stages as illustrated schematically in Fig. 4. 'Stage I' shear crack growth is generally confined to a depth of only a few grain diameters into the material. This stage (together with the initiation stage) may be absent altogether where a sufficiently large defect exists to produce crack propagation in the 'Stage II' mode (perpendicular to the tensile axis) at the start of the fatigue process. Examination of a fracture surface produced during the growth of a fatigue crack, Fig. 5, shows it to consist of striation markings each of which correspond to one fatigue cycle. These markings are on quite a different dimensional scale to the macro 'clam-shell' markings which are visible by eye (Fig. 2) and often characteristic of service fatigue failures. (These are typically produced as arrest markings indicating changes in environmental or load conditions during the in-service propagation of a fatigue crack.) As the crack grows and accelerates (under constant cyclic load) the 'micro' striations coarsen and often terminate in rapid fracture with a relatively coarse (ductile or brittle) fracture surface (see Fig. 2). These striations, in addition to providing proof that the failure is a result of fatigue, can sometimes permit useful estimates of the rate of fatigue crack growth during failure investigations owing to the one-to-one correspondence between the number of striations and the number of fatigue cycles.

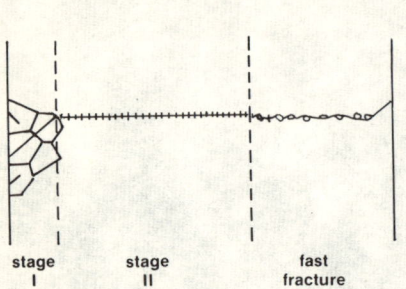

Fig.4: Fatigue crack propagation across a specimen section

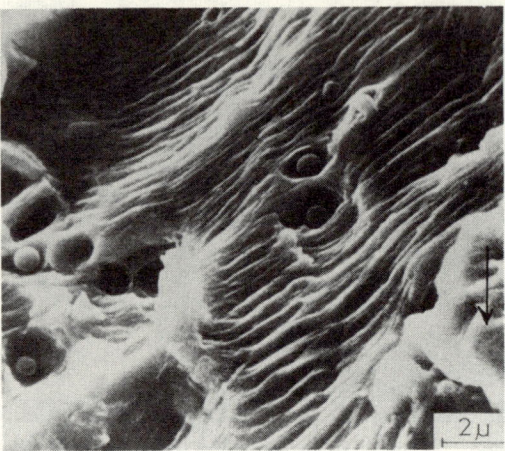

Fig.5: Scanning electron micrograph of fatigue striations in an alloy steel (courtesy R O Ritchie)

ANALYSING FATIGUE

Lifetime Determinations

In the total life determination of simple specimens, the distinction between crack initiation and propagation is not made, at least at the design stage, and it is conventional practice to measure fatigue strength in terms of the number of cycles to cause ultimate failure, N_f, as a function of either the peak applied stress, σ_p, or the applied stress range (i.e. amplitude), $\Delta\sigma$. It is most important to realise, however, that the resulting '$S-N$' curve may be significantly affected by a number of factors, some of which will be discussed in a following section.

First and foremost, we must note that the wide scatter observed in experimental test results, Fig. 6, is an important characteristic of fatigue, and survival probability estimates, Fig. 7 - based on statistical distributions of N_f for a given $\Delta\sigma$ or vice-versa - are essential components of any reliable design analysis. This variability is intrinsically related to the mechanism of fatigue failure and is in no way a result of errors or difficulties associated with the mechanical testing procedure. Thus, fatigue cracks develop from microscopic surface 'damage' in unfavourably oriented grains - because of their random orientation these may well not correspond to regions of highest stress concentration.

Because of the many factors affecting fatigue strength, it has often been found necessary to derive $S-N$ curves for each particular design situation. Under certain conditions, however, on the basis of the large body of available experimental data, it is possible to derive an $S-N$ curve from static tensile data without the need to perform dynamic fatigue tests (Wirsching & Kempert, 1976a). This is particularly the case for steels which, unlike most non-ferrous metals, produce $S-N$ curves that level off after a large number of cycles ($\sim 10^6$) giving a 'fatigue limit' cyclic stress below which fatigue failure does not occur. For steels having U.T.S. values < 1500 MPa, a fatigue limit of U.T.S./2 is often appropriate. A further 'fatigue strength reduction coefficient' may then be applied to relate this value to real-life design conditions to accommodate factors known to affect fatigue life such as: mean stress, corrosion environment, surface finish, stress concentrations, size variations, frequency of loading, temperature, statistical scatter and quality of workmanship, fabrication or welding (Juvinall, 1967). Allowances for such factors may combine in certain instances to reduce the allowable design stress by as much as a factor of 10 below the average smooth specimen stress for a given fatigue lifetime.

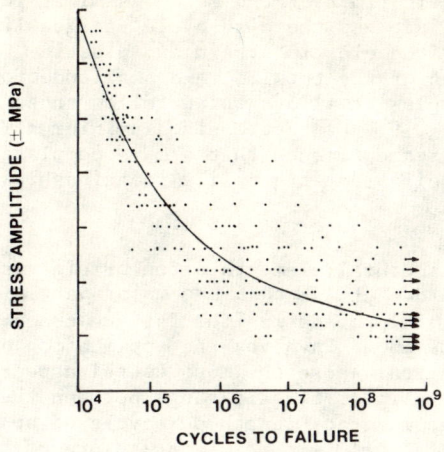

Fig.6: Experimental scatter associated with fatigue testing (aluminium alloy, after Gatto,1956)

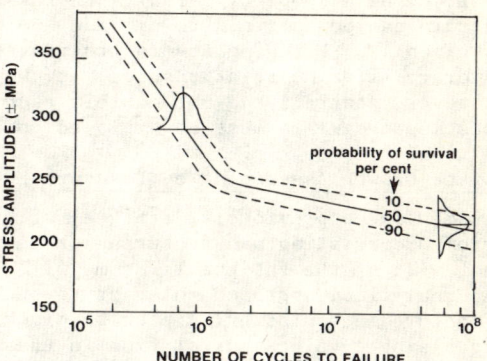

Fig.7: Stress amplitude—fatigue life—probability plot for an aluminium alloy (after Hardrath,1967)

The Strain Range Approach

In the low-cycle range where inelastic deformation is dominant, and in components subject to a controlled cyclic strain amplitude, it is common to relate the lifetime to a strain range rather than a stress range because relatively simple relationships have been found to exist between strain range and fatigue lifetime for fully reversed strain. These relationships have been developed from the so-called 'hysteresis' loop' - or stress-strain behaviour under cyclic loading - for a ductile metal, Fig. 8. Under 'steady-state' conditions, the cycles to failure (N_f) can be related to the amount of plastic strain by

$$2N_f = (\Delta\varepsilon_p/2\varepsilon'_f)^{1/c} \qquad (1)$$

This empirical relationship can be used to predict low-cycle fatigue behaviour. In this equation $\Delta\varepsilon_p$ is the plastic strain range (Fig. 8); ε'_f is the fatigue ductility coefficient (the true strain at fracture on the first reversal: $\varepsilon'_f \simeq \ln\{100/(100-RA)\}$ where RA = reduction in area from a tensile test); and c is the fatigue ductility exponent ($\simeq -0.6$ for metals). When the strain ampli-

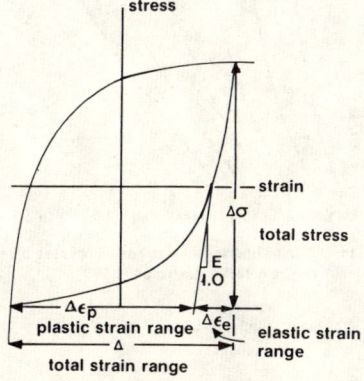

Fig.8: Cyclic stress-strain hysteresis loop

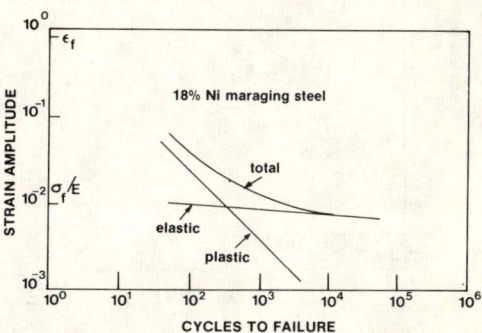

Fig.9: Low cycle fatigue of a maraging steel as a function of elastic, plastic and total strain amplitudes (after Landgraf,1970)

tudes are low enough to remain within the elastic limit of the metal, similar relationships exist between elastic strain amplitude and the high cycle fatigue life (Wirsching & Kempert, 1976a). Expressions for the elastic strain and plastic strain can be combined to provide an indication of the total strain amplitude for a given N_f. Maraging steels provide good examples of alloys which follow these average lifetime relationships, as shown in Fig. 9. High strength aluminium alloys, however, generally do not exhibit such good agreement, such that a more complete testing programme must be employed to obtain reliable fatigue life relationships.

Crack Growth Assessment by Fracture Mechanics

As mentioned previously, fatigue failure in structural elements often originates from pre-existing manufacturing or fabrication defects so that the major part, if not all, of the fatigue lifetime will be spent in propagating the fatigue crack. An analytical approach relevant to such circumstances involves the application of fracture mechanics principles to crack propagation. Here the most useful aspect is a single-valued correlation in the nominally linear elastic range between the rate of fatigue crack growth, da/dN (the change in crack length per cycle of applied load), and the cyclic 'stress intensity factor', ΔK. ΔK is a measure of the magnitude of the stress in the vicinity of the crack tip and is a function of the crack size (a), nominal stress range ($\Delta \sigma$), and the crack and component geometry (through a factor, α), and is of the form

$$\Delta K = \Delta \sigma \sqrt{\alpha \pi a} \qquad (2)$$

Fig. 10 indicates schematically the characteristic dependence of fatigue crack growth rate on the stress intensity factor. There are two asymptotic limits to the curve, the upper limit being set by the 'fracture toughness' of the material, K_C, while the lower limit defines the threshold, ΔK_{TH}, below which crack growth from pre-existing defects is not possible. Although the form of curve shown in

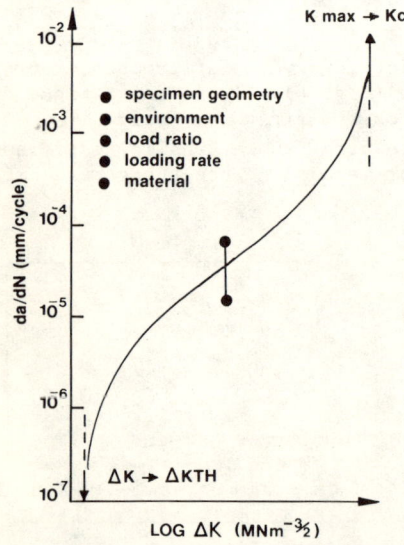

Fig.10: Schematic dependence of the rate of fatigue crack growth, da/dN, on the cyclic stress intensity factor, ΔK, and some of the factors affecting this relationship.

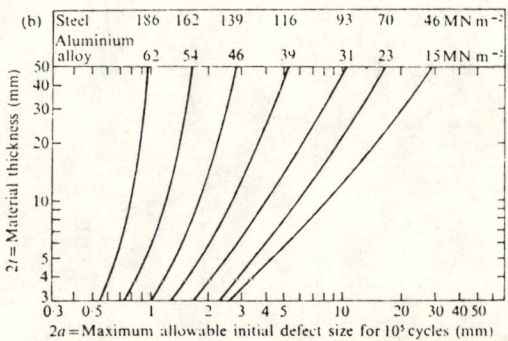

Fig.11: Allowable defect size for 10⁵ cycles of repeated tension (after Harrison, 1969).

Fig. 10 is essentially sigmoidal, for most crack growth rates of practical interest the curve may be approximated to a straight line described by an equation of the form

$$da/dN = A\Delta K^m$$
$$= A(\Delta_\sigma \sqrt{\alpha \pi a})^m \qquad (3)$$

where A is a material constant, and m (the slope of the log da/dN vs log ΔK curve) is a constant for the particular material, geometry and environment under test (Fig. 10). Equations such as (3) may be readily integrated to provide an assessment of the number of fatigue cycles required to grow a fatigue crack from an initial length, a_i (possibly determined by non-destructive inspection) to a final length, a_f (e.g. that required to initiate catastrophic failure). Appropriate re-inspection intervals may then be readily defined.

Under conditions where the fracture mechanics approach applies, experimental results may be used to construct curves giving maximum allowable defect sizes for various applied stress levels, component geometry and initial defect sizes (Fig. 11). Similarly, suitable calculation can reduce crack growth rate data of the form shown in Fig. 10 to the $S-N$ type more familiar to designers (7), Fig.12.

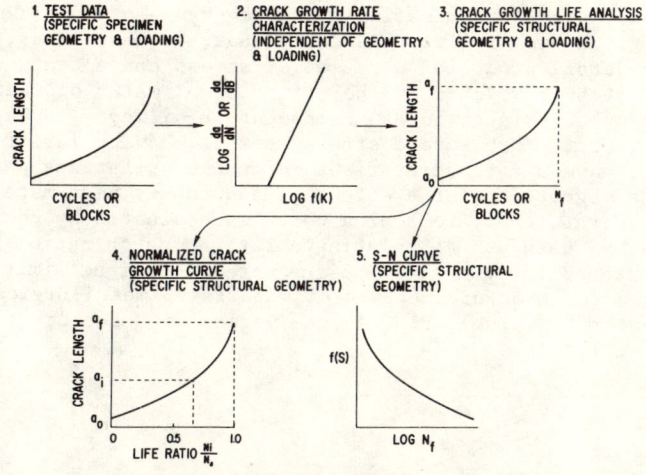

Fig.12: Schematic illustration of the engineering methodology for applying fracture mechanics technology to fatigue crack propagation (after Crooker, 1978)

FACTORS AFFECTING FATIGUE

An indication has already been given of the nature and diversity of the factors which may affect fatigue behaviour. In many instances, the problem is severely complicated when a number of factors act in combination. As a result, design against fatigue may only be possible through a broad-based awareness of the manner in which fatigue strength can vary, coupled with full- or semi-scale simulation tests to refine first stage designs. There follows, then, a summary of some of the major variables affecting fatigue performance. For convenience, these can be divided into 3 general categories, viz: specimen/component/structure condition, loading conditions and environmental conditions.

Specimen Condition

(a) <u>Surface finish/residual stresses</u>. Since the development of fatigue cracking is very much a surface-related phenomenon, it follows that fatigue strength is very dependent on surface conditions such as finish, hardness and the presence of residual stresses. As illustrated in Fig. 13, the magnitude of the effect is a function of the alloy strength level. The following section indicates how the fatigue limit of a specimen containing a surface notch decreases as both the depth and sharpness of a notch increase. However, changes in surface hardness and the introduction of residual surface stresses introduced by the finishing process employed may override geometric effects of surface roughness. Surface hardening inhibits those micro-plastic deformation processes leading to crack initiation, while residual compressive stresses offset the applied tensile stresses and prevent crack opening. Thus, surface hardening treatments, such as carburising or nitriding, can improve fatigue life as can techniques such as shot-peening, sand blasting, surface rolling or quenching, by producing a plastically deformed compressive skin over a relatively undeformed core. (It should be noted, however, that significant improvements in the fatigue lives of plain specimens will only be recorded in bending or torsion (Frost et al, 1974), and not in reversed direct stress tests). Conversely, treatments that produce surface softening (e.g. decarburisation during forging) or tensile residual surface stresses (e.g. electroplating) can seriously reduce fatigue life, irrespective of whether tests are carried out in bending or under direct stress.

(b) <u>Stress concentration</u>. Examination of a wide range of components which have failed in service from fatigue indicates that, in many cases, cracks initiate from a geometric detail producing a region of stress concentration, generally referred to as a "notch". Notches can be cracks, tool marks or metallurgical defects, or they can be designed into a component as oilways, keyways, fillet radii, threads, splines, etc. More severe stress raisers, having larger stress concentration factors, K_t, have a more deleterious influence on fatigue performance (Fig. 14), although the magnitude of the effect of a notch is very material dependent (Fig. 15). In general, the effect of a notch is evaluated by comparing notched versus unnotched S-N data and calculating a fatigue notch ratio, K_f, given by the ratio of the unnotched fatigue limit to the notched fatigue limit ($K_f \geqslant 1$). As well as being a function of the type of materials, metallurgical condition and notch dimensions (or K_t value), Fig. 16, K_f varies with factors such as section size and stress amplitude.

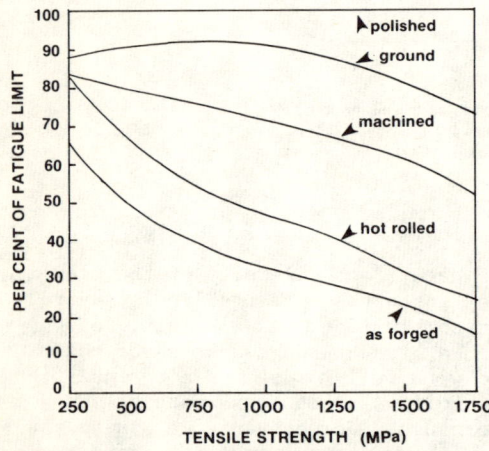

Fig.13: Effect of surface on fatigue as a function of tensile strength of steel (after Lipson and Juvinall)

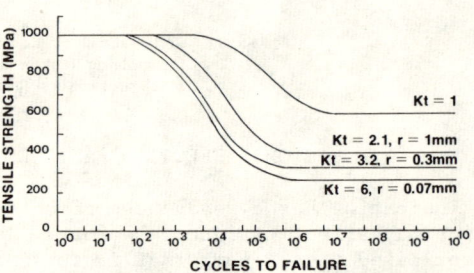

Fig.14: The influence of root radii on fatigue strength of Inconel 718 (Ni,19% Cr, 18% Fe, 5% Nb&Ta, 1% Ti 0,5% Al) at 600°C (after W J Harris, 1976)

Failure by Fatigue

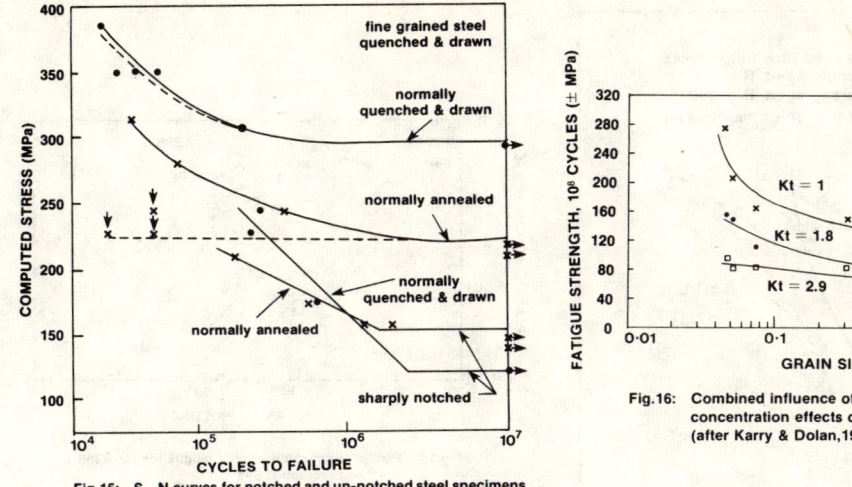

Fig.15: S—N curves for notched and un-notched steel specimens in two heat treated conditions (after Brophy, 1936)

Fig.16: Combined influence of metallurgical and stress concentration effects on fatigue life (after Karry & Dolan, 1953)

Fatigue Loading Conditions

(a) <u>Mean stress</u>. Design situations often involve a steady mean stress superimposed on an alternating stress; much fatigue data, however, particularly that obtained in the low-cycle range, has been obtained under fully-reversed cycling. Various methods are available to extrapolate the allowable stress-amplitude, $\Delta\sigma$, for a given lifetime at any non-zero mean stress level, σ_m, Fig. 17 (Wirsching & Kempert, 1976b). It is readily evident that, for a given mean stress, the allowable maximum alternating stress can vary over a very wide range. In practice, it has been found that the Gerber curve best fits the behaviour of ductile steels, the Goodman & Soderberg curves being too conservative (Wirsching & Kempert, 1976b), whereas the Goodman model describes better the fatigue response of the more brittle steels. However, the general uncertainty of such empirical relationships presents one of the major problems in fatigue life prediction.

The material dependence of the mean stress effect on fatigue crack growth is illustrated in Fig. 18, indicating that mean stress effects are much more pronounced in the case of lower toughness alloys. This may be associated with the occurrence

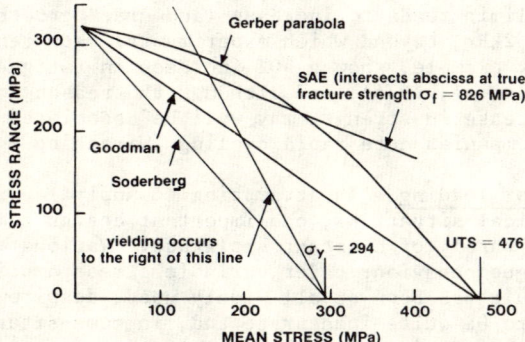

Fig.17: Four models which attempt to predict the effect of mean stress on fatigue life

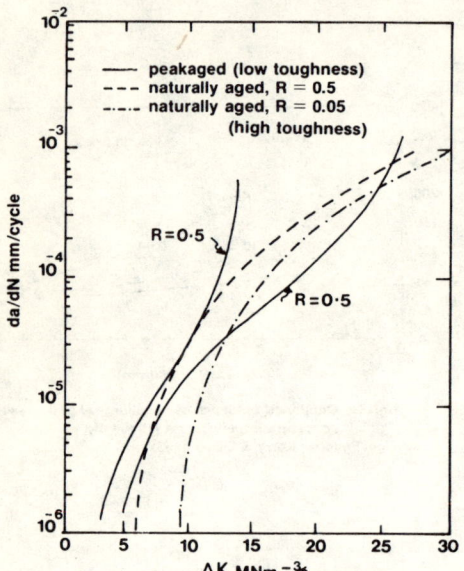

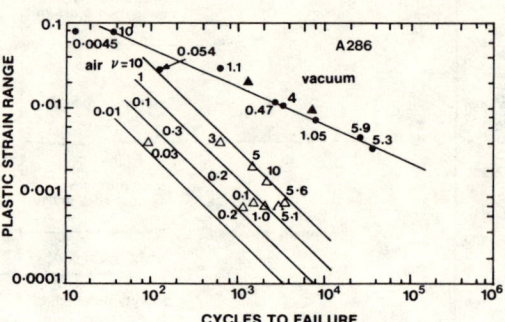

Fig.19: Plastic strain range versus fatigue life for A286 in air and vacuum at 593°C. Numbers adjacent to test points indicate frequency in cycle min.(After Coffin 1972)

Fig.18: The influence of material toughness on the mean stress effect in fatigue crack growth (after Garrett & Knott, 1975)

of 'static' modes of cracking (e.g. void coalescence, or incremental cleavage) in combination with 'classical' fatigue crack growth by the formation of striations (Garrett & Knott, 1975). Although high toughness alloys are much less sensitive to variations in mean stress at intermediate and high crack growth rates, they too depend on mean stress towards the threshold level (Fig. 18). In fact, considerable research is currently aimed at understanding the factors affecting fatigue crack growth in the threshold regime (e.g. see Metals Society, 1977).

(b) <u>Frequency</u>. Fatigue can obviously take place over a wide range of cycling speeds. Over the range 1-200 Hz the bulk of experimental evidence suggests that the fatigue limit remains essentially constant (Frost et al, 1974), provided there is no adiabatic heating effect (which can become serious at higher frequencies) or no environmental attack (which can cause time-dependent damage and therefore reduce fatigue life at low frequencies, Fig. 19). At higher testing frequencies the fatigue limit tends to increase (and crack growth rate decrease) with frequency up to about 2kHz, beyond which experimental data tend to be conflicting. Low carbon steels, for example, show a 50% increase in fatigue limit between 200 and 100 000 Hz (Frost et al, 1974), and although the reason for this is still rather unclear, the increase in strength may well be associated with the increased strain rate which accompanies more rapid cycling (Wirsching & Kempert, 1976b).

(c) <u>Variable and random loading</u>. In attempting to apply laboratory-derived fatigue data to practical situations, one important design complication is that in-service loading is rarely of constant amplitude. Various models have been proposed to predict fatigue behaviour under variable stress amplitudes, but only the Palmgren-Miner (PM) rule has been at all widely used, despite the fact that it is generally considered to be quite inaccurate and, in some situations, dangerous. According to the PM rule, fatigue damage is cumulative in proportion to the magnitude of each stress cycle. Thus, for the range of stress amplitudes shown in Fig. 20 (a), the fraction of the life consumed due to n_1 cycles at a stress range

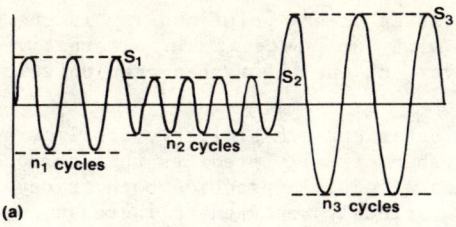
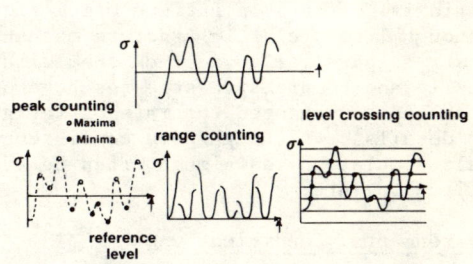
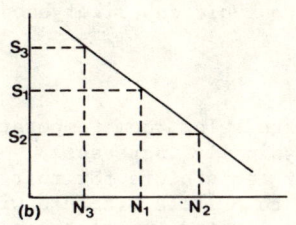

Fig.21: Example of three basic types of counting methods in the analysis of variable amplitude loading (After Buxbaum,1973)

Fig.20: Variable amplitude stresses (a) cause cumulative damage in proportion to their effect on the constant amplitude fatigue life,(b)

S_1 is n_1/N_1 where N_1 is the cycles required to cause failure at S_1, taken from the constant amplitude S-N curve (Fig. 20 (b)). The fraction of life consumed, D, is then given by:

$$D = n_1 / N_1 + n_2 / N_3 + n_3 / N_3 + \ldots \ldots \quad (4)$$

where failure occurs for $D = 1$. In practice D at failure usually varies in the range 0,7 to 2,2 suggesting that design values for D of 0,5 may be more appropriate, although variations between 0,18 and 23 have been reported (Wirsching & Kempert, 1976b). (In general, D is smaller if the higher stresses are applied first, and vice-versa).

Random stresses are a special case of variable-amplitude loading and here a first problem is the selection of a statistical counting method, assuming that adequate load data is available. Current statistical counting methods are modifications of the three basic types indicated in Fig. 21, and methods known as 'rain-flow' and 'range-pair' cycle counting have been used with some success in fatigue life predictions (Dowling, 1972). Thus, the Palmgren-Miner rule, despite it inadequacies, still serves a useful function at the early stages of design. However, in order to achieve optimum design for strength, reliability or safety, there is at present little alternative to full-scale testing of components or structures (Fig. 1) under realistic simulations of service-loading conditions, with the associated cost penalty that this incurs.

(d) <u>Combined loading</u>. In many components and structures the state of stress may be two- or three dimensional - pressure vessels, pipelines and aircraft skin structures, for example, operate under biaxial stresses - and this has presented considerable difficulties when attempts are made to extrapolate from the more usual and straightforward, laboratory-derived tests. One design procedure used for evaluating fatigue behaviour under such conditions, based on the so-called 'distorsion energy' theory of failure, calculates an 'equivalent' alternating bending stress, $\Delta\sigma_e$, for use with a convential S-N curve to predict fatigue life. $\Delta\sigma_e$ is given by

$$\Delta\sigma_e = (S_x^2 + S_y^2 - S_x S_y + 3\tau_{xy})^{\frac{1}{2}} \quad (5)$$

where S_x and S_y are the peak $x-y$ normal stresses at any point and τ_{xy} is the peak shear stress. Although this has been used with some success, many alternative theories have been proposed. As yet, however, no one theory has been universally proved and accepted.

Multiaxial effects on fatigue crack growth behaviour are similarly conflicting. Although linear elastic fracture mechanics theoretically predicts that components of stress parallel to a crack should have no effect, in practice both transverse tensile and compressive stresses have been variously reported to increase, decrease or leave unchanged the rate of fatigue crack growth (see Garrett et al, 1979 for details). Although some consistency is gradually emerging, once again full-scale simulation tests must often be adopted in order to avoid any likelyhood of fatigue failure.

Environmental Conditions

(a) <u>Temperature</u>. When components are required to operate below room temperature, although the fatigue strength of plain specimens typically increases (~ 1,5 - 2,5 over the range from ambient to liquid air temperature for most metals and alloys) there is an increasing susceptibility to brittle fracture with metals failing at shorter crack sizes. Conversely, at elevated temperatures fatigue strength is generally reduced and factors which may be relatively unimportant at room temperature, such as test frequency or shape of the cyclic waveform, may be dominant. In addition, if a mean stress is present, creep mechanisms may contribute significantly to accelerating failure. Furthermore, poor elevated temperature fatigue properties may not so much be associated with softening at operating temperatures, but rather due to surface oxidation or chemical attack by the pervading atmosphere; in these situations, protective surface treatments may result in considerable improvements.

Fluctuating temperatures may themselves lead to thermal fatigue, a process whereby thermal stresses (either due to thermal gradients or if free expansion/contraction is constrained) may induce failure even in the absence of externally applied loads. It is not necessary to subject a complete component to temperature variations to cause thermal fatigue (Wirsching & Kempert, 1976b); one face exposed to ambient temperatures and another face heated or cooled may produce a sufficiently large thermal gradient to cause thermal fatigue - for example in heavy-walled gun barrels.

(b) <u>Corrosion fatigue</u>. The influence of a corrosive environment in causing a reduction in fatigue strength is well illustrated in Fig. 19 - indicating that in many cases air must be classified as a corrosive environment - and Fig. 22, showing that the fatigue limit for steel is essentially eliminated in salt water.

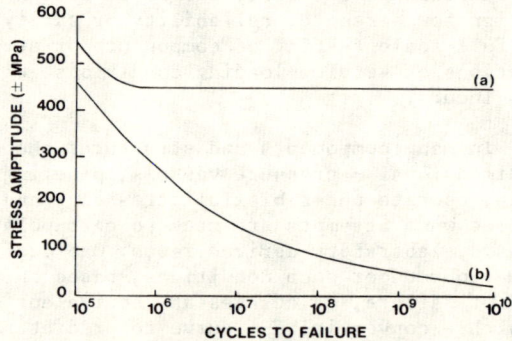

Fig.22: The rotating bend fatigue behaviour of a 2,5% Ni—Cr—Mo steel tested (a) in air, (b) in 3% NaCl (after W J Harris, 1976)

It is now reasonably well established that the simultaneous application of dynamic stress and corrosive environment leads to lower strength values than the simple effects of corrosion and fatigue considered separately would suggest.

In practical situations, corrosive agents are unavoidable ingredients in fuels, lubricants and hydraulic fluids. Also, while metallic components or structures in static stress situations can often be well protected against corrosion by paints, electrodeposited metals or protective films, these methods are usually less effective under fatigue conditions since the protective layer must meet a rheological criterion as well as a chemical one.

DESIGNING AGAINST FATIGUE

Preceeding sections have attempted to convey the diversity with which a number of different factors can affect fatigue behaviour. When several of these factors act in combination, as may frequently occur in realistic operational situations, the problems of extrapolating design information from laboratory data can readily be appreciated; historically, therefore, design codes have developed from a conservative base. While existing codes are under some pressure for amendment to create lighter, more efficient designs, the overriding requirement for absolute structural reliability and safety means that code writers are generally somewhat reticent in changing the status quo.

Today there are really two main approaches to design against fatigue. The first is based on designing for a 'safe life' in terms of the $S-N$ properties of the material. This requires estimation or measurement (e.g. by strain gauges, accelerometers, etc.) of the anticipated load spectrum, followed by laboratory testing or analysis of this loading history to predict the fatigue life. The safe life - at which the component in service should be replaced - then incorporates a suitable safety factor on the predicted fatigue life. The consequences associated with underestimating this factor of safety have led designers to move from a modest 3 or 4, to 10 or even 15, with corresponding severe weight penalties or, in many situations, economically unviable safe lives.

Such restrictions have led to the development of the second main approach to fatigue design which requires damage tolerant ('fail-safe') structures so that in the event that fatigue cracks do develop, their presence will not be catastrophic. In practice this philosophy is characterised by designers having to accept that crack-free structures are rare and therefore having to 'live with defects', and designing to ensure the capability of the structure to withstand operating loads should any one element fail. Using this approach, the designer must consider: crack propagation rates; residual strength; cumulative damage estimates; critical defect sizes based on fracture mechanics estimations, coupled with appropriate non-destructive inspection procedures; and design details which act as crack propagation barriers (Harris, 1976).

CONCLUDING REMARKS

This brief review has served to introduce the problem of fatigue and mechanisms leading to it, together with the various factors affecting in-service fatigue performance and the analytical approaches currently available to predict fatigue life. It is important to realise that this paper, like much of the early work in fatigue, has been very much a surface study. The voluminous information currently appearing in the literature on the topic of fatigue, perhaps characterised by the emergence in 1979 of two new international technical journals dealing with the subject, precludes any in-depth review at this level. However, much of the nature of current information, if hardly any of the specific detail, has been described.

In recent years, improvements in analysis and understanding rather than improvements in materials' resistance to fatigue have been obtained. While it is still to be hoped that material improvements will be forthcoming, so far this goal has been a difficult one to achieve. Perhaps more modest goals such as reduction in scatter, in notch sensitivity, and in environmental sensitivity may be realisable and more useful than improvements in average properties. In any event, a wider dissemination of information already available about factors affecting fatigue and new principles of fatigue would constitute a very positive step towards the elimination of fatigue failures.

ACKNOWLEDGEMENT

The author is grateful to Professor A.J. McEvily for permission to use material previously published by him, and for constructive comments on the manuscript.

REFERENCES

Brophy, G.R. (1936). Damping capacity, a factor in fatigue. Trans. Am. Soc. Metals, 24, 154-185

Buxbaum, O. (1973). Methods of stress-measurement analysis for fatigue life evaluation. In Fatigue Life Prediction for Aircraft Structures and Materials, AGARD-LS-62

Coffin, L.F. (1972). The effect of vacuum on the high-temperature, low cycle fatigue behaviour of structural metals. In Corrosion Fatigue, NACE-2 Nat. Assoc. of Corrosion Engineers, Houston, p. 590

Crooker, T.W. (1978). Subcritical crack growth in high-strength alloys. In N. Perrone, H. Liebowitz, D. Mulville and W. Pilkey (Eds.), Fracture Mechanics, University Press of Virginia, pp. 333-346

Dowling, N.E. (1972). Fatigue failure predictions for complicated stress-strain histories. J. Mater. ASM 7, p. 71.

Freudenthal, A.M. (1970). Fatigue mechanisms, fatigue performance and structural integrity, Proc. Air Force Conference on Fatigue & Fracture of Aircraft Structures & Materials, AFFDL TR 70-144, p.9

Frost, N.E., Marsh, K.J. and Pook, L.P. (1974). Metal Fatigue. Clarendon Press, Oxford

Garrett, G.G. and Knott, J.F. (1975). On the influence of fracture mechanics on fatigue crack propagation in aluminium alloys. Met. Trans., 6A, 1663-1665

Garrett, G.G., Anderson, P.R.G. and Charvat, I.M.H. (1979). The influence of biaxial stresses on high cycle fatigue crack propagation. Exp. Mech. (in press)

Gatto, F. (1956). Colloquium on Fatigue. Springer-Verlag, Berlin, p. 66

Hardrath, H.F. (1963). A guide for fatigue testing and statistical analysis of fatigue data. ASTM STP, 91-A

Harris, W.J. (1976). Significance of Fatigue. Engineering Design Guide No. 14, Oxford University Press

Harrison, J.D. (1969) in P.L. Pratt (Ed.), Fracture, Chapman & Hall, London

Juvinall, R.C. (1967). Engineering Considerations of Stress, Strain & Strength, McGraw-Hill

Karry, R.W. and Dolan, T.J. (1953). Proc. Am. Soc. Test. Mater., 53 p. 789

Landgraf, R. (1970). The resistance of metals to cyclic deformation. ASTM STP 467, p. 3

Lipson, C. and Juvinall, R.C. (19xx). Handbook of Stress and Strength, Macmillan, New York

Metals Society 1977). Special issue on Fatigue. Proc. Cambridge Conf. 'Fatigue 77', 11

Sorkin, G., Pohler, C.H., Stavory, A.B. and Borriello, F.F. (1973). An overview of fatigue and fracture for design and certification of advanced high performance ships. Eng. Fr. Mech., 5, 307-352

Wirsching, P.H. and Kempert, J.E. (1976a). A fresh look at fatigue. Machine Design, (May) pp. 120-123

Wirsching, P.H. and Kempert, J.E. (1976 b). Fatigue failure in the real world. Machine Design (Aug.) pp. 86-90

BIBLIOGRAPHY

see above: Frost et al (1974); Harris (1976); Metals Society (1977); Wirsching and Kempert (1976); and

Rolfe, S.T. and Barsom, J.M. (1977). Fracture and Fatigue Control in Structures. Prentice-Hall, Inc., New Jersey

Schijve, J. (1979). Four lectures on fatigue crack growth. Eng. Fracture Mech., 11, pp. 167-221

ON THE MICROSTRUCTURAL CONTROL OF THE FRACTURE PROCESSES INVOLVED IN WEAR

C. J. Heathcock*, C. Allen*, B. E. Protheroe** and A. Ball*

*Department of Metallurgy and Materials Science, University of
Cape Town, R.S.A.
**Mining Technology Laboratory, Chamber of Mines,
Johannesburg, R.S.A.

ABSTRACT

The microstructural control of the plastic deformation and microfracturing processes involved in wear is examined. Results obtained by a laboratory study of cavitation erosion on a range of materials has verified the validity of this approach and predictions can now be made concerning the suitability of a material and its microstructure for a given wear situation.

Keywords: Wear, cavitation erosion, microstructures, metals, polymers, fracture mechanisms.

INTRODUCTION

Wear occurs when material is lost from surfaces that are in relative motion or when component tolerances are affected through surface deformation. Wear represents an enormous financial loss, particularly to the mining and associated industries, through component replacement, repair, loss of production and reduced efficiency of operation. It is clearly important therefore to understand the nature of the wear process and the precise effect of the operating variables.

Numerous attempts have been made to classify wear in terms of abrasion, erosion and fretting on the basis of observed phenomenological processes. However this does not define adequately the mechanisms of material loss nor the important 'metallurgical' parameters that influence these processes. In the majority of situations both plastic deformation and fracture are important rate controlling processes. Despite the extensive amount of literature concerning the mechanical behaviour of solids very little attempt has been made to utilise this knowledge in the analysis of the parameters which affect wear. Rather, reliance has been placed on in-house practical experience and tenuous correlations with the bulk properties of materials when assessing the ability of solids to resist wear in industrial situations. Such an approach relied heavily on shrewd deductions and good judgement on the part of technologists who are involved in combating wear.

Although it is obviously essential to consider bulk mechanical properties such as strength when selecting materials for wear resistance, it can be misleading to rely too heavily on such data. Hardness has been universally accepted as one of the most important tests in the selection process, yet there is no simple relationship between material hardness and wear resistance. It is well known, for example, that the abrasive wear resistance of steels heat treated to the same hardness level depends upon carbon content. Furthermore, some 'soft' polymeric materials have better abrasive wear resistance in certain situations than hard steels. Clearly such discrepances must be avoided, since they will have serious consequences in practice.

The important factors causing wear are mechanical, thermal or electrochemical in nature. In the majority of situations they do not operate singly but in combination which makes the wear analysis somewhat complex. Furthermore slight changes in operating conditions can often lead to large changes in wear rates. Thus the use of the title 'Wear Resistant Alloy' based solely on uniform and controlled laboratory testing is ambitious since performance will be affected by the actual operating conditions. For example Hadfields manganese steel performs well in high stress-impact abrasion conditions but behaves poorly when exposed to a corrosive environment. It is most important therefore to identify the precise conditions under which wear occurs and to systematically evaluate the microstructures that will resist such conditions. It should then be possible to make valued judgements, on the basis of existing knowledge, or the type of surface structure and properties of wear resistant materials for particular industrial situations.

The present work is an attempt to examine the value of this approach and is based on the results of cavitation erosion experiments carried out at Cape Town University together with in-situ examples collated by the Chamber of Mines in Johannesburg.

REQUIREMENTS FOR CAVITATION EROSION RESISTANT MATERIALS - A RATIONALE

Cavitation is the process of repeated growth and subsequent violent collapse of cavities in a liquid. The high strain rate shock waves and liquid microjets produced on collapse of the cavities can cause severe surface damage to engineering materials. Cavitation occurs in systems having high flow rates and pressures (eg. hydraulic turbines and pumps) and in systems having low flow rates and pressures (eg. diesel engines). Materials subjected to such conditions should have a high elastic resilience, in order to accommodate the shock loading, with a good fatigue resistance to the oscillating stress pattern. They should not be strain rate or notch sensitive and in addition should not be susceptible to corrosion in the particular environment. High elastic resilience can be achieved by either increasing the yield stress or yield strain through alloying, heat treatment or through cold working of metallic materials. It can also be achieved by utilising polymeric materials which have the ability to absorb and return energy without permanent deformation.

In general, good fatigue resistance is exhibited by materials which have high yield strengths coupled with fine homogenous structures. Surface compressive stresses are also known to be most beneficial in combating the effects of applied tensile stresses. Thus surface deformation or strain induced phase transformations such as the austenite to martensite reaction may well be beneficial. This phase transformation not only increases the strain hardening rate of the material but changes the residual stress pattern in the surface from tensile to compressive and thus hinders the formation and propagation of cracks.

It might also be expected that materials with high strain rate sensitivity and notch sensitivity such as body centered cubic metals and alloys should be avoided. The relative importance of all of these parameters will obviously depend on in-service conditions and in particular the local strain rate.

OBSERVED MODES OF MICROFRACTURE

A large number of materials have been eroded using a vibratory cavitation apparatus (Heathcock, Protheroe and Ball 1979). Fractographic studies of the damaged surfaces have revealed a number of distinct fracture mechanisms, viz:

a) plastic deformation and rupture in ductile metals and alloys.
b) brittle transgranular and intergranular cleavage in strain rate sensitive metals and alloys.
c) fibrous tearing and 'flow like' deformation of polymeric materials.

The damaged surface of Monel 400 (Fig.1) is typical of the ductile mode of failure. Impinging shock waves and jets produced by bubble collapse deform the ductile phases in the surface region and indentations and pits are formed. Asperities and lips which build up at the edges of the pits are subsequently removed by ductile fracture processes. A section through 304 type stainless steel shows the nature of these caviation pits and lips (Fig.2).

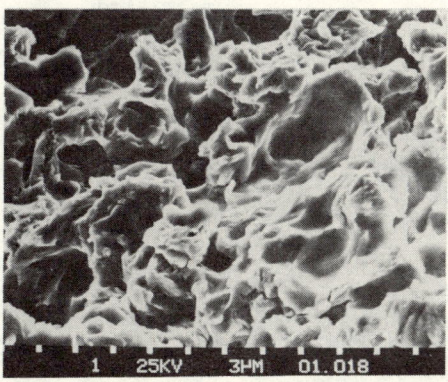
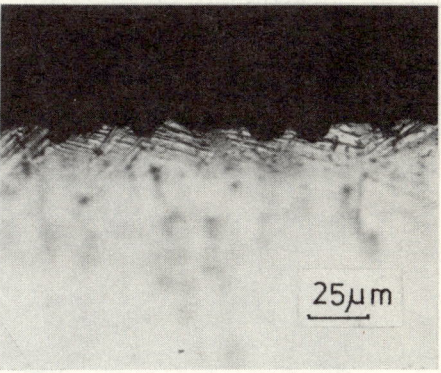

Fig.1. Ductile mode of failure in Monel 400.

Fig.2. Cavitation pits in the eroded surface of 304 SS.

The second mode of failure was found to occur in a number of strain rate sensitive body centered cubic metals and alloys such as pure iron and the high chromium ferritic stainless steels. In these ferritic stainless steels the amount of brittle cleavage increased as the chromium content was raised. The fracture surface of type 409 stainless steel (Fig.3) is typical of this mode of failure.

The fibrous tearing mode of fracture occurred in polymers such as high density polyethylene, polyacetal copolymer and nylon (Fig. 4). Polytetrafluorethylene eroded by a process of flow-like deformation of the polymer-structure. (Fig. 5)

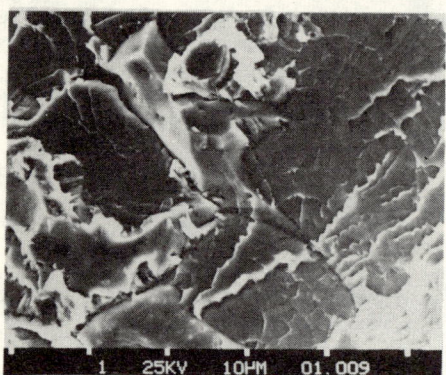

Fig. 3. Brittle cleavage mode of failure in type 409 stainless steel.

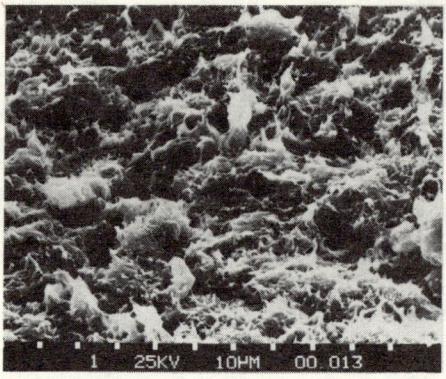

Fig. 4. Fibrous tearing mode of failure in Nylatron.

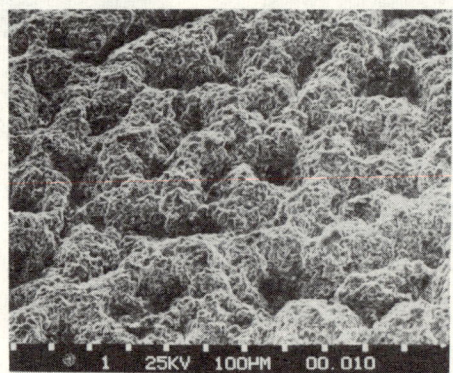

Fig. 5. 'Flow like' damage in PTFE.

RELATIONSHIP OF MICROSTRUCTURE AND PROPERTIES TO EROSION RESISTANCE

The relationship between erosion resistance and elastic resilience as measured by the Shore Scleroscope is shown in Fig. 6 for a wide range of materials. Elastically resilient materials such as sintered carbides, quenched and lightly tempered steels and pure polymers such as nylon and high density polyethylene show excellent erosion resistance. Less resilient materials such as PTFE, Monel 400 and type 316 austenitic stainless steel have poor erosion resistance. Metals which have been classified as ductile by virtue of their observed erosion mechanism show a clear correlation between high wear loss and low resilience.

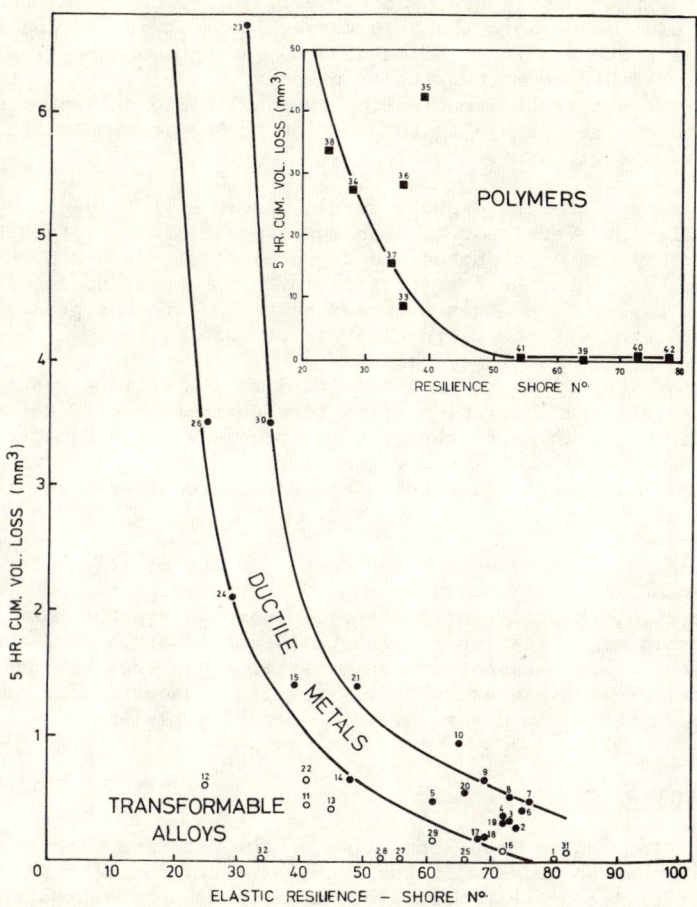

1.	EN24	200	11.	EN57	200	21. 316 SS	32. NiTi
2.		300	12.		300	22. 304 SS	33. PTFE
3.		400	13.		400	23. S.G. 80	34. PTFE - 60% Bronze
4.		500	14.		500	24. Incalloy 825	35. PTFE - 25% Carbon
5.		600	15.		600	25. Stellite 3	36. PTFE - C + Graph.
6.	EN30B	200	16.	DIN4112	200	26. Monel 400	37. PTFE - 15% Glass
7.		300	17.		300	27. WC - 8MC	38. PTFE - 25% Glass
8.		400	18.		400	28. WC - 8LC	39. Nylatron
9.		500	19.		500	29. WC - CNO2	40. Nylon 66
10.		600	20.		600	30. 3Cr12Ni	41. Flurodur
						31. EN24 - A.Q.	42. Polyacetal

Fig. 6. Relationship between erosion loss and elastic resilience as measured by a Shore Scleroscope.

The transformable alloys, which also fail in a ductile manner, have resistances which deviate from the ductile group of alloys. This improved resistance is due to the presence of a metastable phase in these alloys which, on loading, transforms to a stable state and in so doing absorbs cavitation energy. The stress induced face centred cubic to martensite transformation in type 304 stainless steel, the face centered cubic transformation to hexagonal close packed structure in cobalt alloys and the reversible transformation in the intermetallic alloy NiTi all improve erosion resistance.

Transgranular cleavage in single phase ferritic steels is a low energy process and hence these steels have very poor erosion resistance. It is interesting to note that the eroded surfaces of quenched and tempered steels show a more ductile mechanism of fracture than the ferritic steels. This apparent anomalous behaviour is partly due to a very fine grain size and partly due to the presence of laths of martensite which interrupt the micro-cleavage process.

Addition of fillers to polymeric materials assists the failure process by providing elastic inhomogeniety and initiation sites for debris removal. Removal of particles results in holes which further aid the progression of damage. In the case of carbides embedded in a matrix the situation is more complex; removal of the matrix leaves the hard particles exposed and these are afterwards dislodged by cavitation action.

Thus high elastic resilience, an ability to accumulate plastic deformation, either through dislocation movement or stress induced transformations, and good toughness with respect to the propagation of surface microcracks are the most important factors in determining cavitation erosion resistance. Often more than one factor contributes to the overall response of a material to an erosive environment but fine, tough metallurgical microstructures in which phase transformation can occur would appear to be the optimum for cavitation erosion resistance.

COMMENTS

This paper has attempted to focus attention on the need for more systematic microstructural approaches to be made in the study of wear. Only by such investigations can we hope to formulate the composition and microstructural characteristics of the 'ideal' wear resistant material for items of industrial equipment. The paper has also attempted to show that general predictions on the surface properties and metallurgical characteristics of wear resistant materials for cavitation erosion conditions can be made and, in principle, we can now formulate similar predictive arguments for the 'design' of materials for other wear situations. However, it is important to realise that wear occurs over a very wide range of local strain rate conditions and that cavitation erosion represents a very high strain rate situation. It can be postulated for example, that the body centred cubic metals will have improved performance in abrasive-slow strain rate deformations and similarly that a material's resistance to slow ductile microtearing will become more important than the ability of a fine microstructure to prevent microcleavage. Also the slow strain rates imposed by abrasive wear situations will accentuate the role of corrosion. Corrosive action will have a greater opportunity to assist the removal of stressed material when the strain is applied slowly. Thus much could be learnt from the results of corrosion fatigue experiments and the influence of microstructure. It is believed that this type of approach will provide a sound basis for the development of wear resistant materials.

ACKNOWLEDGEMENT

The Chamber of Mines of South Africa and the University of Cape Town are acknowledged for their generous provision of financial support and laboratory facilities. Mrs. G. Perez and Mrs. J. Lomberg are gratefully thanked for their assistance with the laboratory work and the preparation of the manuscript.

REFERENCES

Heathcock, C.J., Protheroe, B.E., and Ball, A. (1979). The influence of external variables and microstructure on the cavitation erosion of materials. To be published in Proc. 5th Int. Conf. on strength of metals and alloys, Aachen, 1979.

BIBLIOGRAPHY

Glaeser, W.A., Ludema, K.C., and Rhee, S.K. (Ed.)1977. Wear of Materials. Proc. Int. Conf. on wear of materials, St. Louis, Missouri. The American Society of Mechanical Engineers, N.Y.

THE GENERAL CHARACTERISTICS AND EVALUATION OF STRESS CORROSION CRACKING

D. Twigg

South African Bureau of Standards, Private Bag X191, Pretoria 0001, R.S.A.

ABSTRACT

The general characteristics and findings on stress corrosion cracking behaviour are discussed and illustrated with appropriate examples. The caution required in applying methods to prevent SCC and the various test methods used to evaluate the resistance of materials to SCC then receives attention. The paper concludes with a description of the activities of the ISO SCC Technical Committee responsible for developing internationally accepted standard test procedures.

INTRODUCTION

The SABS is involved in the stress corrosion cracking problem in two areas:

1) The diagnosis of stress corrosion problems encountered in industry.
2) Development of standard stress corrosion cracking test procedures.

Experience has shown that many South African engineers still lack an appreciation of the subject. In fact, in many existing chemical plant constructions, it is doubtful whether the possibility of stress corrosion cracks occurring in service was ever considered at the design stage.

Stress corrosion cracking is defined as non-ductile failure resulting from the simultaneous presence of a static surface tensile stress and a specific corrodent. In a simple equation form this can be expressed as follows:

STATIC SURFACE TENSILE STRESS – often residual only

+

SMALL AMOUNT OF CORROSION – usually localized, i.e. pitting

+

TIME
↓
STRESS CORROSION CRACKING

In other words a crack-free metal or alloy can be put into service and after some time stress corrosion cracks can develop if tensile stress is present and localized corrosion occurs. Corrosion in the absence of surface tensile stress does not cause stress corrosion cracking. Prevention of stress corrosion cracking falls under one or other of the following methods:

i) Removal of surface tensile stress or reducing the stresses below a so-called threshold value.
ii) Changing the surface tensile stresses to compressive.
iii) Prevention of corrosion (inhibitors, cathodic protection, etc.)
iv) Elimination of the specific environment.
v) Changing to another metal composition not susceptible to stress corrosion cracking in the particular environment.

Engineers cannot be expected to be experts in the technology involved in all of these preventative methods. All that is required is that the engineer should appreciate that stress corrosion cracking could be a problem and then the appropriate expert in the appropriate discipline should be consulted for advice. Ideally advice should be sought at the design stage rather than after a crack has developed in service. Prevention is better than cure.

This paper considers examples and characteristics of stress corrosion cracking to enlighten the engineer so that the present situation can be improved upon.

SOME EXAMPLES OF STRESS CORROSION CRACKING WHICH CAN OCCUR IN METAL SURFACES THAT ARE IN TENSION

a) 20 mg/ℓ of chloride in drinking water can cause the cracking of thick stainless steel sections.
b) A slight amount of ammonia in the air can crack copper alloys.
c) Titanium alloys at one time thought to be immune to SCC are now found to be susceptible in such diverse environments as nitrogen tetroxide (N_2O_4), methanol, and salt water.
d) Inconel-600, an often used alternative for stainless steel in chloride environments, may crack in high purity water in the 300 °C range with a few mg/ℓ of oxygen or lead contamination.
e) Caustic cracking of mild steel.
f) Nitrate cracking of mild steel.

The alarming feature of SCC is that it is most prevalent in alloys considered the passive and non-corroding alloys which in many environments corrode uniformly at rates of less than 0,1 mm per year.

Many major equipment failures have occurred as a result of SCC, consequently design engineers should no longer be satisfied with information on tensile properties or fatigue properties. Environmental affected variations on usable strengths should be made accessible to designers.

Presenting reliable information to the engineer is not as simple as it might appear as shown by the following examples:

1. Solutions containing nitrates (i.e. the nitrate ion) crack mild steel but do not crack stainless steel.
2. Solutions containing chlorides (i.e. the chloride ion) crack austenitic stainless steels but not plain carbon steel.
3. Ammonia cracks brasses but not austenitic stainless steels or plain carbon steels.

General Characteristics of Stress Corrosion Cracking   105

In each case there is only a small amount of corrosion with negligible metal loss.

<u>Research has failed to give a foolproof explanation of the need for a specific corrodent to produce cracking in a given alloy.</u>

The engineer should always refer to the literature to find out if the environment under consideration is one reported to cause cracking in the alloy being used.

In Table 1 Johnson lists a series of common alloys in which SCC has been studied, and the compound or anion believed to be responsible for the corrosive action.

TABLE 1* Alloy-Corrodent Combinations in which Stress Corrosion Cracking has been Reported

Alloy	Environment
<u>Aluminium base</u>	
Al-Zn	Air
Al-Mg	$NaCl + H_2O_2$, NaCl solutions air
Al-Mg)	
Al-Cu-Mg)	Sea water
Al-Mg-Zn)	
Al-Zn-Cu	NaCl, $NaCl + H_2O_2$ solutions
Al-Zn-Mg-Mn)	Sea water
Al-Zn-Mg-Cu-Mn)	
Al-Cu-Mg-Mn	$NaCl + H_2O_2$ solution
Al-Cu)	$NaCl + H_2O_2$ solution
Al-Cu)	NaCl, $NaCl + NaHCO_3$, KCl, $MgCl_2$
Al-Mg	$CaCl_2$, NH_4Cl, $CoCl_2$ solutions
<u>Magnesium base</u>	
Mg-Al	(a) HNO_3, NaOH, HF solutions
	(b) Distilled water
Mg-Al-Zn-Mn	(a) $NaCl + H_2O_2$ solution
	(b) Coastal atmosphere; $NaCl + K_2CrO_4$ solution
	(c) Moist air + SO_2 + CO_2
Mg	KHF_2 solution
<u>Copper base</u>	
Cu-Zn)	
Cu-Zn-Sn)	NH_3 vapors and solutions
Cu-Zn-Pb)	
Cu-Sn-P	Conc. NH_4OH
Cu-Zn	Amines
Cu-Zn-Ni)	NH_3 vapors and solutions
Cu-Sn)	
Cu-Sn-P)	Air
Cu-As)	
Cu-P, -As, -Sb, -Ni) -Al, -Si, -Zn)	Moist NH_3 atmosphere
Cu-Si-Mn	
Cu-Zn-Si	Water vapor

Alloy	Environment
Cu-Zn-Sn-Mn	Water
Cu-Au	NH_4OH, $FeCl_3$, NHO_3 solutions
Cu-Zn ⎫ Cu-Zn-Mn ⎭	Moist SO_2; $Cu(NO_2)_2$ solutions
Cu-Mn	Moist SO_2; $Cu(NO_2)_2$, HCl, HNO_3 solutions
Cu-Zn plus minor amounts of Al, As, Be, B, Cd, Co, Au, Pb, Mn, Ni, Pd, Ag, Sr, Tl, Sn, Sb, Ba, Bi, Ca, Ce, Cr, Fe, Mg, P, Si, Te, Ti, Zr, Li, Nb, Mo, K, Se, Na, S, Ta	Moist NH_3 atmosphere
Cu-Ni-Si	Moist NH_3 atmosphere
Cu-Al-Fe	Steam
Cu-Be	Moist NH_3 atmosphere

Iron base

Alloy	Environment
Mild steel	(a) $NaOH + Na_2SiO_3$
	(b) $Ca(NO_3)_2$, NH_4NO_3 and $NaNO_3$ solutions
	(c) $HCN + SnCl_2 + AsCl_2 + CHCl_3$
	(d) Na_3PO_4 solution
	(e) Pure $NaOH$ solution
	(f) $NH_3 + CO_2 + H_2S + HCN$
	(g) $NaOH$, KOH solutions; Monoethanolamine solution + $H_2S + CO_2$, $Fe(AlO_2)_3 + Al_2O_3 + CaO$ solution
	(h) $HNO_3 + H_2SO_4$
	(i) $MgCl_2 + NaF$ solution
	(j) Anhydrous liquid NH_3
	(k) H_2S media
	(l) $FeCl_3$ solution
Fe-Cr-C	(a) NH_4Cl, $MgCl_2$, $(NH_4)H_2PO_4$, Na_2HPO_4 solutions
	(b) $H_2SO_4 + NaCl$ solution
	(c) $NaCl + H_2O_2$ solution, sea water
	(d) H_2S solutions
Fe-Cr-Ni-C	(a) $NaCl + H_2O_2$ solution, sea water
	(b) $H_2SO_4 + CuSO_4$ solution
	(c) $MgCl_2$, $CoCl_2$, $NaCl$, $BaCl_2$ solutions
	(d) CH_3CH_2Cl + water
	(e) $LiCl$, $ZnCl_2$, $CaCl_2$, NH_4Cl solutions
	(f) $(NH_4)_2CO_3$ solutions
	(g) $NaCl$, NaF, $NaBr$, NaI, NaH_2PO_4, Na_3PO_4, Na_2SO_4, $NaNO_3$, Na_2SO_3, $NaClO_3$, $NaC_2H_3O_2$ solutions
	(h) Steam + chlorides

Alloy	Environment
	(i) H$_2$S solutions
	(j) NaCl + NH$_4$NO$_2$ solution
	NaCl + NaNO$_2$ solution
	(k) Polythionic acids
Nickel base	
Ni	NaOH, KOH solutions; fused NaOH
Ni-Cr-Fe	NaOH + sulfide solution, steam
Ni-Cu	Fused NaOH, H$_2$SiF$_6$ solution, chromic acid, sulfonated oil, steam
Ni-Cu-Al)	
Ni-Cu)	
Ni-Al)	HF acid vapor
Ni-Cr-Fe)	
Ni-Cr-Fe-Ti)	
Miscellaneous alloys	
Au-Cu-Ag	FeCl$_2$ solutions
Cu-Au	HNO$_3$ + HCl, HNO$_3$, FeCl$_3$, NH$_4$OH solutions
Ag-Au	HNO$_3$ + HCl, HNO$_3$, FeCl solutions
Ag-Pt	FeCl$_3$ solutions
Pb	(a) Pb(OAc)$_2$ + HNO$_3$ solutions
	(b) Air
Ti alloys	Solid NaCl, Temperature > 550 °F or 290 °C
	Electrolytes, ambient temperatures
Ti-6Al-4V	Liquid N$_2$O$_4$
Zr	FeCl$_3$ solutions

*This table, except for minor additions, is taken from H E Johnson, Thesis, University of Alberta, Edmonton, Alberta, 1964.

GENERAL FINDINGS ON STRESS CORROSION CRACKING

1. The application of stress and corrodent must be simultaneous for failure by SCC to occur. Alternate application does not give rise to SCC. Therefore to prevent SCC, eliminate stresses or eliminate corrosion.

2. Surface or sub-surface tensile stresses must be present. These may be acting or residual. SCC's often propagate with no externally applied stress. Those from welding, forming, or machining are sufficient. These dangerous residual tensile stresses should be eliminated by stress relief anneals or by putting surface layers of material in compression by rolling, shot-peening or other means.

3. The corrosive environment may be removed or rendered less harmful. This may be possible by changing the pH of the fluid, elimination of oxygen or chloride from a solution, or blanketing a part of a system with an inert gas during the time it is not in use. Oxygen and chloride contents in parts per billion is prescribed for water used in stainless steel reactor components, Fig. 1.

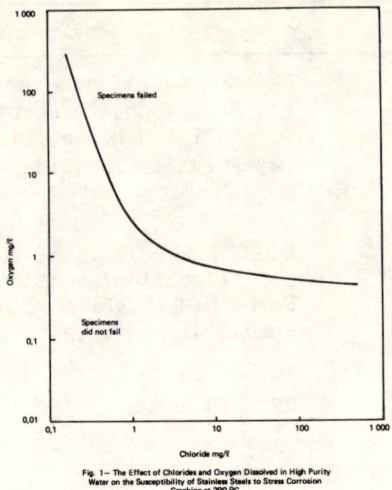

Fig. 1 — The Effect of Chlorides and Oxygen Dissolved in High Purity Water on the Susceptibility of Stainless Steels to Stress Corrosion Cracking at 290 °C

4. Inhibitors, cathodic protection or design changes may render the environment less harmful, e.g. avoiding crevices where chlorides may concentrate in stainless steel heat exchanger units. Finally if other means fail, it may be necessary to replace the metal in the particular application with another material that does not fail in the specific environment.

5. There have been numerous attempts to establish threshold stresses below which SCC's would not occur for various alloys susceptible to SCC. Residual stresses above the yield strength of the material are often responsible for failure, e.g. a component cold worked but not stress relieved. Laboratory tests have revealed threshold stresses in 42 % boiling magnesium chloride for stainless steel, NH_3 - air - CO_2, brass, $3\frac{1}{2}$ % NaCl solution aluminium alloys, etc., Fig. 2.

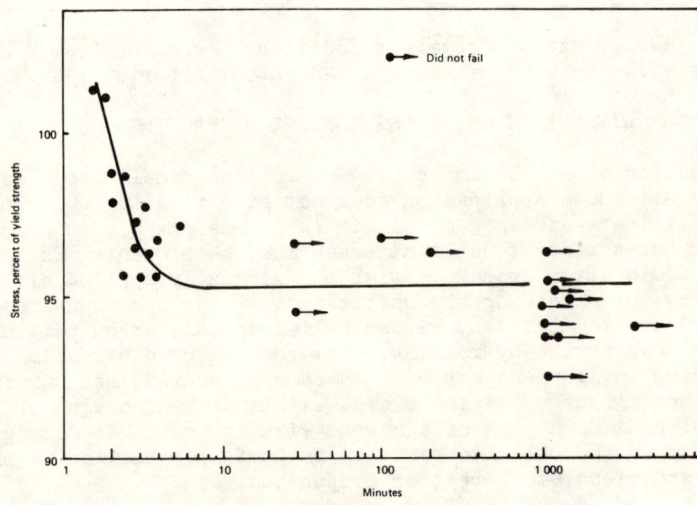

Fig. 2 — Stress-Time to Failure Curve for Annealed Specimens of Magnesium Alloy

6. Pure metals are generally considered immune to SCC. Impurities in the p.p.m. range can however cause susceptibility, e.g. Cu + 0,004 % P .
7. Heat treatment and hence microstructure affects susceptibility.
8. High temperature promotes SCC. Some systems appear to have a limiting temperature and chloride cracking of austenitic stainless steel is not reported below 70 °C .
9. There is a time lag between exposure and the development of faults. This is complex. Stainless steel cracks within 24 h of immersion in hot concentrated magnesium chloride solutions, but nitrate cracking of mild steel has occurred in weeks, months, and even years.
10. The rate of crack propogation is high, 1 to 4 mm per hour which is several orders of magnitude less than brittle fracture.
11. The features of SCC are that the cracks create the impression of brittleness but the uncracked regions are normally ductile in a tensile test.
12. Cracking is inter- or trans-granular or mixed. The mode of cracking can frequently be changed, e.g. mild steel suffers intergranular failure in nitrate solutions but transgranular failure in carbon monoxide/carbon dioxide/water mixtures. When the cracking is intergranular it is usually possible to detect structural or compositional features at the grain boundaries, e.g. segregation or precipitates with different electrochemical properties. Either the precipitate is anodic and dissolves, or it is cathodic and dissolution occurs in the adjacent matrix.

A Theory of Intergranular SCC

1. Corrosion at the grain boundary forming a deep point pit as opposed to saucer shaped.
2. Tearing along the grain boundaries due to stress build up at the crack tip. This relieves the stresses and corrodent diffuses into the crack.
3. Corrosion occurs again and there is stress build up at the crack tip in part due to the products of corrosion which occupy a larger volume than the metal destroyed. When the residual stress is high enough tearing occurs again.

The process is repeated over and over again. Note the grain boundaries are sometimes mechanically weaker than the grain because of a lower alloy content, e.g. a depletion in copper atoms in aluminium alloys (Al plus Cu, Mg, Mn, Si) under certain conditions of heat treatment.

A Theory of Transgranular SCC

When the cracking is transgranular the microstructure has less influence.

1. Stresses destroy the passive film or polarized condition at narrow regions normal to the applied stress.
2. Corrosion pitting penetrates to a vicinity of localized embrittlement within the material.
3. A mechanical burst follows.
4. Crack slowed down and brought to a stop by the presence of ductile regions in the material.
5. No further crack propagation until the slower electrochemical attack makes possible another rapid mechanical burst.

Metal Physicists inform us that SCC may follow intercrystalline or transcrystalline paths and give an explanation in terms of electron microstructure. In materials having complex slip systems or high stacking fault energies, cracks will most probably follow intercrystalline paths.

In materials having low stacking fault energies, short range order, and having planar arrays of dislocations after plastic deformation cracks may follow transcrystalline or intercrystalline paths depending on the corrodent and the extent of plastic deformation.

<u>Hydrogen embrittlement</u> cannot be divorced from SCC. Some authorities believe that hydrogen is absorbed at the tip of the crack at a rate faster than it can disperse from that region by diffusing into the bulk of the specimen. In one of several possible ways it causes embrittlement of a small volume of metal which cracks under the influence of tensile stress.

CAUTION IN APPLYING METHODS TO PREVENT SCC

Numerous examples are given in the technical literature of methods reported to have been used to eliminate SCC under a particular set of materials and service conditions. The engineer should view recommendations given in these papers with extreme caution since a cure under one set of conditions can be a disaster under another. Small changes in the chemical composition of the metal or electrolyte and the service conditions can cause big differences in SCC susceptibility or create other problems. An example experienced recently was a stress relief anneal at 850 °C (slow cool) applied to an austenitic stainless steel. This would have been quite safe on an extra low carbon grade (< 0,03 % C) but was applied to a normal carbon content grade (0,07 % C) and was sensitized and thus made susceptible to intergranular corrosion.

Another method often recommended to prevent SCC is annealing after fabrication. This method can hardly be depended upon as a means of control, except in borderline cases involving very mild cracking conditions. Major difficulties are often associated with thermal stressing during cooling of complicated equipment and accidental stresses in handling after the stress relief anneal.

Inhibitors are capable of eliminating stress corrosion cracking in particular environments, for instance alkaline phosphates in boiler water. The problem remains of ensuring the required concentration of the inhibitor in contact with all critical surfaces such as those exposed only to occasional splashing above the usual liquid level. In this same general class are chemicals that serve as oxygen scavengers and avoid SCC by elimination of oxygen from environments, as in boiler water where cracking does not occur in the complete absence of oxygen.

Since there is ample evidence that SCC is an electrochemical phenomenon, it should be possible to control it by the application of protective currents. However there are practical difficulties in locating anodes in complex equipment so as to ensure that an adequate current will reach all surfaces that require this protection.

The risk for SCC in steam generator tubing can be minimized by introducing compressive stresses in the tube surface. This is achieved by shot-peening. Essential parameters such as particle size, air pressure, and tube feed rate have to be chosen to give an optimum production rate, uniformity in treatment, and penetration depth of the shot-peening.

The residual stress level in weldments is controlled by such factors as coefficient of expansion, melting point, phase changes, volume change during solidification, difference in thermal expansion between dissimilar metals, energy input, preheat, annealing or softening temperature of the material, and the welding technique. Prevention of SCC susceptibility in weldments is practised by eliminating or minimizing the level of residual stress by control of the aforementioned variables and post-welding stress relieving treatments such as peening and heat treatment.

External stress corrosion cracking of austenitic stainless steel used in chemical process plant has caused leakage in numerous plants. The concentration of chloride ions on the surface of hot stainless steel is the most common cause of external SCC and most procedures to prevent failures are designed to avoid this concentration. Chloride ion concentration usually results from the evaporation of rain water, fire water, or process leaks which contain or have become contaminated with chloride. In coastal regions airborne salts are frequently the cause of external SCC. Wrapping the equipment with aluminium foil before applying lagging will reduce the risk of SCC. An adequate overlap (50 mm) should be provided at joints. Painting the outside of stainless steel before application of insulation provides a barrier to prevent water reaching the metal surface. Heat-resistant aluminium silicone paint (40 micron film thickness) or zinc-free silicone alkyd paint and clear silicone lacquers have been used successfully in the range of 80-200 °C.

Evaluating the Resistance of Materials to SCC

1. A constant load is applied that will produce tensile stresses over the cross-section in an axially loaded specimen or in the outer fibres of a bend specimen.
2. Stresses are induced by subjecting a specimen to a predetermined and constant deflection. The effective stresses in a specimen subject to a constant load will increase as cracks penetrate into the material. The specimen eventually fails from overload. In a specimen subject to constant deflection, cracks <u>may</u> penetrate into the specimen until the stresses are no longer sufficient to propagate the crack, and will stop before final failure of the material. The test should be chosen to simulate service conditions.
3. <u>Fracture mechanics approach</u>. This approach asks the question: 'What is the largest defect size that will not propagate under the expected stress distribution in the component under its worst working condition?'

The duration of conventional tests is governed primarily by the time taken for a corrosion pit to form and grow to a sufficient size to initiate a stress corrosion crack. Fracture mechanics, on the other hand, deals only with the growth of cracks and does not therefore consider the initiation phase of SCC. Typical apparatus is the Brown test apparatus, Fig. 3. Typical results are given in Fig. 4.

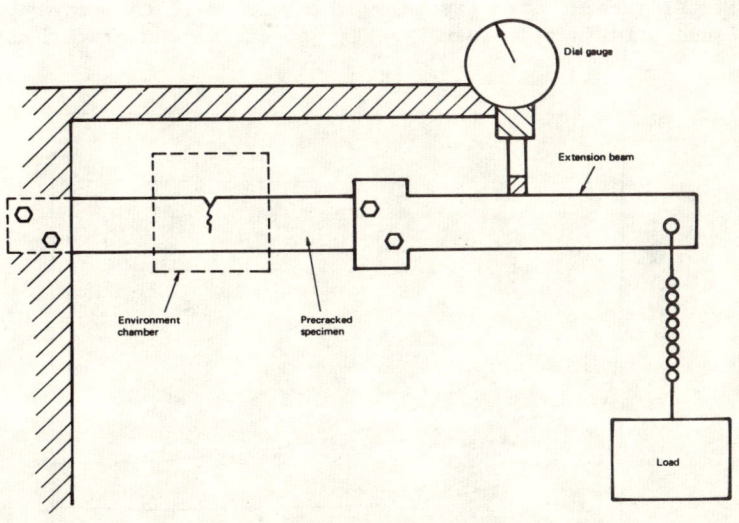

Fig. 3 — Brown Test Apparatus

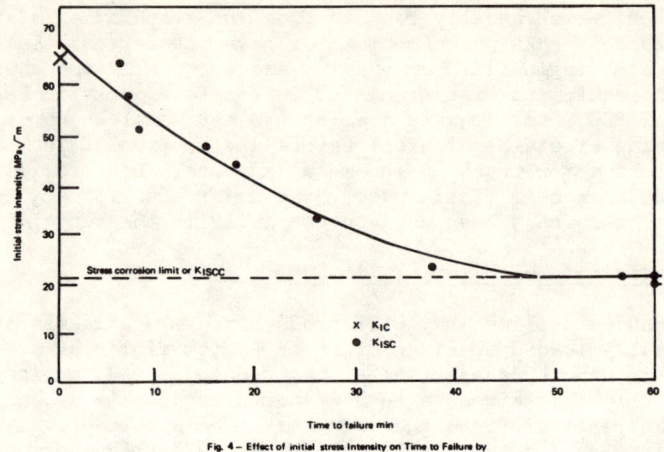

Fig. 4 – Effect of Initial Stress Intensity on Time to Failure by Stress Corrosion for a 4½ % Ni-Cr-Mo Steel

K_{IC} measures the resistance to crack extension under plain strain conditions at the crack front, these being the most favourable for crack propagation. K_{IC} is an intrinsic property of a material.

$$K_{IC} = \frac{P_F \, Y}{B \, W}$$

P_F = fracture load
Y = constant, allowing for the shape of the specimen
B = breadth of specimen
W = thickness of specimen

K_{ISC} is a materials x environment parameter and in this respect differs from K_{IC}.

K_{ISC} is a critical stress intensity below which SCC will not propagate. A very useful relationship is that between K_{IC}, K_{ISCC}, stress and crack length, Fig. 5.

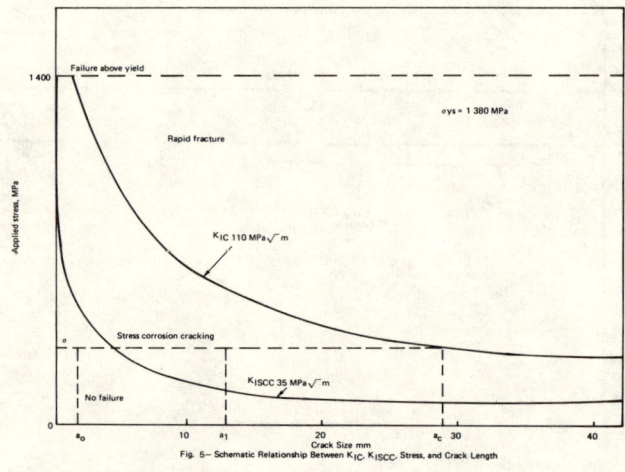

Fig. 5 – Schematic Relationship Between K_{IC}, K_{ISCC}, Stress, and Crack Length

From Fig. 5

a) a component containing a crack of length a_c loaded with the stress shown will fail on initial loading,
b) a component containing a crack of length a_1 loaded under the same stress will not fail on inital loading,
c) in a corrosive environment if stress corrosion cracking leads to an increase in length to a_c failure occurs,
d) if the stress intensity is below the stress corrosion cracking limit, K_{ISCC} as would be at a stress of σ and a flaw size a_0, crack growth will not occur.

To summarize, the designer needs information on four parameters:

1) The threshold stress for crack propagation.
2) The critical crack length or flaw size that can be tolerated without leading to fast fracture.
3) The time required for a crack to propagate through a given section thickness.
4) The time taken to initiate a crack or flaw in the corrosive environment.

In practice, it is usually necessary to design on the basis of data obtained in dry conditions and make an allowance for stress corrosion rather than design on the basis of K_{ISCC} values, because environments encountered in service are often quite different in their effect on materials from those used in the test laboratory. Those responsible for furnishing designers with data on materials would do well in future to concentrate on measuring crack growth rates which are less subject to variation than K_{ISCC}.

It should also be noted that materials have to be sufficiently high strength to be studied by current K_{ISCC} test procedures. Of the three classes of stainless steels, viz. austenitic, ferritic, and martensitic, it is only martensitic that are of sufficiently high strength.

THE SABS AND STRESS CORROSION CRACKING

The main objectives of the SABS can be summarized as follows:

- The promotion of standardization
- The preparation of specifications and codes of practice
- The setting up of testing facilities
- The administration of quality certification, e.g. the SABS mark on products
- The rendering of assistance to the State, public bodies, and private enterprises on standardization matters

In its effort to achieve these objectives, the SABS has a staff of nearly 1 200 of whom more than 700 are highly qualified scientists and technicians. The Metallurgy Division is mainly responsible for corrosion and corrosion control, and can offer the engineer a range of standard test procedures. The importance of standard test procedures cannot be over-emphasized since unfortunately a new test procedure for studying the stress corrosion resistance of materials appears to be invented by each investigator.

Internationally accepted test methods are obviously required to enable valid comparison of results to be made on a world-wide basis. Assistance in this respect is now under way by means of ISO Technical Committee 156: Corrosion of Metals (Secretariat USSR). Working Group 2: Stress Corrosion Cracking held their first meeting in London in November 1977. At a subsequent meeting on 30 May 1978, this Working Group agreed to recommend that four ASTM specifications be circulated as draft ISO specifications for comment:

ASTM G35-73   Testing stainless steel in polythionic acid
ASTM G38-73   Preparation and use of 'C' ring s.c. specimens
ASTM G39-73   Preparation and use of bent beam s.c. specimens
ASTM G44-75   Alternate immersion s.c. testing in $3\frac{1}{2}$ % sodium chloride

ASTM G49-76 'Preparation and use of direct tension s.c. specimens' is to be compared with existing National Standards and any ISO Standard for tensile test specimens.

New standards were considered desirable to cover

a) constant strain rate s.c. testing,
b) drop evaporation s.c. testing for stainless steels,
c) s.c. testing with pre-cracked specimens,
d) s.c. and hydrogen cracking of steel in sour oil or gas containing hydrogen sulphide.

ASTM is preparing a standard covering (a) and will liaise with laboratories in Germany, Sweden, and UK with special interests in the same method. Sweden, Finland, Germany, and Czechoslovakia are interested in (b) and will exchange experience. (c) involves correlation with standards for plain strain fracture toughness test methods and national standards are to be studied. (d) is important commercially as well as technically, and interested parties in various countries have been requested for their views.

The object of ISO (International Organization for Standardization) is to promote the development of standards in the world with a view to facilitating international exchange of goods and services, and to developing mutual co-operation in the spheres of intellectual, scientific, technological, and economic activity. The results of ISO technical work are published as International Standards or Technical Reports.

The member bodies of ISO are the bodies most representative of standardization in their respective countries. Only one body in each country may be admitted to membership and, of course, the SABS represents South Africa.

ISO technical work is undertaken by technical committees and their subcommittees and working groups, subject to the general authority of the ISO Council. It is planned and co-ordinated by the Planning Committee (PLACO), the technical divisions, and the Central Secretariat. Technical Committees for clearly defined fields of activity are created by Council, e.g. TC 156 'Corrosion of metals', to undertake the preparation of International Standards in a new field. The member countries of Technical Committees are the member bodies which have expressed their willingness to participate actively (P-members) or their desire to be kept informed of the progress of work (O-members). South Africa has P-membership of TC 156 and documents are received by the SABS from this Committee for close study, making comments and proposals where necessary, bringing the documents to the notice of South African industry, etc. and eliciting their comments where possible, and finally voting on proposed international standards (specifications).

The Metallurgy Division is more than willing to circulate titles of documents in the first instance to Companies etc. wishing to actively participate in the drawing up of standards such as SCC test procedures. Photostat copies of documents would then be sent on if required for comment.

The Technical Committee Secretariat is responsible to ISO Council and to the member countries of the Committee for all activities of the Technical Committee, including its subcommittees and working groups. The Technical Committee Secretariat should act in all respects as an international secretariat and must not be influenced by national considerations in the pursuit of its work. Working as a Secretariat, a member body shall maintain strict neutrality and distinguish sharply between proposals it makes as a member country body and proposals made in its capacity as Secretariat. The Secretariat for TC 156 is held by Russia, but working groups are active in both the East and the West.

CONCLUDING REMARKS

Stress corrosion still requires considerable research, particularly to explain the need for a specific corrodent to produce cracking in a given alloy. Whilst research has not yet produced all the answers, worthwhile advances are continuously taking place. Many alloys are presently available which can be used in place of susceptible ones. Construction and maintenance procedures based on a fundamental general knowledge of the mechanism of SCC would prevent many failures. Properly planned corrosion tests in anticipated environments would forecast otherwise imminent difficulties. Standardized SCC test procedures must be developed to enable world-wide comparative data to be available to the engineer. South African industry, research organizations, etc. are invited to participate via the SABS to ISO in achieving this end.

ACKNOWLEDGEMENT

Particular thanks are due to Mr J van Heerden of the Metallurgy Division of the South African Bureau of Standards for making useful comments.

REFERENCES

Notes and figures were compiled by reference to the following literature:

Advances in Corrosion Science and Technology (1973). Vol. 3. Plenum.
ASTM special publication (1966). Stress corrosion testing.
ASTM special technical publication No. 264 (1960). Stress corrosion cracking of austenitic chromium-nickel stainless steels.
Barer, R.D., and B.F. Peters (1970). Why Metals Fail. Gordon and Breach.
Bosich, J.F. (1970). Corrosion Prevention for Practicing Engineers. Barnes and Noble.
ISI Publication 121 (1968). Fracture toughness.
ISO/TC 156. Documents.
Logan, H.L. (1966). The Stress Corrosion of Metals. John Wiley.
National Association of Corrosion Engineers (1969). Fundamental aspects of stress corrosion cracking. Proceedings of Conference.
Scully, J.C. (1975). The Fundamentals of Corrosion. Pergamon.
The Institute of Chemical Engineers (1978). Guide notes on the safe use of stainless steel in chemical process plant.
The Institution of Corrosion Technology, England (1970-77). Newsletters.

FACTORS CONTROLLING HAZ AND WELD METAL TOUGHNESS IN C-Mn STEELS

R. E. Dolby

The Welding Institute, Abington, Cambridge, England

ABSTRACT

Recent information on metallurgical, welding procedural and other factors which control the toughness of heat affected zones (HAZs) and weld metals is reviewed. The paper deals with C-Mn and C-Mn microalloyed steel weldments in the as-welded condition.

PART 1. HEAT AFFECTED ZONES

INTRODUCTION

The risk of a brittle fracture developing, for example, from HAZ hydrogen-induced cracks, reheat cracks, fatigue cracks, or lack of root or sidewall fusion will depend on the properties of the HAZ microstructure which locally surrounds the tips of these defects. These properties, including toughness, are mainly determined by steel composition and welding procedure since these control the HAZ microstructure.

To study factors influencing HAZ toughness two experimental approaches are possible, tests being carried out either on specimens containing thermally simulated microstructures or on specimens extracted from welded joints. Because of the difficulties associated with the thermal simulation technique (Dolby & Widgery, 1971), emphasis at The Welding Institute is normally placed on testing welded joints. Also the preferred test method is a fracture initiation test, e.g. the K_{IC} or COD test, rather than the Charpy V test, because the toughness measurement can be more easily related to the local microstructure at the tip of the specimen crack or notch and to the micromechanism of fracture as seen from a fractographic examination of failed specimens.

TOUGHNESS OBSERVATIONS

Embrittlement can be found either in the transformed (visible) or in the sub-critical HAZ. In general, the transformed HAZ shows greatest embrittlement in the grain-coarsened region where peak temperatures during welding exceed $\sim 1200^\circ$C. A distinction must be made between HAZ regions which are not reheated, e.g. those situated at the toes of weld, and those regions which have been reheated and grain refined by subsequent passes. The latter generally have higher toughness.

A further observation is that toughness of the HAZ can be similar, better, or worse than that of the parent steel. In grain refined C-Mn steels, it is generally worse. Finally, it is important to distinguish between resistance to the various micro-mechanisms of fracture, i.e. cleavage, microvoid coalescence, and intergranular fracture. The metallurgical factors controlling resistance to these mechanisms can be different. Intergranular fracture is encountered only in one or two restricted situations, e.g. in the stress relieved HAZs of certain low alloy steels, and the following discussion concentrates on the other two mechanisms.

THE TRANSFORMED (OR VISIBLE) HAZ

The prior austenite grain size close to the fusion boundary coarsens with increasing heat input, being around 100μm at ∿3kJ/mm and over 300μm for high heat input HAZs (>20kJ/mm). The presence of grain boundary pinning agents in the steel, e.g. AlN and NbCN does not usually alter the grain size close to the fusion boundary, but AlN, in particular, narrows the width of the coarse-grained region by restricting austenite grain growth at lower peak temperatures. Within the austenite grains the transformed microstructure in the HAZ is usually a complex mixture of two or more of the following constituents, placed in approximate decreasing order of transformation temperature: primary ferrite associated with austenite boundaries or in transgranular Widmanstatten plate form; areas of high carbide content ranging from pearlite to forms in which the carbides have precipitated as rods or spheroids; bainitic colonies in which the ferrite plates have grown in a side-by-side manner resembling upper bainite; lower bainite; martensite. Some of these microstructures are illustrated in Fig 1. The constituents, their colony size, and their proportion depend on steel composition and welding procedure.

Typical HAZ transition curves obtained from slowly loaded COD tests on C-Mn and low alloy steels in the as-welded condition are shown in Fig 2. The shape of each curve depends on two main properties of the HAZ:

    a.   resistance to cleavage, and

    b.   resistance to microvoid coalescence.

The transition from cleavage to microvoid coalescence with increase in test temperature reflects the greater difficulty of initiating cleavage fractures at the higher temperatures. Because the metallurgical factors controlling these fracture mechanisms are different, the curves for different steel compositions can cross, as seen in Fig 2, and it is quite possible for compositional variations which improve cleavage resistance to cause a lower resistance to microvid coalescence. When optimising steel compositions for HAZ toughness, a high cleavage resistance must usually be sought in the first instance, since fracture initiation by this mechanism can lead to catastrophic failure, an event only rarely associated with fracture by microvoid coalescence.

Improving Resistance to Cleavage

C-Mn Steels. Figure 3 shows schematically that with increasing heat input (expressed as a cooling time from 800°C to 500°C) the HAZ cleavage resistance of a typical C-Mn steel, say 0.17%C-1.3%Mn, passes through a maximum at $\Delta t_{800-500}$ ∿10sec, equivalent to a heat input of about 2.5kJ/mm on 25mm plate. At lower heat inputs, the increasing proportion of martensite and associated higher HAZ hardness results in a lower cleavage resistance (higher transition temperature) and, at higher heat inputs, coarse ferrite structures derived from coarse austenite grains, also show a lower resistance to

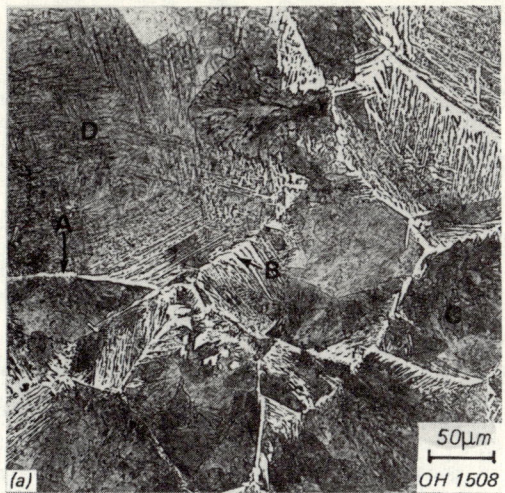
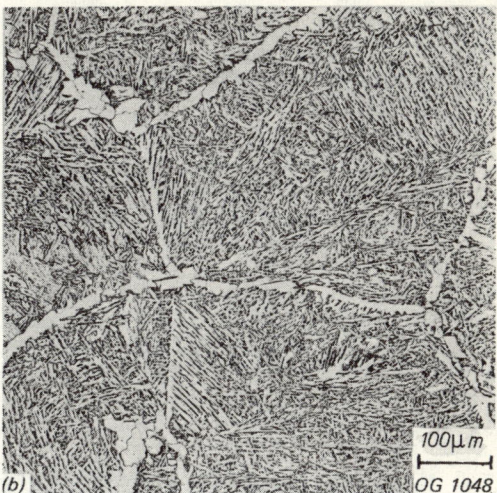

Fig. 1. HAZ Microstructures of: a) MMA weld in 0.21C-0.07Mn steel at 2kJ/mm showing grain boundary ferrite A, ferrite side plates B, ferrite with interphase carbide C, martensite D; b) electroslag weld in 0.17C-1.27Mn-0.02Nb steel at 25kJ/mm showing grain boundary ferrite and upper bainite colonies.

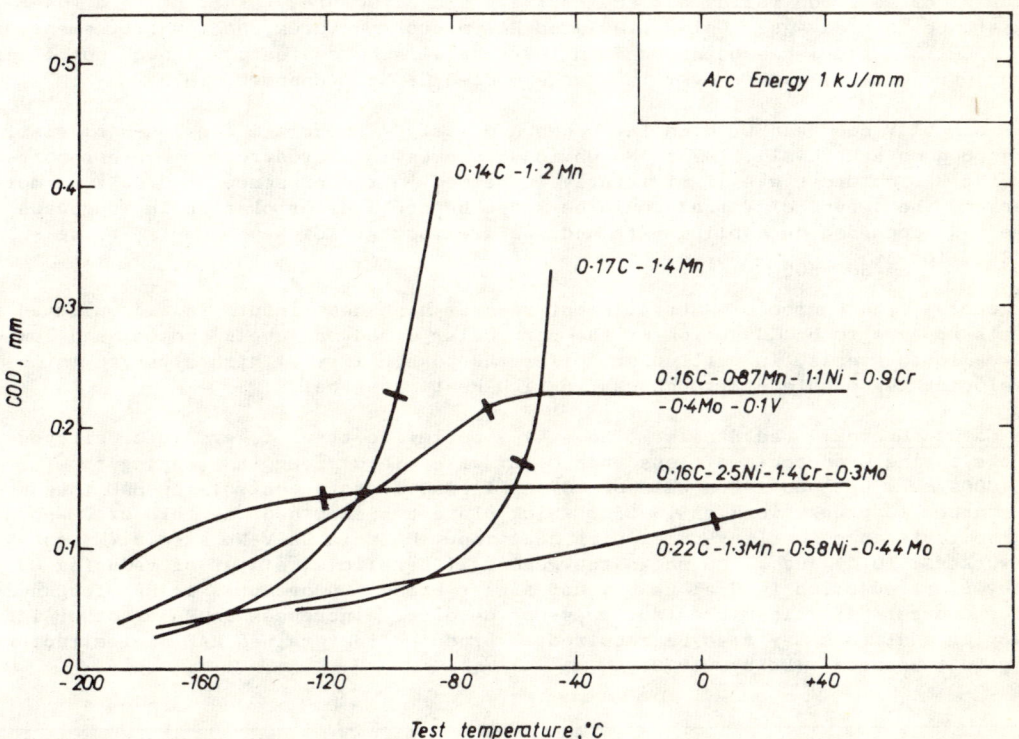

Fig. 2. COD test data on grain coarsened HAZ regions of various steels. Specimen size 10 x 10mm section. Bars on lines indicate cleavage initiation below this temperature

cleavage despite the general fall in HAZ hardness. Since the IIW carbon equivalent (CE) for hydrogen cracking[1] is, to a large extent, indicative of the chances of forming martensite, high CE steels tend to show a lower resistance to cleavage at heat inputs equivalent to $\Delta t_{800-500} < 10$ sec, particularly if the C level itself is high. Steels having high C contents also tend to show lower toughness at high heat inputs. However, the compositional factors controlling the HAZ cleavage resistance of C-Mn steels are not well understood and HAZ toughness cannot yet be readily predicted from knowledge of composition and welding procedure, although some progress has been made recently. (Dolby, 1979)

No general correlation between parent plate toughness and HAZ toughness can be expected. For example, two steels of similar chemistry but of different grain size and toughness caused by a different rolling or heat treatment practice, would be expected to have similar HAZ toughness when welded under identical procedures, since the HAZ microstructures should be similar. However, limited correlations may be possible if specific steel types only are considered.

C-Mn Microalloyed Steels. Steels containing microalloying elements such as Nb and V show similar if not improved toughness at low heat inputs ($\Delta t_{800-500} <$ sec) compared with non-microalloyed steels of comparable C and Mn content. (Hannerz, 1975; Bonomo & Rothwell, 1971). However, with increasing heat input, tests on simulated and weld HAZ microstructures show that resistance to cleavage of the microalloyed steels falls more rapidly than for C-Mn steels, Fig 4. This has been explained by Nb depressing the austenite transformation temperature at slow cooling rates and forming fine bainitic structures of high hardness (Dolby, 1976). The increase in hardness offsets the refinement in ferrite grain structure, resulting in a lower resistance to cleavage. Using simulated HAZ microstructures, HAZ embrittlement of a 0.19C-1.3Mn steel has been found to increase with Nb content, although not linearly, being particularly marked over the range 0.01-0.04%Nb. (Hannerz, 1975)

The role of V has been studied in Sweden and Italy. (Hannerz & Jonsson-Holmquist, 1974; Bonomo & Rothwell, 1971) No obvious changes in microstructure were reported but the HAZ hardness was significantly increased in the presence of 0.03%V or more. However, the amount of V that could be added before a deterioration in toughness was observed, depended on cooling rate and was greatest at low heat inputs, being $\sim 0.15\%$V for $\Delta t_{800-500}$ = 33secs.

In general, the degree of embrittlement seen at high heat inputs in microalloyed steels appears to be a function of the particular C and Mn levels chosen, and low C microalloyed steels, in particular, offer the possibility of high strength while developing reasonable HAZ toughness in high heat input welds.

Low Carbon Microalloyed Steels. There is a worldwide trend towards microalloyed steels having C contents of less than 0.12% with yield strengths ranging from 400-600N/mm^2, but published data on compositional factors controlling HAZ toughness is scarce. Japanese work on 16-25mm thick plate has examined the role of C and Ni on the resistance to cleavage of high heat input HAZs in Cu-V-Nb steels,(Myoshi & co-workers, 1974) and Fig 5 shows the generally beneficial effect of reducing C. However, a reduction in C alone may not always bring improvements in HAZ toughness since coarser bainitic microstructures may develop. Increases in Mn or other lean alloying additions may also be required to produce fine grained HAZ microstructures of good toughness. (Dolby, 1979)

---

[1] $CE = C + \dfrac{Mn}{6} + \dfrac{Ni + Cu}{15} + \dfrac{Cr + Mo + V}{5}$

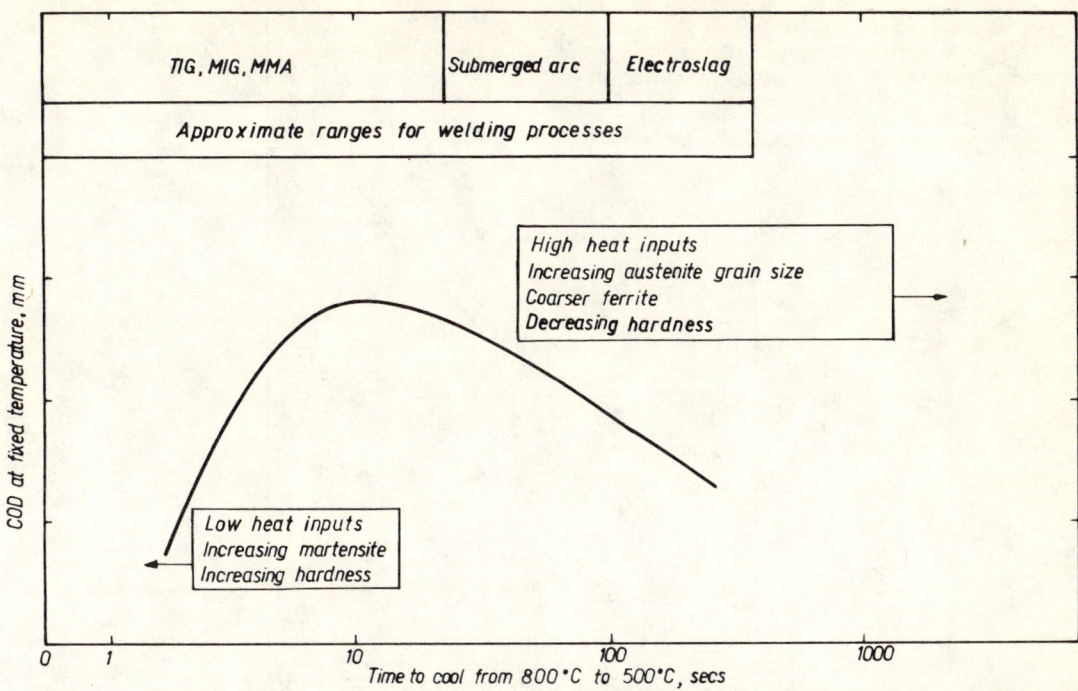

Fig. 3. Effect of cooling time and heat input on HAZ cleavage resistance of a C-Mn steel. Schematic diagram.

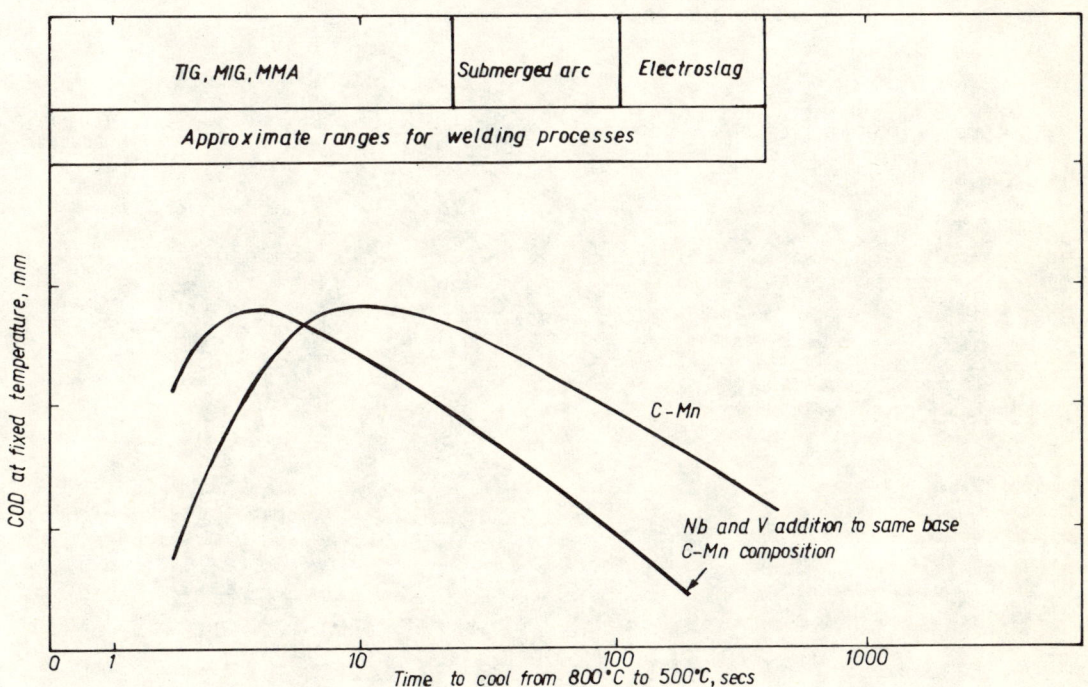

Fig. 4. Schematic diagram showing effect of Nb and V on HAZ cleavage resistance.

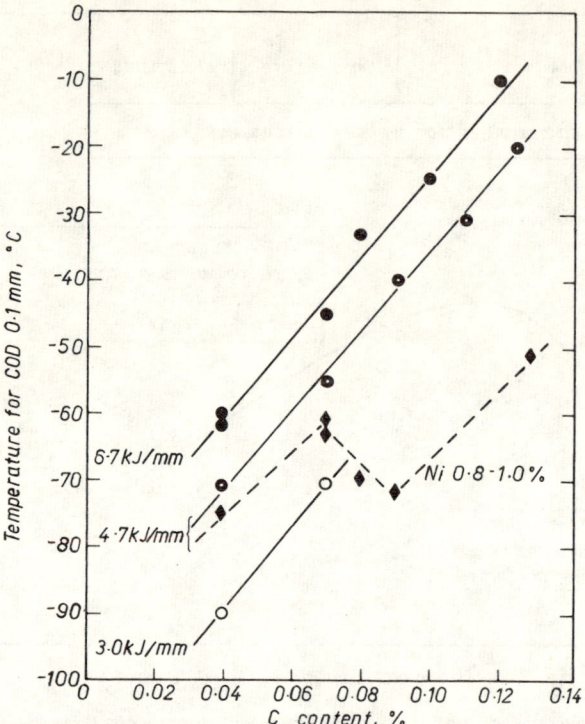

Fig. 5. The effect of carbon content on the HAZ toughness of a Cu-V-Nb and Cu-Ni-V-Nb steel (after Myoshi and co-workers, 1974).

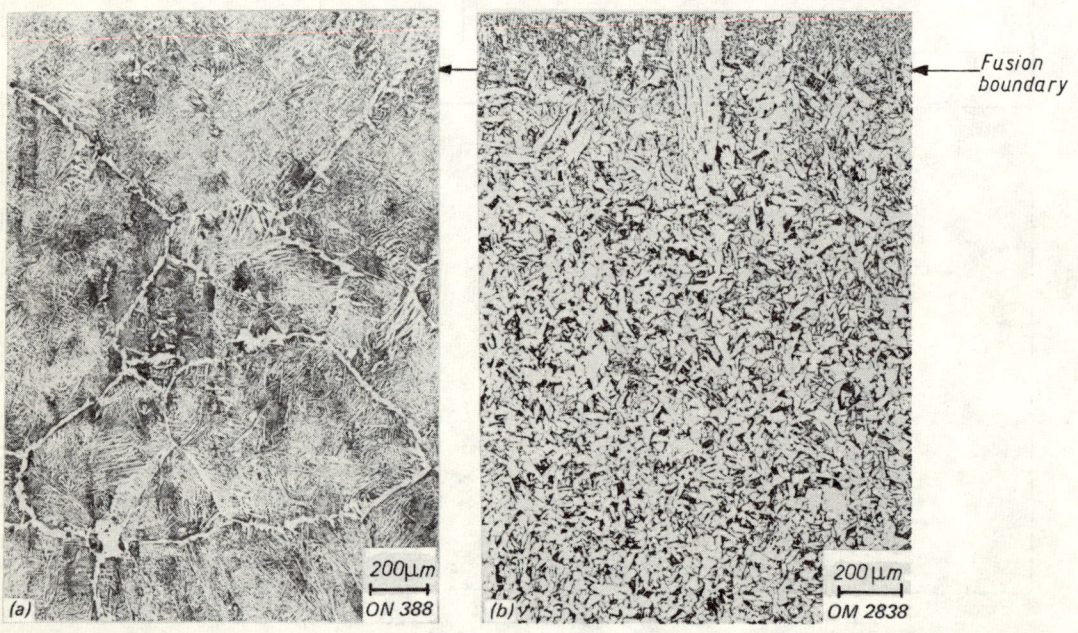

Fig. 6. Micrographs showing the effect of Ti in restricting HAZ austenite grain growth in electroslag welding: a) 0.18C-1.36Mn steel; b) 0.08C-1.39Mn-Ti steel.

Several special steel developments are now taking place to attempt to improve HAZ toughness at high heat inputs. Most of these are low C microalloyed steels and do not use Nb for reasons outlined earlier. Of particular interest are the developments on Ti-containing steels. TiN particles are extremely stable and in a certain particle size range can considerably restrict HAZ austenite grain growth as seen in Fig 6. Markedly improved HAZ toughness is reported and several commercial steels are now available around the world.

Improving Resistance to Microvoid Coalescence

The principal controlling metallurgical factor for all steel grades is the cleanness of the steel with respect to non-metallic inclusions. In steels which are Al-treated, the expected inclusion types will be Type II or Type III MnS with some alumina inclusions. Improvements in HAZ toughness in these steels can be brought about by lowering the S level, this directly reducing the volume fraction of MnS inclusions. Improvements can also be achieved through shape control of the MnS inclusions, e.g. by rare earth metal additions.

For steels which are not Al-treated the expected inclusion types are silicates or duplex sulphide-silicates, and here improvements in toughness can be achieved only through a vacuum degassing treatment which should eliminate silicate inclusions. Control of S would improve toughness still further.

THE SUBCRITICAL HAZ

Embrittlement at pre-existing defects sited in the subcritical HAZ where peak temperatures are less than about $700°C$ has been widely studied because of its relevance to Wells wide plate test results, Fig 7. It has been established that embrittlement is a reduction in cleavage resistance and that it is important mainly in C and C-Mn steels, the mechanism being one of strain concentration with the hardness at the defect tip enhanced by dynamic strain ageing.(Dolby & Saunders, 1972)

The two principal metallurgical factors which need to be controlled are :
1. the ferrite grain size of the parent material, and
2. the interstitial nitrogen content.

For a high resistance to subcritical HAZ embrittlement steels should have low interstitial nitrogen levels and a small ferrite grain size. These requirements are generally met by normalised fine-grained C-Mn and C-Mn microalloyed steels containing Al, although not necessarily by controlled rolled steels where much of the nitrogen may not be combined with Al.

OTHER FACTORS

Nitrogen

Work on electroslag HAZ microstructures has shown that AlN particles in the parent steel are taken into solution in the coarse grained transformed HAZ. (Dolby, 1976) Assuming this occurs in lower heat input processes although to a smaller extent, interstitial nitrogen could act over a wide heat input range to lower cleavage resistance of HAZ regions. The total nitrogen in the parent steel therefore assumes some importance in controlling HAZ toughness. Nitrogen ageing at ambient temperatures is also a possible source of embrittlement of the HAZ, and the extent of any embrittlement should be assessed by ageing HAZ test specimens at $100°C$ before testing.

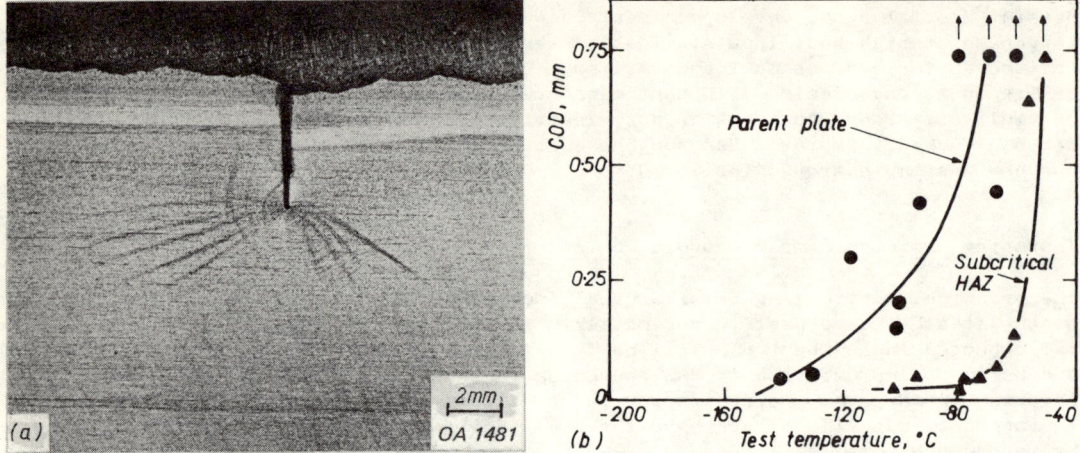

Fig. 7. a) Strain field patterns associated with welding over a notch sited in the subcritical HAZ; b) embrittlement in the sub-critical HAZ as seen from COD tests on a 0.15C-1.0Mn steel. Initial notch depth 5mm. Specimen size 10 x 10mm section.

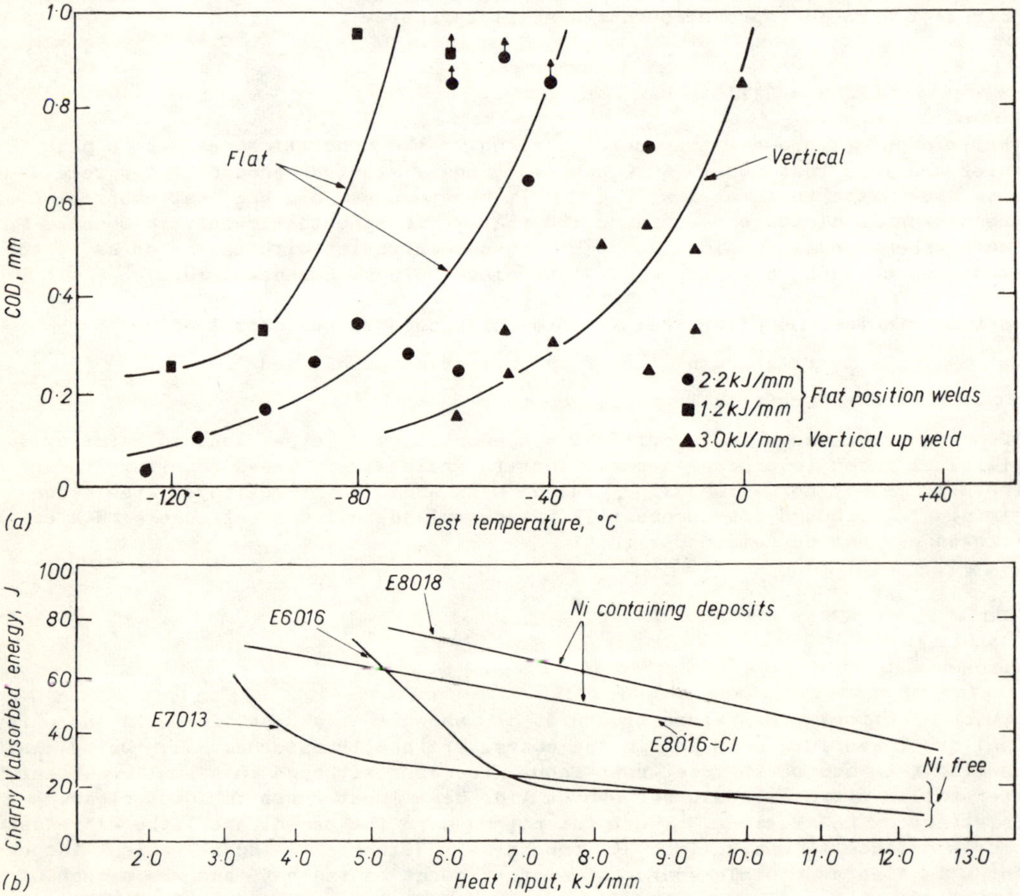

Fig. 8. Effect of heat input on the toughness of MMA welds: a) C-Mn deposits (Dawes 1972); b) C-Mn and C-Mn-Ni vertical-up deposits (Nicholson and Rogers 1974).

Hydrogen

The presence of hydrogen in as-welded HAZs can lower measured toughness values in slowly loaded tests (but not impact tests). The elapsed time between welding and testing is important and a decision is generally required when testing HAZs as to whether a low temperature hydrogen diffusion treatment is required ($\sim 150^\circ C$) or whether the hydrogen is left in the specimen to simulate conditions in practical structures where first loading could occur soon after welding.

WELDING PROCEDURAL AND TESTING FACTORS

Generally, improvements in toughness will be obtained by minimising the heat input, although this will usually reduce the deposition rate. Particular care should be taken to avoid very small weld beads at the toes of joints since high hardness, low toughness microstructures may result in these regions.

Due to the beneficial effect of grain refinement by succeeding passes, closely overlapping beads in fillets or butt joints should be helpful in improving toughness. The maximum amount of refinement can be obtained when the bead size and the angle of attack, i.e. the angle between the electrode and the joint preparation, are both small. Welding in the flat and overhead position gives the lowest angles of attack, whilst horizontal-vertical welding usually needs a high angle of attack. Closely overlapping beads, whilst ensuring good toughness in the depth of the weld, can still leave untempered HAZ regions at the toes. In critical situations, temper bead techniques can be used to soften the transformed HAZ but in practice this is an extremely difficult technique to control since the location of the bead needs to be close enough to the toe to adequately soften the HAZ without creating a new HAZ.

Measured values of HAZ toughness are affected by the method of extraction of specimens from welded joints. The difficulty in taking Charpy specimens from two-pass submerged-arc seam welds in line pipe is an example of the general problem that the specimen notch and propagation path cannot usually be sited completely in the HAZ. The toughness measured can therefore be influenced by both weld metal and parent steel microstructures. More realistic HAZ toughness assessments can be made using fracture initiation tests, and it is recommended that the method of extraction of specimens and their design should be such that the specimen notch and location matches the orientation of typical HAZ and weld defects. (Dolby & Archer, 1971)

PART 2. WELD METALS

INTRODUCTION

Brittle fractures in weld metal can initiate from solidification cracks, hydrogen induced cracks, reheat cracks, or other weld defects such as lack of fusion. The tips of such defects may be sited in weld metal microstructures not significantly affected by reheating from succeeding passes, or in regions which have been grain refined by reheating above the Ac_3 temperature.

A complicating factor in multipass deposits, not present in the HAZ situation, is that the weld composition, and therefore microstructure, can vary from one run to another through the joint thickness because of varying dilution. Regions near the root in a multipass butt joint can incorporate $\sim 70\%$ of the parent steel composition whereas, adjacent to the surface, the dilution by parent steel is usually less than 5%.

TOUGHNESS OBSERVATIONS

The toughness of weld metals can vary considerably, even when consumables of nominally the same type are assessed, and it is often difficult to predict for a given plate thickness whether the toughness of the weld metal is better or worse than the HAZ or parent steel.

Generally, basic coated manual metal arc (MMA) electrodes or basic fluxes for submerged arc or electroslag welding give better toughness than acid types, although basicity as such is not considered the controlling factor.

It is general experience that low heat inputs lead to the highest toughness, and the greatest problems in meeting specifications usually arise in the high heat input processes such as submerged arc, electrogas and electroslag welding. However, vertical-up welding can be classified as a moderately high heat input technique when significant weaving is employed, and this welding position can give low toughness deposits. Stringer beads deposited at low heat inputs usually give the highest toughness.

Figure 8a reproduces some MMA COD toughness data (Dawes, 1972) illustrating the detrimental effect of increasigg heat input for C-Mn deposits and Fig 8b shows that alloyed MMA deposits, in this instance made in the vertical-up position and containing Ni, are more tolerant to increases in heat input than C-Mn deposits. (Nicholson & Rogers, 1974) Deposits alloyed with Mo and Ni are often used when welding C-Mn steels with high heat input processes to develop the required toughness.

The level of interpass temperature is another procedural factor which can result in marked changes in weld toughness, (Garland & Kirkwood, 1975; Evans, 1978) this factor influencing the deposit microstructure, and the proportion of grain refined/unrefined microstructure.

It is also observed that the toughness of root regions in multipass MMA butt joints of C-Mn steels can be lower than in surface regions. (Robinson, 1978) Yield strengths are usually higher in the root region, but the properties of root regions can be improved by back gouging the first side root. Most national specifications require Charpy specimen extraction away from the root area, but some companies are asking for test specimens to sample weld root regions.

IMPROVING RESISTANCE TO CLEAVAGE FRACTURE

The important microstructural constituents of weld metals placed in order of decreasing transformation temperature, some of which are shown in Fig 9, are: primary ferrite nucleated in austenite grain boundaries, usually existing as thin bands outlining a columnar grain structure; ferrite side plates growing from the ferrite at the austenite boundaries; intragranular ferrite plates known as acicular ferrite, plates typically being 1-2μm in width and interlocking; pearlite. Small areas of retained austenite can be found with the electron microscope at low heat inputs. At higher heat inputs retained phase regions, ∿0.5μm in size, comprising martensite or other lower temperature transformation products can comprise up to 10% of the weld microstructure, these forming from austenite entrapped between ferrite plates. (Garland & Kirkwood, 1975). The various constituent microstructures, their colony size, and their proportion in the deposit are determined by the weld metal composition, the welding conditions, and the plate thickness used, and they control the resistance to cleavage and the Charpy V transition temperature.

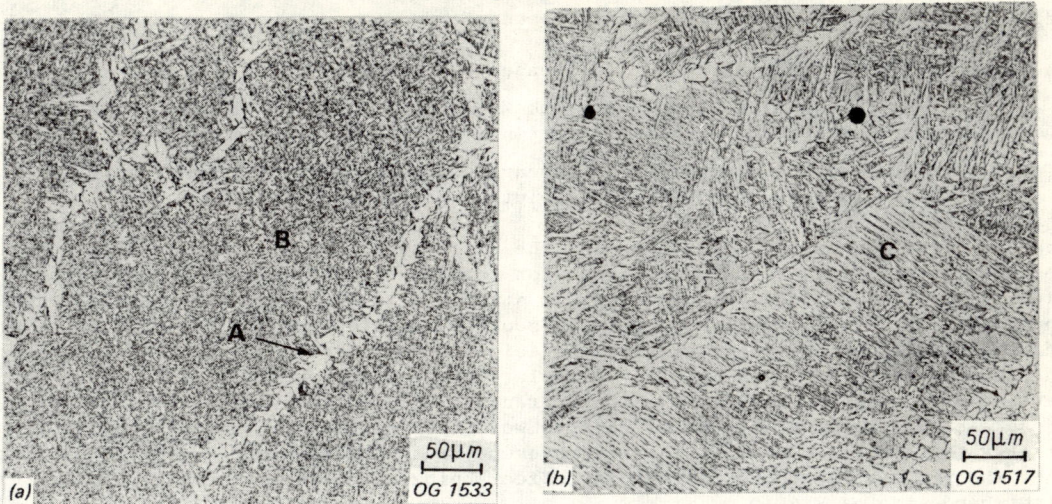

Fig. 9. Weld metal microstructures showing: a) grain boundary ferrite A and acicular ferrite B; b) lamellar structure including ferrite side plates C.

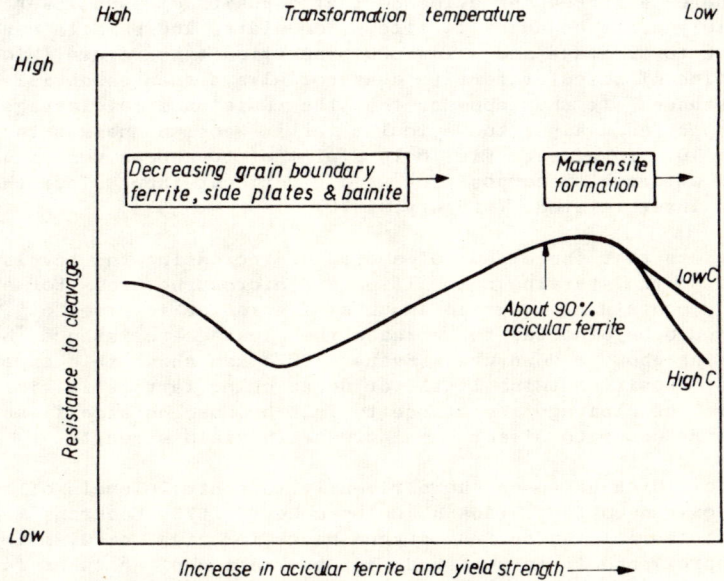

Fig. 10. Schematic diagram showing relationship between microstructure and cleavage resistance.

Deposits not Containing Microalloying Elements

The elements Mn, Mo, Ni, Si and Cr influence toughness in two main ways:
1. They modify the microstructure, with Mn, Mo and Ni in particular, promoting acicular ferrite and eventually martensite at high alloy contents.
2. They increase the deposit yield strength as a result of the changes in microstructure and through solid solution hardening effects.

Dealing first with plain C-Mn deposits, the literature indicates that a high resistance to cleavage and a low transition temperature will result if the proportion of the fine grained acicular ferrite structure is kept high and that of grain boundary ferrite and lamellar structures such as ferrite side plates kept low. In general, the finer the plate size of the acicular ferrite the better the toughness in C-Mn and low alloy weld metals. Acicular ferrite is promoted by Mn, Mo and Ni particularly and, for heat inputs in the range 2-4kJ/mm, increasing the level of these elements is generally beneficial to weld toughness where yield strengths are greater than about $500N/mm^2$. This improvement in toughness occurs <u>despite the fact</u> that the yield strength increases, the effect being shown schematically towards the right hand side of Fig 10.

It is not possible to continually improve toughness by progressively raising the alloy content. Once a high level of acicular ferrite is achieved, say 95%, further increases in an alloying element such as Mn or Mo will serve only to increase the yield strength by solid solution hardening and to promote brittle martensitic regions which may be segregated. The cleavage resistance then starts to fall as indicated in Fig 10.

Although the importance of acicular ferrite for high toughness is now widely recognised, there is increasing evidence that deposits of much lower yield strength, containing mainly grain boundary ferrite, side plates and bainite can also show a high resistance to cleavage and a low COD transition temperature.(Widgery, 1976) A high proportion of acicular ferrite does not always seem essential for good cleavage resistance. It thus appears that the resistance to cleavage of C-Mn deposits when plotted against the acicular ferrite content passes through a minimum as seen in Fig 10. In effect, Fig 10 is a plot of toughness versus acicular ferrite, yield strength or composition at a given heat input since these latter factors can be inter-related. (Widgery, 1976)

Figure 10 predicts that the effect of adding or increasing the levels of Mn, Mo or Ni will depend on the starting composition and microstructure. For example, for a deposit already containing moderate amounts of acicular ferrite, e.g. 50%, further Mn additions would be expected to increase the cleavage resistance, but too high a level would bring about a downturn. Evans (1977) has shown this experimentally. However if the deposit contains little or no acicular ferrite initially, a higher Mn level may cause the cleavage resistance to fall because no significant improvements in microstructure occur to offset the increase in yield strength.

A further factor which has been shown recently to control levels of acicular ferrite is weld metal oxygen content. (Abson, Dolby & Hart, 1978; Cochrane & Kirkwood, 1978). It appears that there is an optimum oxygen range for high acicular ferrite contents (0.02-0.04% approx) and it is believed that nucleation of acicular ferrite plates is directly controlled by non-metallic inclusions. (Abson, Dolby & Hart, 1978) Control of weld oxygen by suitable choice of flux or coating formulation is assuming increasing importance in consumable design.

Diagrams of the type shown in Fig 10 need confirmation and quantification, but it is clear that the consumables and deposit compositions giving the highest cleavage resistance will be different at different heat inputs. Also, unless heat inputs are comparable, investigators are certain to draw different conclusions concerning the effect of changing levels of the alloying elements, Mn, Ni and Mo.

Deposits Containing Microalloying Elements

Elements such as Nb, V, Ti and B appear in the deposit either as deliberate additions made through the consumables or from dilution with the parent steel. Like the major alloying elements they can have two main effects on cleavage resistance:

1. The deposit microstructure may be modified, for better or worse.
2. The deposit yield strength may increase as a result of precipitation hardening, and this effect is generally damaging to toughness.

Some elements, e.g. Ti & V tend to reduce the interstitial N content and improve cleavage resistance.

Niobium. Although Nb has been reported to give marked embrittlement in as-welded deposits, (Hannerz, Valland, Easterling, 1972) other work on submerged arc welds has shown that this is not always so. Garland and Kirkwood (1975) found that for deposits containing between 1.4 and 2.0%Mn (50-70% acicular ferrite), Nb inhibited the formation of grain boundary ferrite and increased the amount of acicular ferrite. The yield strength was substantially increased and it was noted that small lath martensite regions were segregated at solidification boundaries when Nb was present. Both these factors would offset the beneficial effect of the higher proportion of acicular ferrite but the net effect on toughness of 0.025%Nb in the deposits studied was found to be very small.

The literature suggests that the conditions under which small amounts of Nb can be diluted into the weld deposit without markedly lowering toughness are those which result in a deposit of high acicular ferrite content in the absence of Nb. Microstructures of low acicular ferrite content on the other hand tend to show a significant fall in toughness with Nb additions. Figure 11 shows a modified Garland and Kirkwood model detailing the various effects of Nb.

Vanadium. Hannerz and Johnsson-Holmquist (1974) found an increasingly detrimental effect on Charpy V transition temperature above about 0.05%V, when the welding conditions corresponded to a high heat input (7kJ/mm, $\Delta t_{800-500°C}$ = 220sec). The deposit yield strength was observed to increase progressively with V above 0.05%, although no changes in the metallographic structure were reported except for some fine scale precipitation of V_4C_3 observed in the high V content weld deposits. Garland and Kirkwood (1975) noted an increase in acicular ferrite content with increase in V, for C-Mn-V-N submerged arc welds made at 3.3kJ/mm and generally a deterioration in toughness occurred for V up to 0.12%. However Sawhill and Wada (1975) working with 0.8%Ni-0.4%Mo-0.05%Nb weld deposits showed that V can be beneficial to toughness at the 0.05-0.1% level, the optimum addition depending on whether Ti was present. Some changes in microstructure were detected including a decrease in the proportion of upper bainite.

It appears that V inhibits ferrite side plate formation to a greater degree than Nb and the latter inhibits grain boundary ferrite more strongly than V, possibly due to the different precipitation kinetics of the carbides. In general, most tolerance to V would be expected where the addition increased the proportion of acicular ferrite without excessively increasing the yield strength or creating a high proportion of small segregated martensitic constituents. A maximum addition would be postulated beyond which toughness fell, and this maximum level will be a function of heat input and initial weld composition. This is in accord with experimental data and explains differences in findings between investigators as to tolerable V levels.

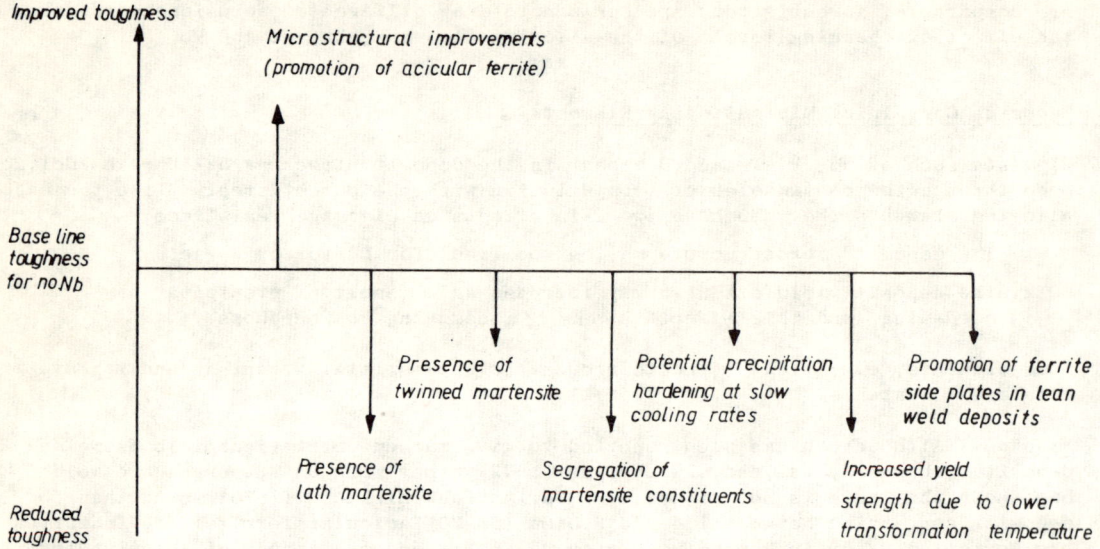

Fig.11. Schematic diagram indicating the different metallurgical effects of Nb — the arrows indicate direction of effect but are not vectors.

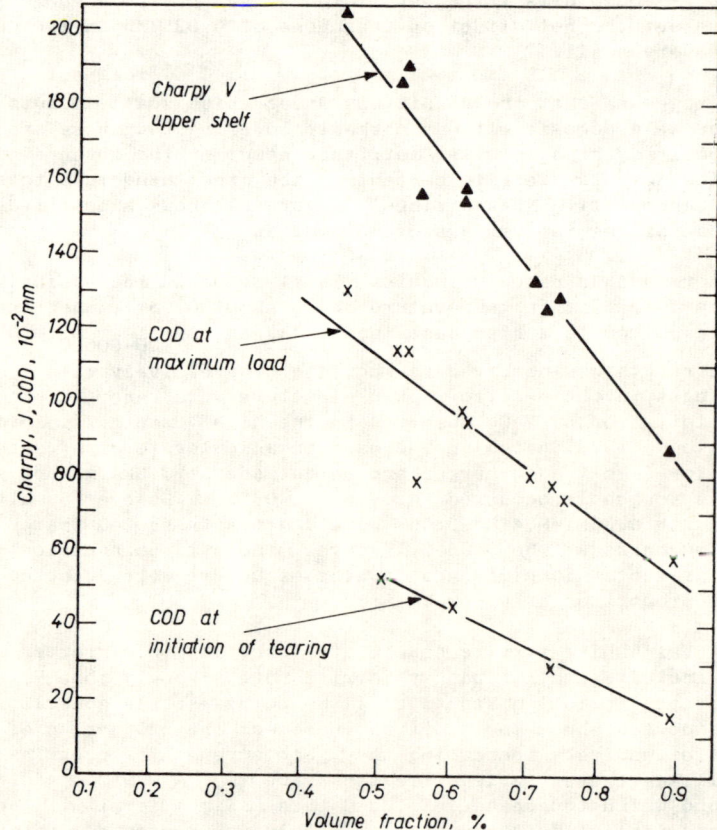

Fig. 12. The effect of volume fraction of inclusions on weld metal COD and Charpy V test results (Steel, 1972).

Titanium. A number of investigators have found that small additions of Ti are beneficial to Charpy V toughness of multipass welds, 0.03-0.04%Ti being quoted as the optimum deposit content for CO_2 welding of C-Mn steels, (Boniszewski, 1969) Ti was found to increasingly refine the microstructure and promote more acicular ferrite, but above the optimum level, despite the microstructure being progressively refined, the yield strength increased markedly and the toughness fell. In more alloyed weld metals deposited by the MMA, submerged arc or MIG processes, similar beneficial effects of Ti have been found, although often the optimum level has been found to be 0.015% or less. (Nakano, Shiga & Tsuboi , 1975)

The common feature of the above investigations is that the welding conditions and alloy content were such as to develop a relatively high acicular ferrite proportion in the absence of Ti. Titanium inhibits side plate formation and promotes more acicular ferrite, eventually eliminating grain boundary ferrite at very high levels (∿0.10%). (Nakano, Shiga & Tsuboi , 1975) As with V, additions above a critical level would be expected to cause a downturn in cleavage resistance, due to hardening by precipitation and martensite segregation offsetting the increased acicular ferrite fraction.

Boron. Boron is being increasingly studied as a potential alloying element for weld metals. (Garland & Kirkwood, 1975; Homma & co-workers, 1978) Its principal advantage is that it can eliminate grain boundary ferrite and promote fine grained structures without the penalty of significant solid solution hardening as seen with elements such as Mn, Mo. Large increases in submerged arc toughness using Ti-B wires have been reported (Garland & Kirkwood, 1975) and commercial wires are now available in both solid and cored form although a full understanding of the weld metal metallurgy may take some time. Several commercial submerged arc fluxes transfer boron to the deposit and substantial improvements in toughness are claimed for this approach. (Shinmyo & co-workers, 1978)

IMPROVING RESISTANCE TO MICROVOID COALESCENCE

As with HAZs the principal controlling factor is the non-metallic inclusion population. Low volume fractions and a high mean free spacing of the inclusions are needed for good resistance to microvoid coalescence and Fig 12 shows this effect for both COD and Charpy results. (Steel,1972)

Low inclusion volume fractions result principally from control of oxygen and sulphur levels in the deposit. Oxygen levels can vary markedly between processes, typical analyses being for MMA welding 0.03-0.07%, for submerged arc welding 0.02-0.15%, for MIG welding 0.01-0.02% and for CO_2-shielded welding 0.04-0.06%. In submerged arc welding oxygen levels are related particularly to SiO_2 activity in the slag, and low SiO_2 fluxes are beneficial to toughness. It should be noted that oxygen content is not linearly related to basicity.

For gas shielded welding a low oxygen potential in the gas shield gives low oxygen levels in the deposit, and so argon based gases containing small amounts of O_2 or CO_2 are used when looking for the highest toughness, together with vacuum deoxidised wires. A high oxygen process will tend to reduce Mn, Si and C levels and this, in itself, will affect the transition curve because these elements can change the transformation microstructure and resistance to cleavage.

Manganese and Si can directly affect the inclusion volume fraction because of their influence on the melting range of the inclusions and the ease with which inclusions are removed from liquid metal by the welding slag. By plotting the ratio $\frac{Mn}{Si}$ against inclusion volume fraction Widgery (1976) confirmed that a high $\frac{Mn}{Si}$ ratio leads to cleaner weld metal.

The impurity element S contributes directly to the inclusion volume fraction by forming sulphide inclusions, and tight control of S in wires, fluxes and plate is needed for highest toughness. Basic fluxes can reduce the weld S level, but in most applications this effect on inclusion volume fraction is small. It should be noted that fluxes of low basicity can transfer S to the deposit.

OTHER FACTORS INCLUDING WELDING PROCEDURE

Similar comments apply to the effect of nitrogen and hydrogen on weld metal toughness as discussed in relation to HAZs. A decision on whether to carry out post weld hydrogen diffusion treatment or an ageing treatment is always required when assessing weld metal toughness in the as-welded condition.

As regards welding procedural controls, improvements in toughness will usually be obtained by reducing the heat input in both single and multi-pass welds. This applies to Charpy V data generally but, in multipass MMA welds, the effect of changing electrode diameter on K_{IC}/COD test results has not been systematically investigated. Interpass temperature is another variable which needs further study.

As seen in Fig 8a, electrode weaving can produce lower COD results in MMA weldments compared to a stringer technique, but it is interesting to note that Charpy V values can often be improved by weaving provided specimens are extracted away from root regions. (Robinson 1978) The improvement is due to an increase in the ratio of refined to unrefined weld metal.

Finally as regards toughness testing of weld metal, both surface notched and through thickness notched specimens can be used when carrying out fracture toughness tests of the K_{IC}/COD type. The former being of BxB configuration (B = plate thickness) is of lower constraint than the 2BxB through thickness notched specimen but can sometimes give lower results due to the fatigue crack front being sited in a roughly uniform microstructure. Precompresssion of the through thickness notched specimen to produce a uniform fatigue crack front is usually necessary in multipass welds. (Dawes 1976)

ACKNOWLEDGEMENTS

The author would like to thank his colleagues at The Welding Institute for their helpful comments in preparing this review.

REFERENCES

Abson, D.J. R.E.Dolby, and P.H.M.Hart (1978). The role of non-metallic inclusions in ferrite nucleation in carbon steel weld metals. Procs of Trends in Steels and Consumables for Welding. London Nov 1978. The Welding Institute.

Boniszewski, T. (1969). Titanium in steel wires for CO_2 welding. Metal Constr 1 225-229.

Bonomo, F. and A.B.Rothwell (1971) The weldability of steels and its implications for products development criteria. International Welding Conference, Bratislava Czechoslovakia.

Cochrane, R.C. and P.R. Kirkwood (1978). The effect of oxygen on weld metal microstructure. Procs of Trends in Steels and Consumables for Welding. London, Nov 1978. The Welding Institute.

Dawes, M.G. (1972). Fracture initiation in weld metals. Weld and Metal Fabr. 40 95-104

Dawes, M.G. (1976). Contemporary measurements of weld metal fracture toughness. Welding Journal 55, 1052-1057

Dolby, R.E. and D.J. Widgery. (1971). The simulation of HAZ microstructures. Weld Research Int'l. 1 (3)

Dolby, R.E. and G.G. Saunders (1972). Subcritical HAZ fracture toughness of C-Mn steels. Metal Constr., 4, 185-190

Dolby, R.E. and G.L. Archer (1971) The assessment of HAZ fracture toughness. Conf. on Practical applications of fracture mechanics to pressure vessel technology. London, May 1971. Institution of Mechanical Engineers.

Dolby, R.E. (1976). The effect of Nb on the HAZ toughness of high heat input welds in C-Mn steels. Procs of Conf on Welding of HSLA (microalloyed) structural steels. Rome, Nov 1976, American Society for Metals.

Dolby, R.E. (1979). HAZ toughness of structural and pressure vessel steels - improvement and prediction. To be published in The Welding Journal.

Evans, G.M. (1977). The effect of Mn on the microstructure and properties of all-weld-metal deposits. IIW Doc. II-A-432-77.

Evans, G.M. (1978). Effect of interpass temperature on the microstructure and properties of C-Mn all weld metal deposits. IIW Doc. II-A-460-78.

Garland, J.G. and P. Kirkwood (1975). Towards improved submerged arc weld metal. Metal Constr. 7. 275-283 & 320-330.

Hannerz, N.E. (1975). Effect of Cb on HAZ ductility in constructional HT steels. Welding Journal 54. 162s-168s

Hannerz, N.E. and B. Johnsson-Holmquist (1974). Influence of vanadium on the HAZ properties of mild steel. Metal Science J. 8. 228-234

Hannerz, N.E., G. Valland and K.E. Easterling (1972). Influence of Nb on the microstructure and mechanical properties of submerged arc weld metals for C-Mn steels. IIW Doc. IX-798-72.

Hannerz, N.E. and B. Jonsson-Holmquist (1974). Vanadium in mild steel weld metal. Metal Constr. 6. 64-67

Homma, H., N.Mori, S.Saito and K.Shinmyo (1978). Effects of Ti-B and Nb additions on the mechanical properties of submerged arc weld metals. IIW Doc IX-1072-78

Kirkwood, P.R. and J.G.Garland (1975). The influence of vanadium on submerged arc weld metal toughness. Weld & Metal Fab. 45. 17-28 & 23-99

Myoshi, E., S. Hasebe, K Bessyo and Y.Yamaguchi (1974). Fracture initiation characteristics of HAZ assessed by COD test. IIW Doc. IX-878-74.

Nakano, S., A.Shiga and J.Tsuboi (1975). Optimising the Ti effect on weld metal toughness. IIW Doc. XII-B-182-75.

Nicholson, S. and P.F.Rogers (1974). Controlling weld metal toughness in manual metal arc welds. Conf. on Welding in offshore structures. Newcastle. The Welding Institute.

Robinson, J.L. (1978) Through thickness toughness variations in multipass arc welds. Procs of Trends in steels and consumables for welding. London, Nov 1978. The Welding Institute.

Sawhill, J.M. and T Wada. (1975). Properties of welds in low carbon Mn-Mo-Cb line pipe steels. Welding Journal 54 1s-11s

Steel, A.C. (1972). The effects of sulphur and phosphorus on the toughness of mild steel weld metal. Weld Res Int'l. 2 (3)

Shinmyo, K., S. Saito, N.Mori, T.Takami, and R Kono (1978). A new submerged arc welding flux for offshore structure in arctic use. IIW Doc. IX-1073-78.

Widgery, D.J. (1976). Deoxidation practice for mild steel weld metal. Welding Journal. 55. 57s-68s

SECTION 3A

Solving Fracture Problems— The Tools Available

DESIGN AND THE PREVENTION OF METALLURGICAL FAILURES

P. J. Eccleston

TUV-Rheinland South Africa (Pty.) Limited, R.S.A.

ABSTRACT

Design philosophy is examined and the "designer's" role is placed into perspective against a back cloth of quality by "design".

Structures for the containment of pressure are selected to demonstrate the wide spectrum of metallurgical failure modes which need to be considered at the initial design stage.

Major Codes of Practice are reviewed to reveal that an overall design concept has been found to be necessary in order to maintain pressure vessel integrity, and that it works with good effect.

Nevertheless, it is dared to suggest that the designer does not fully appreciate nor fulfil his responsibilities and that changes in education and management philosophy are required to improve the situation. Furthermore it is emphasised that a generation and communication gap has been created inadvertently by the Codes, leading to the stifling of common sense, and the resolving of old problems with economic shortcomings.

KEY WORDS

Design, Metallurgical-Failure, Codes of Practice, Pressure Vessels, Modes of Failure, Quality Control.

INTRODUCTION

Bitter experience has taught us that structures for the containment of pressure are indeed a complex design problem. Economics have placed ever increasing demands on the pressure vessel to sustain the effects of greater pressure - temperature ranges and environments both hostile to vessel and habitat alike.

The rate of change of development in this field has on occasions outstripped the current technological resources. Engineering history is littered with the records of our failures, whilst our successses have tended to be taken for granted and overlooked. Metallurgical failures in pressure vessels cover a wide spectrum, and much research has been conducted in recent years toward eliminating these occurrences and gaining a better understanding of their nature. If we want to review the solutions to prevent certain failure problems in heavy engineering where better to start than with the problems associated with pressure containment, a historical and topical subject.

It took a long time to dispel the "Iron Boiler Syndrome" from the academic fraternity, such that some respectability was bestowed on the pressure vessel and the necessary cross-fertilization and research work could be undertaken on a coordinated basis to improve reliability. In fact, it was not until the age of nuclear power that things began to really improve, the consequences of radioactive contamination providing the stimulant. How is it possible to blend experience and research in a manner that may be used by "ordinary engineering folk"? The codes of practice that have evolved over the years provide some of the answers.

Pressure vessel design is indirectly monitored by law, such that minimum standards are maintained. It is usual that certain codes of practice are referenced or implied. Codes provide a useful tool in the prevention of metallurgical and other failure.

This paper deals with the prevention of failure by "design". The design of the non-nuclear pressure vessel is examined and it is attempted to put the design philosophy outlined in the codes into perspective as a total concept. It is not intended to give specific design examples. However, a review is made of American, British and German Codes to highlight their similarities and differences and to demonstrate that CODE may may also be described as a four-letter word.

ENGINEERING DESIGN

Design Philosophy

The design process may be seen as the formulation of a need into a specificiation and the translation of that specification into information (by way of drawings, etc.) to enable the preparation of components, to fulfil the original need.

Design activity occurs in the following areas :-

1. TECHNOLOGICAL - Functional : e.g. performance, reliability, safety, maintenance, etc.

    - Constructional : e.g. production technology

2. ECONOMIC   - Marketing
                 - Economics

3. HUMAN FACTORS - Ergonomic
                 - Aesthetic
                 - Safety

Design quality may be described as the measure by which design meets and optimises all identified requirements in a product to the satisfaction of both user and producer.

Technological Sphere

It is in the technological area where the design process can be used with effect to minimise metallurgical failure, however, to understand the real meaning of this statement one has to appreciate the total concept of design.

In the technological area the "designer" has two roles: firstly, to translate a specification into information for others to use in creating a real product. This includes ensuring that the specification itself defines the need; that the conceived ideas and their detailed execution make it possible to provide the performance, safety requirements and the level of quality specified; and that the design approach adopted make it possible to maintain the equipment effectively in service. The second role concerns the style of the manufacturing information issued by the "designer". The ability of the manufacturing function to implement the design requirements, the ability of the procurement function to acquire the materials and components required, the ability of the inspection function to check that the product embodies the design requirements, the ability of the construction function to correctly assemble on site, the ability of the maintenance function to maintain the product in service, etc. etc., quite clearly depend upon the clarity and quality of these instructions.

Weakness in the Design Process

Reliability of product and hence safety, are of a statistical nature.

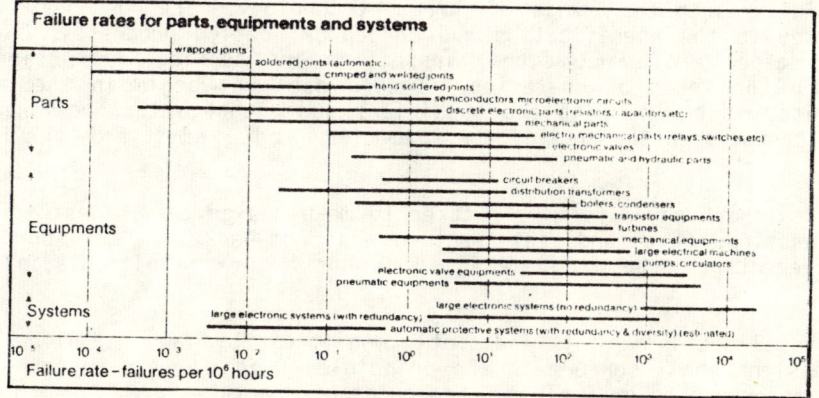

Fig. 1   Failure Rates Histogram
(After UKAEA)

It is possible to approach optimum reliability by careful design, choice of materials, the application of fail-safe techniques and the implementation of quality procedures, but it is impossible to attain infallibility.

It follows, therefore, that failures will occur in service, even in basically sound design. These may be as a result of not meeting the performance requirements, which may have been difficult to predict with accuracy, or they may be due to weaknesses in design not understood at the time the design was made. However, a high proportion of failures arise from known causes that could have been avoided by adequate study of the technology available when the design was carried out. If the available information sometimes fails to be embodied in a design, this may also be due to weaknesses in the the technical organisation and procedures, rather than weaknesses in the academically based design technologies. Furthermore, a technical generation gap is created between engineers of different ages, due to different methods of education and datum points such that cetain key information is often not passed on. There is also a communication gap between these engineers, which is also true for engineers who work in different departments or countries due to lack of common vocabulary throughout industry. This may be likened to radio transmission on VHF and reception on a crystal set.

This has resulted in failures occurring in time cycles and old problems having to be resolved by new hands. It would appear that in most cases, old design knowledge, especially that empirically collected, is being dumped for the newer, more sophisticated analytical approaches. Otherwise it is discarded due to the production techniques it represented falling into natural disuse, e.g. riveting of steam boilers. In either case, no effort has been made to collect the key data and modes of failure from former engineering know-how.

Quality by Design

Thus, with the above in mind and the increasing complexity of design, systems, the narrower design margins, the higher the risks consequential on failure, the greater precision needed in manufacture and the decline in trade skill have all created a need for a more identifiable approach to quality in design supported by formal codes of practice and a senior management objective to ensure all identified criteria are met. Only with this kind of philosophy is it possible to reduce failures to the limits of current technology. The "designer" not only holds the key to the specification and conceptual design processes, but beyond influencing aspects of manufacture, inspection and testing, commissioning and further into the modes of operation. Design does not begin in the so-called design office, it begins with the specifier, and although the "designer" holds the key he cannot open the door if the specifier has barred it from the inside.

The term "designer" nowadays may be taken to mean design team. Design is not an isolated function, it demands inter-action and feed-back from all sources, such that improvements may be made, design work has become a multi-discipline team effort.

Pressure vessel design work is indirectly monitored, by law, and it is required that the designs shall conform to the principles laid down in certain statutes which make reference to or imply certain codes of practice.

CODE A TOOL FOR SAFETY

The fundamental principle of Codes is to set <u>minimum</u> standards of safety. The codes are drafted by a balanced proportion of users, manufacturers and inspectors (including in some countries government departments). They provide technical rules to afford adequate protection of life and property and a margin for deterioration in service so as to give a reasonably long safe period of usefulness.

Codes are based on sound engineering principles, they reflect the current state of the art and enable a legal means of procedure. However, because they are written by engineers for engineers they may be interpreted too literally by the legal profession in the case of an accident enquiry.

Codes can only set minimum standards and offer general guidance and thus allow continual development in engineering, material, fabrication and testing towards the goal of safety and econonmy.

All technologically advanced countries have their own pressure vessel codes each with its own nationalistic idiosyncrasy. Leaders in this field must be the American, British and Germans, all other Codes are apparently influenced by the Big 3.

Pressure vessels may be divided into two categories: -

- fired (Boilers - Steam Generators)
- unfired

Traditionally the boiler has represented the keystone in the harnessing of power and for some reason boilers and pressure vessels have developed separate sets of rules.

Codes have been distilled from long years of experience in the operation and failure of pressure vessels and have reduced explosions which were considered as inevitable or "Acts of God" during the last and the earlier part of this century.

History has shown (Green '55) that a violent boiler explosion which occurred on the 20 March 1905 in Brockton, Massachusetts in a shoe factory gave stimulus to the state to enact the first legal (US) code of rules for steam boilers in 1907. The explosion killed 58 people and injured 117 others causing $ 250 000 of property damage.

Fig. 2 Brockton Mass. Shoe Factory

Fig. 3 Shoe Factory after Boiler Explosion

Afterwards, it soon became apparent to other states and cities in which explosions had taken place, that accidents of this nature could perhaps be prevented by the proper design, construction and inspection of boilers and pressure vessels. The regulations drawn up differed from state to state and often conflicted with one another, manufacturers found it increasingly difficult to produce vessels for the use in one state which would be acceptable to another. The ASME rationalised the situation and published "Rules of Construction of Stationary Boilers and for Allowable Working Pressures", known as the 1914 edition. Fig. 4 serves to indicate the steady decline in boiler explosion after this date.

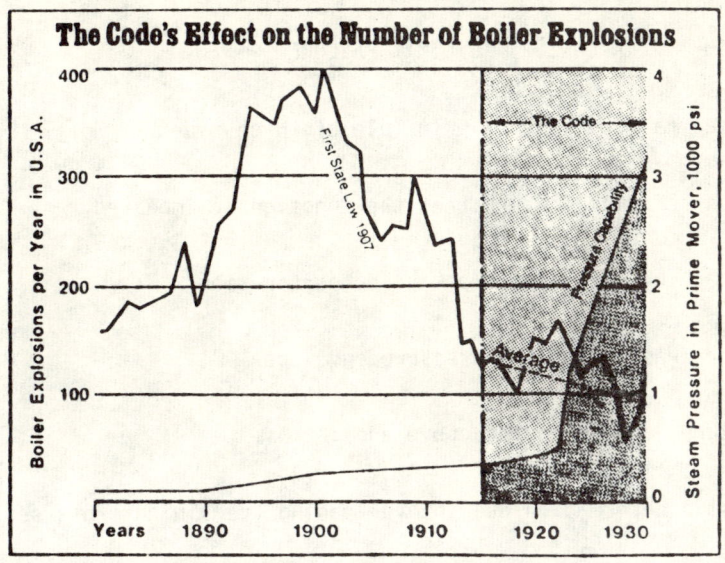

Fig. 4 ASME Code Effect on the number of Boiler Explosions

A similar state of affairs existed in Europe, although somewhat earlier (Eccleston '79) which led to the formation of the Inspection Authorities and Engineering Insurance Companies we know today.

Although explosions nowadays attributable to faulty design and construction are few and far between, one notable disaster (BSI News, Financial Times, HMSO, Insurance Monitor) occurred in 1974 at Flixborough, England. Figure 5. The Nypro Caprolactam Plant was almost completely destroyed by a blast of explosive gas, escaping from a site modified pressure vessel connection which had failed. It caused 28 deaths and injuries to more than 100 people, leaving losses of the order of 20 million Pounds Sterling in material damage together with a national interruption of business related to the processing of Caprolactam into Nylon-6. This serves to indicate implications not considered at the specification and design stage of the pressure vessels and the plant as a whole, this point will be further amplified later.

Fig. 5 The Shadow of Flixborough

CODE - A CONCEPT TO LIMIT FAILURE

Pressure vessel failures are in principle of three types :

(a) Gross deformation such that the function is impaired.

(b) Leakage by localized cracking or perforation.

(c) Major fracture of the pressure envelope.

Modes of failure leading to the above are :-

(a) Plastic deformation due to overloading (remaining cross-section)-(shear)

(b) Instability-(buckling)

(c) Fatigue - mechanical and thermal

(d) Creep

(e) Corrosion

(f) Erosion

(g) Brittle fracture-(cleavage)

(h) Interaction of modes (a) to (g).

Fig. 6 Collapse of Road Tanker
(Gross deformation)

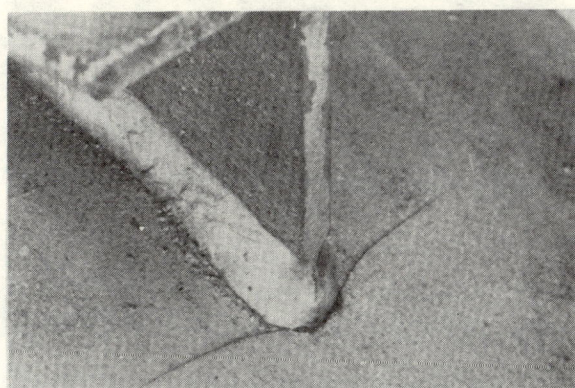

Fig. 7 Fatigue Crack at Fillet Weld
(localized cracking)

Fig. 8 Cracking due to Corrosion
(localized or major?)

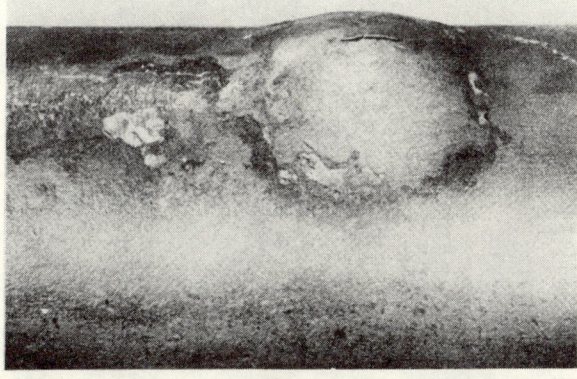

Fig. 9 Crack due to Creep
(localized or major?)

Fig. 10 Boiler Furnace Collapse
(Major Rupture)

Fig. 11 Brittle Fracture of a PV under Test
(Major Rupture)

Fig. 12 Steam Generator Explosion - due to over pressure
(Major Rupture)

Four factors affect pressure vessel integrity (Eccleston '79), they are :- material properties, state of stress, material discontinuities and operating environment. Interaction of these factors leads to the modes and types of failure described above.

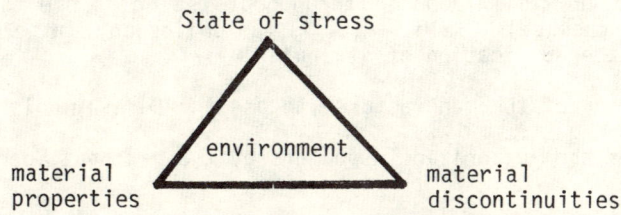

Fig. 13 Triangle of Integrity

The codes attempt to systematically balance these factors.

CODES - IN CLOSE-UP

A discussion will focus on the pressure vessel rules developed by the Big 3, the boiler rules; whilst being similar; will only serve to complicate the issue at this stage. The Pressure vessel codes taken for review, in alphabetical order of country of origin:-

US America

ASME Boiler and Pressure Vessel Code -   American National Standards
Section VIII   Division 1   Institute ANSI
Section VIII   Division 2   American Society of Mechanical Engineers - ASME

Britain

BS 5500   - British Standards Institution
Specification for unfired fusion welded   BSI
pressure vessels

W-Germany

A-D Merkblaetter   - Arbeitsgemeinschaft
Technical Rules for Pressure Vessels   Druckbehaelter (A-D) Vd TUV

A review is made of the Pressure Vessels (PV) Codes in Appendix I. This review is not intended to be exhaustive, but merely informative to give one an insight into the manner the codes attempt to achieve safety by balancing the factors affecting integrity.

ASME Section VIII Division 1

The ASME Section VIII Div. 1 Code represents the traditional approach to pressure vessel design and construction.

The 1914 edition was implemented in 1915 and over the years the Code has kept itself up to date by constant revision. In recent years the Code has been completely revised every three years and during this period addenda are published twice a year.

The Code has been widely adopted throughout the world due to American influence in the petro-chemical field. It is a philosophy of the Code to ensure uniformity in the application of its rules by:-

(a) Authorization of the manufacturer to use the Code symbols of the ASME.

(b) Approval of the authorized inspectors by the National Board.

True conformance to the Code requires implementation of a quality control system by the manufacturer to cover all aspects of design, purchasing, fabrication and testing. Additionally, a third party authorised inspector is required to vet the quality control system and ensure that the design and construction meet the requirements of the Code.

This, however, has met with economic resistance outside of the U.S., because the laws governing Code usage do not apply. Obviously, the ASME is concerned, for a number of reasons, regarding the improper application of the Code and it has recently taken legal steps to prevent its name being associated with unauthorized constructions.

Appendix I, outlines the rules governing PV design, construction and testing. It will be seen that the Code bases its design requirements on the ultimate tensile strength of the material (UTS). The UTS is factored for safety (ignorance) and further factored for fabrication quality (casting, welding, etc.). This approach is a natural progression from the rivetted construction, with only minor modifications to Code rules, e.g. rivetted joint efficiency becoming weld joint efficiency.

The Code basis is semi-empirical, extrapolations beyond conventionally accepted practice may result in an unreliable construction. Code rules do not consider many of the modes of failure encountered today, partly because of the relatively high factor of safety afforded by the Code has shown that reliable service usually follows. However, it is a misleading assumption to believe that the Code rules will provide adequate safety and reliability in every case. It is expected that the manufacturer shall be sufficiently competent to realise when accepted practice is exceeded and that the necessary compensation is provided to ensure a margin of safety and adequate reliability is maintained. It must be stated here that the latter statement is a subtle Code change which has been introduced due to recent advances in technology. Such subtlety has not been fully recognized throughout the industry.

An interesting point in the Code rules is the mandatory provision for inspection openings. Provision for in-service inspection has been found to be a powerful tool in increasing the safety and reliability of plant. This simple fact is often overlooked by the designer.

Pressure testing was not always a requirement in the manufacture of PV before implementation of the Code. The over-pressure test has always provided evidence of integrity against leakage and plastic deformation. It has only been appreciated in recent years the real benefit of an over-pressure test with the re-distribution of discontinuity and residual stresses.

ASME VIII Division 2 (Alternate Rules)

As a consequence of material and technological economics the ASME has published Division 2 of Section VIII of the Code. It permits higher design stresses than those allowed in Division 1, but it is to be noted that Division 2 is not simply an alternative providing identical coverage with an up-grading of design stresses. The rules of Division 2 are more restrictive in the choice of material, more precise design procedures are required and some common design details are prohibited. Permissible fabrication procedures are specifically delineated and more complete examination-testing and inspection are required.

ASME VIII/2 may be described as the grandson of ASME VIII/1 and the son of ASME III, Nuclear Power Plant. It possesses the same characteristics of its grandfather without the hang-ups of its father. However, in order to comply with its rules, the specifier needs to understand the Code and issue a properly defined user specification, certified by a professional engineer (US). Furthermore, the manufacturer is required to produce a fully documented design report also approved by a professional engineer. The onus for design approval is thus taken from the authorized inspector, however, he is required to establish that the design has been properly documented and approved by others.

The revised philosophy and engineering effort enables the construction to be in general approximately 20% thinner than the ASME VIII/1 at ambient and moderate temperatures, with economic advantages. However, additional engineering often outweighs the material savings, thus the economics of the situation need to be established. The effect of this may be appreciated in Figure 14 showing the distribution of Code stamps (Irving '77). It may be seen that few manufacturers have found it necessary to gain authorization to manufacture in accordance with ASME VIII/2 so far.

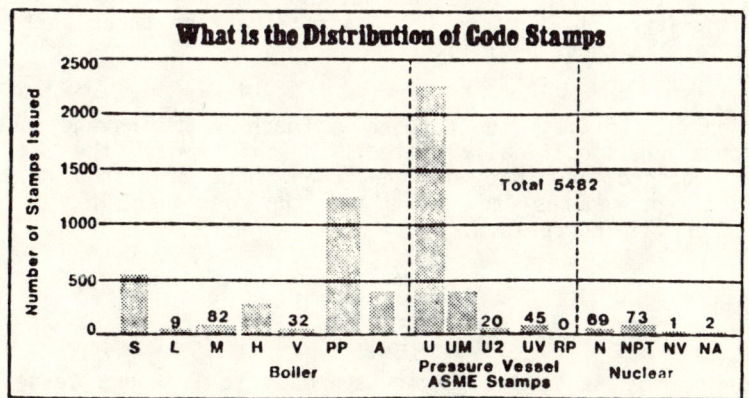

Fig. 14 ASME Code Stamps Distribution - 1975

It is obvious that the ASME recognizes the significance of designing around the yield strength of the material and the need to make manufacturers aware of their added responsibilities by the stringent nature of the rules. This section of the Code was the first of its type to provide a stress analysis failure criteria based on the maximum shear stress. Figure 20 (Appendix I). It is an intention to prevent primary stresses from exceeding the yield stress. Further, total stresses which include secondary stresses are limited to twice the yield stress so that "shake-down" occurs with repeated loading.

BS5500

Designated as a master Pressure Vessel specification it supersedes BS1500 Pt 1, BS1515 Pts 1 and 2 and will eventually replace standards which cover air receivers, vessels in other materials and nuclear reactor vessels etc. on the publication of appropriate supplements.

It has been recognized, in the past, the British system of Code revisions has not been effectively communicated and as such an effort has been made to revise the format to provide for continuous updating. This is not on the same scale as the systematic revision system provided by the ASME.

BS 5500 follows the philosophy adopted by ASME VIII/2, i.e., via the ISO and many of the rules are similar. The British rules vary significantly where material and welding are concerned. There are also differences in design stress, caused by the adoption of different proof stress levels and different factors of safety. Mechanical properties based on hot testing do not feature in the application of the ASME rules except in so far as they have been statistically determined at the material acceptance stage. British material specifications provide for hot certified properties and the Code allows increased design stresses over the materials which are not hot certified.

No mandatory manufacturing quality control system has been specified, but it is implicit that some system needs to be applied to ensure adequacy of product. The purchaser and manufacturer are free to set the limits of many requirements only outlined by the Code, providing that they meet with the approval of the inspection authority. It is taken for granted that the three parties concerned are competent to make these decisions.

The application of a fitness for purpose approach to design and construction is implied although not specifically stated. In this regard the use of fracture mechanics is broached, but not enlarged upon. The British Code has included many analytical procedures to stimulate the designer in a standardised application of certain concepts.

AD-Merkblaetter

A-D Merkblaetter represents the German approach to Pressure Vessel design and construction; the rules are closely linked to the German Law: - Accident Prevention Regulation - Pressure Vessels (UVV - Druckbehaelter VBG 17).

It is a fundamental principle in Germany to utilise the expertise of its engineers to provide economy in design and construction coupled with the overall goals of safety and reliability. The AD-Merkblaetter provide a general framework within which to work and are presented in the form of separate instruction sheets which are continuously updated. The VdTUV (Association of Inspection Authorities) takes a similar role to that of the ASME in the U.S., in that it is the Code steering and drafting body. In Germany the law recognises the VdTUV as its authorized inspector and although the individual TUV's are independent they are regulated through the Association.

The philosophy of the rules are similar to ASME VIII/2 and BS 5500 again with the influence of the ISO. German practice with regard to quality control is through the expertise of its engineers rather than with the formal quality control procedures required by the ASME. It is presumed that individuals will occupy certain positions within the industry will be experienced and possess the relevant academic qualifications. Due to this philosophy there is scope for the application of fitness for purpose reviews during manufacture with economic advantage.

The requirements for materials of construction and welding consumables are strictly controlled by the inspection authorities.

It is a requirement of the Code that the vessels, as installed, are acceptance tested in conjunction with overall plant and safety installations together with periodic examinations in service.

The International Theme

It will be appreciated from Appendix I and the foregoing that there exists a theme common to the PV Codes which may be summarised as follows:-

1. Policy statement of SAFETY
2. Material selection
3. Design and Design Approval
4. Manufacture and Workmanship
5. Inspection, Testing and Certification

In general, the purchaser shall be responsible for furnishing the manufacturer with sufficient information to enable a pressure vessel to be designed and constructed to the full requirements of the code. The manufacturer shall be responsible for the completeness and accuracy of all design calculations and for compliance with all applicable requirements of the code. An independent Inspection Authority shall be responsible for checking that the design of the vessel complies with the requirements of the code and that all inspections and test required during manufacture are conducted and are properly certified. The examinations carried out by the Inspecting Authority do not absolve the manufacturer of his responsibilities.

CODE - 'A FOUR LETTER WORD'

Reliability of present PV

The reliability of PV is in part a function of the Code rules. This may best be illustrated by considering some statistics. The National Board of Boiler and Pressure Vessel Inspectors, published the following:-

TABLE 1 Accident Report Details (US) 1975

Power Boilers	Accidents	Injuries	Deaths
Rupture of drums during hydro test	+ 20	0	0
Shell rupture	+ 10	0	0
Furnace explosions	+ 82	2	0
Flarebacks	+ 18	0	0
Low water	- 510	0	0
Overheating	- 106	0	0
Piping failures	+ 58	0	0
Poor maintenance - controls	- 86	0	0
Unsafe practice	- 52	0	0
Code violation - welds	- 14	0	0
Dry fired	- 25	3	0
Tube rupture or failure	+ 8	5	0
	Σ 489	10	0
Heating Boilers			
Dry Fired	- 85	0	0
Gas Explosion	+ 45	0	0
Shell rupture or cracks	+ 45	0	0
Furnace explosions	+ 63	1	0
Flarebacks	+ 34	0	0
Low water	-1339	0	0
Runaway burner	+ 2	0	0
Inadequate safety valve	- 27	0	0
Code violation	- 1	0	0
C.I. Heating Boilers	+ 192	0	0
Domestic Hot Water Heater	+ 12	2	0
Overheating	- 6	0	0
	Σ 1851	3	0
Unfired Pressure Vessels			
LPG & NH_3 Cargo Tanks	+ 3	0	0
Piping & Hoses LPG, NH_3, H_2	+ 2	3	1
Auxiliary Piping	+ 23	0	0
Air Receiver	+ 3	0	0
Air & Water (or oil)	+ 24	0	0
Jacketed Kettle	+ 2	0	0
Misc. Failures	+ 147	0	0
Safety valve failure	+ 5	0	0
Paper machine drier	+ 3	2	0
Flash vessel	+ 1	0	1
	Σ 213	5	2
	ΣΣ 3053	18	2

A survey based on the records of several British engineering insurance companies during the period 1967 to 1972 (Smith and Warwick) revealed several interesting technical facts:

TABLE 2 Some statistics of PV failure (U.K.) 1967-'72

Type of vessels	-	Class 1	- BS 1500, BS 1515 or Similar
			- Some pipework of same standard
			- Tubing excluded

No. of failures recorded	-	16	- catastrophic
	-	123	- potentially dangerous
		139	

Causes		117 (84%) -	cracking - fillet welds or nozzles
		22 (16%) -	other defects

Causes of cracking:
Pre-existing from manufacture	-	48 (41%)
Fatigue - mech/thermal	-	20 (17%) pre-existing?
Corrosion	-	4 (3%)
Creep, mal-operation, not ascertained	-	45 (38%)
		117

Method of Discovery:
Explosion	-	16 (12%)
Leakage	-	49 (35%)
Periodic Examination (visual/NDE)	-	74 (53%)
		139

The statistics reveal that there are nowadays few injuries and deaths resulting from reportable accidental occurrences to PV and related equipment. Many failures are due to abnormal conditions and can be attributed to human fallibility, e.g. low water in boiler, unsafe practice, pre-existing defects, etc.

Table 1 lists 20 boiler drum ruptures during hydraulic testing. This is an interesting fact since the ASME Code for power boilers is similar to Section VIII Division 1 and gives no recommendations for toughness of the material for the test condition. It must be remembered that steam drums are invariably of thick material containing a large number of stress concentrations.

Table 2 indicates that nozzles and fillet welds are still a cause of major problems in PV. It follows therefore, that there is scope for refinements in both the design and fabrication of these items.

Table 2 also indicates the usefulness of periodic examination as a valuable aid to reliability of PV since 53% of potentially dangerous cases were discovered in this manner before they led to catastrophic failure. This also emphasises the need for more effective NDT, since 41% of the cracks originated from manufacture. Thus, it is desirable to make improvements to the structure as a whole regarding its tolerance to crack-like defects.

At this stage it is necessary to distinguish between Reliability and Safety. Reliability may be regarded as the characteristics of a system expressed as the probability of its performing in the manner desired for a specific period of time under relevant environmental conditions (Eames '68). Whilst, Safety is affording security or not involving danger (a human factor). Reliability embraces safety but not vice-versa. The Codes have tended in the past to concentrate on Safety and because of certain factors of ignorance applied to the design, reliability usually followed.

The ASME Code VIII/1 demonstrates this point adequately, paragraph UHA-105 states, inter alia: "it is a basic principle that the Code rules are intended to provide minimum safety requirements for new construction, not to cover deterioration which may occur in service as a result of corrosion, instability of the material or unusual operating conditions, such as fatigue or shock loading." Which contradicts the foreword to the same code, which now adopts the ISO philosophy of safety coupled with the provision of a margin for deterioration in service, so as to give a reasonably safe period of usefulness. Whilst the statement has been representative of all PV codes, in this case it reflects the heritage of ASME VIII/1, perhaps it could be modified for the future since nowadays the statement should be no longer true.

Code Safety: Equated to Strength Criterion

The Codes concur that the strength of the material as determined in a uniaxial tensile test of some description shall be used as the design parameter by which the pressure bearing parts are proportioned. Whether this strength be the ultimate tensile, yield, proof or creep value, the properties are factored for safety and the resulting figure becomes the "design stress".

Traditionally, engineering design has concentrated on this approach due to its roots having been steeped in elastic stress analysis and partly because we knew no better. With this technique, the number of explosions resulting from plastic over-loading were significantly reduced. However, this was not the final answer. Premature failures depend little on tensile strength values, and strength-based safety factors are a misconception when dealing with modes of failure such as fatigue, brittle fracture and corrosion combinations particularly with the ever increasing demands placed upon the structure.

The strength concept is only suitable for ideal structural behaviour, but all real structures contain flaws, the significance of which is not explained by the traditional theory. This has created a tendency to adopt a no-flaw philosophy in both design and construction. In reality, flaws have been limited to a level commensurate with the best manufacturing techniques, the factors of ignorance being trusted to cater for the unknowns. It would appear that by increasing the factor of safety, reliability towards premature failure is also improved.

This may have been true for mild steels in thicknesses of up to 25mm, but with increasing thickness, restraint has become a problem providing stress concentrations and possible fatigue effects, also leading to plain strain conditions facilitating brittle fracture.

ASME VIII/1 is the only code in this review used by the Big 3 adopting higher factors of safety, the material thickness requirements for ambient and moderate temperature applications are 20% in excess of those required by the higher strength codes. This particular code is the most widely used world-wide, mainly due to American influence in the petro-chemical field. Its wide application and adoption (Refer to Fig. 14) create an inertia to change to the higher strength code. From an economic point of view materials have in the past been less costly in the United States than they have been in Europe.

The restrictions placed on the higher strength ASME VIII/2 indicate an awareness of the lack of attention being paid toward PV design and construction. However, the costs of the extra engineering effort required by ASME VIII/2 at present in 90% of cases outweigh the cost of the extra material required in ASME VIII/1. There is nothing wrong with adopting this approach providing the methods of ASME VIII/1 demand the same levels of safety and reliability as those of the higher strength codes - which is not true of every case. There appears to be a need to rationalize the two separate Codes into one.

The acceptance criteria for defect levels in welded joints are similar for all codes and are empirical, they bear no relationship with real fitness-for-purpose nor safety factors based on strength criteria, but by implication are law. The question must be therefore asked, is strength a redundant engineering concept (Saunders '79) in this sense?

Code - Simplicity - a common misconception

The rules of the Codes are written to a large extent in simple language, to the inexperienced, they give a sense of security tending to suggest that the design of a PV is a very simple matter, even though the most modest pressure vessel presents a structural problem practically impossible of precise stress analysis.

The simplicity of the rules tends to lull the inexperienced into a false sense of security and all too often it is revealed that pressure loading is the only criterion considered with consequent effects on reliability.

Codes have created a generation and communication gap. They are drafted by extremely experienced engineers probably a generation or more apart from the younger engineers who have to apply their rules, certain implications being taken for granted. Communication is hampered by the rules tending to stifle common sense, because in general terms they are empirical and extrapolations based on certain conditions tend to be added on ad nauseam, disguising the true meanings. Thus, younger engineers either end up resolving old problems or over-specifying or being thoroughly confused, wasting money in either case. There is a definite need for a "Manual of Application" as it were, expanding the philosophy and approach of the Code to the various types of design problems.

In the country of Code origin, the rules are applied with a knowledge which may not exist elsewhere. National experience, the application of unwritten rules, the application of guidelines and skills taken for granted within that industry. Playing to these same rules outside the clique alters the odds of reliability unless extreme caution is employed. Examples of this may be :-

(a) American industry is aware of welding problems with certain grades of steels conforming to ASTM standards, which are allowed by the code without restriction. Industry self limits the carbon contents to cope.

(b) British industry acknowledges that certain extrapolation of its code rules are unreliable. Industry adopts other techniques in these circumstances, e.g. Flanges.

(c) German industry realises that the factors of safety proposed for steel based upon the yield strength alone cannot provide the degree of integrity required by a pressure vessel, unless provision is made in the material specifications elsewhere, to confer the ductility required.

Codes do not contain all the answers, they provide the minimum rules for safety. The British Code does not directly acknowledge its safety heritage, although this is implied, a strong statement to this effect would be welcomed, lest we forget the consequences of failure. The Codes assume that pressure vessels will be designed and constructed by the experienced. In recent years the American Code has recognised the decline in trade skills coupled with increasing complexity of applying the rules and has demanded that management implements quality control systems to monitor their production and the adequacy of its operation alike. Whereas the Europeans have not enforced this procedure, it is implicit that some system is required to exist to ensure adequacy of product.

Code - the going gets tough

The concept of toughness is universally accepted. IZOD and Charpy produced tests at the beginning of the century in order to investigate this phenomenon, but nevertheless many failures have occurred due to the lack of it. All the Codes appear now to have adopted the Charpy Vee notch specimen as the standard production test for toughness. However, in application, the necessary toughness requirements vary widely.

All codes recognise that ferritic steels lose toughness with decreasing temperature, the Codes have adopted specific temperature limits to their minimum requirements for toughness. American at $-29^{\circ}C$, British at $0^{\circ}C$ and German at $-10^{\circ}C$. It must also be appreciated that the requirements are empirical, for instance a toughness figure of 20-30J may be acceptable for a mild steel plate of 25mm thick in static loading to prevent the initiation of brittle fracture, but what of greater thicknesses and shock loadings?

The Charpy Vee Notch Specimen is at most 10mm x 10mm x 55mm long. It is therefore not possible with this form of test to simulate the reality of a full size structure since the impact loading uses a high strain rate, the Vee notch does not produce the same effect as a natural crack and the degree of restraint will be different. This type of test has been found to correlate with other forms of testing (Wells '64) and may be used as quality control check, but should not be extrapolated for different materials without experience.

The German Code specifies materials with general impact properties for all pressure vessel applications, and differs from the other Codes where certain exemptions are allowed. The British Code makes reference to fracture mechanics tests providing they are agreed between pruchaser and manufacturer for certain applications. The American Code appears to be quite lenient on toughness aspects as it makes no special references to ASME Sections III or XI which cover the linear elastic fracture mechanics approach in some detail, it assumes that the engineer will make the necessary recourse. After so many encounters with brittle fracture during pressure test, (some writers infer that all major brittle fractures of PV occur at this time) the high strength Codes recommend that certain levels of toughness should be inherent at the test temperature. Toughness is often improved significantly by marginal heating of the test water, this is particularly true of complex high strength materials for high temperature application.

Code - material gets tired

The American and British Codes have adopted a similar fatigue assessment criteria based on peak stresses not exceeding certain master fatigue curve values. The German Code is quite different in this respect and somewhat simpler to apply. It assumes that the design will have minimised the stress concentrations. The maximum allowable working pressure of the critical parts of the system are evaluated and compared with a master fatigue curve, the stress concentrations normally applying have been built-in.

All rules provide only minimum requirements, they may be extended by fatigue testing. The criteria determined for other structures, e.g. BS 153 may be applied to members external to the PV, but are not directly applicable to the pressure envelope.

CONCLUDING REMARKS

The Designer

The designer must realise that he is, primarily, responsible for PV quality from conception into operation. If we are to avoid premature failure in service he must identify the most likely modes in his specification and apply systematically all relevant design methods to prevent such failure.

His starting point must be the relevant Codes of Practice and where these are exceeded he must resort to other methods and where his knowledge is exceeded he must draw upon the help of specialists. The designer must never be afraid to ask.

However, it is considered that most designers do not appreciate the need for nor facilitate their total involvement.

Materials

Probably the most significant difference in the Coded requirements of the Big 3 relates to materials of construction. Material specifications are nationalistic and depend so much on the practices in that country, it is quite dangerous to make comparisons on the basis of chemical analysis alone. The complete spectrum of properties including weldability needs to be reviewed before intermixing specifications.

Code Rules

Engineers consider it is their duty to conserve materials at all costs - even to the extent of mixing Codes to obtain the minimum in all respects - this is a dangerous practice and must not be attempted. The rules are not interchangeable and in fact are not strictly operational outside their country of origin, since each rule is interdependent upon the whole set being fully complied with.

Only the British Code has so far come close to being a truly international PV Standard without the self-protective approaches of the ASME and the VdTUV.

Complete standardisation of code rules on an international basis will not be in the economic interest of all parties, whilst interchangeability and acceptance will follow with an apparent lowering of costs, excessive standardisation may inhibit innovation and lead to stagnation. International competition, works, with good effect.

The Specifier

Design starts with the specifier, all too often the specifier does not have the experience or qualifications to produce the correct information for the "designer". There is a need for improvement in communications between the specifier, the designer/manufacturer and the inspecting authority.

Reliability/Risks

There are three types of pressure vessel possible within the bounds of the Code rules:-

- those we design
- those we build   — any similarity may be purely coincidental!?
- those we operate/maintain

Each type has an effect the reliability of the structure, it is difficult to design a pressure vessel to fail-safe in all circumstances. Failures of PV today are associated with the abnormal condition - we live in a statistical world, the abnormal condition is possible - but such a condition is usually a previously unrelated combination of events, often simply analysed.

Risk in operation is something that the Codes do not take into consideration, other than of an isolated PV constructed in accordance with Code rules. Accidents like Flixborough are caused by a simultaneous combination of events and in the cases of complex dangerous plant an analysis needs to be undertaken to establish weak links, the overall plant reliability and the risks involved. Would a ruptured pipe destroy your plant?

In addition to the above, there is also an economic incentive to improve reliability of lesser faults which are not considered safety hazards, such as leakage and distortions, etc. which cause the plant to be shut-down for inspection and repair.

Education

Further education of engineers is required in the fundamentals of design, coupled with the understanding that in order to prevent failure a whole set of rules needs to be applied in a systematic manner. He must understand his limitations. This calls for cross fertilization and support from many disciplines.

It must be remembered modes of failure are not normally those associated with traditional stress analysis that we spent our youth assimilating, but are more often apparently unrelated results of human fallibility, a stray arc strike, an undiscovered crack, a dimensional error, etc.

The changing face of industry requires engineers with a knowledge of the fundamental principles of how to apply design knowledge from experience, from research and development and how to pass this information on to their successors without loss of meaning or rancour. Unfortunately this knowledge is not easily obtained in the first place and in the second place may fall on deaf ears because the effort required to gain it usually outweighs the rewards. Nevertheless, unless the Nation and Industry in general are prepared to back the education of engineers, in order to stimulate their efforts, future developments are not possible.

In order that a systematic approach be given to the application of design from concept to final product, senior management need to revise their approach in order to make this possible.

THE LAST WORD

Large factors of safety are no substitute for a carefully applied design and construction coupled with material homogeneity, toughness and weldability.

Fig. 15 Failure of Pressure Vessel or Bridge?

ACKNOWLEDGEMENT

The author wishes to thank colleagues at TUV-Rheinland S.A. for their help and useful discussions and Mr. H.F. Prinsloo, Managing Director, for permission to publish this paper. Thanks also to Prof. G.G. Garrett for his extreme patience and encouragement.

The views expressed in this paper are the author's and do not necessarily represent those of TUV-Rheinland S.A.

REFERENCES

ASME	Blr & PV Code Sect. III (Nuclear Power Plant) '77
ASME	Blr & PV Code Sect. IX '77 Welding Qualifications
ASME	Blr & PV Code Sect. XI (Rules for inservice Inspection of Nuclear Plant) '77
Am. Soc. N.D.T.	SNT-TC-1A '75 Personnel Quals & Cert. in NDT
BSI	BS470 '76 Manhole and Inspection openings in chem. plant
BSI	BS1500 '58 Fusion welded Pressure Vessels for general purposes carbon and low alloy steels, Part 1
BSI	BS1515 '65 Fusion welded Pressure Vessels, Part 1, Carbon and Ferritic Alloy Steels
BSI	BS1515 '65 Fusion welded Pressure Vessels, Part 1, Austenitic and Stainless Steels
BSI	BS4870 '74 Appro. Testing of Welding Procedures Pt. 1 - Fusion Welding-Steel
BSI	BS4871 '74 Appro. Testing of Welders working to approved procedures - Pt. 1 - Fusion Welding-Steel
BSI	BS5447 '77 Methods for K_{1C} toughness testing
BSI	DD-19 '72 Methods of COD Testing
BSI News	The Flixborough Disaster - How it happened June '75
DIN	DIN 8560 Approval Testing of Welders (Steel)
DIN	DIN 8561 Approval Testing of Welders (Non-Ferrous)
DIN	DIN 8563 Quality Control - Welded Structures
DIN	DIN 50 049 Certification for Material Testing
Eames A.R.	Current Trends in System Reliability Analysis CIRO Abstracts and Review No. 23 1968
Eccleston P.J.	Indefineability of Defects Pressure Vessel Needs for the Future SAIW '79
Financial Times	The Shadow of Flixborough 14 Sept. '76
Green A.M. Jr.	History of the ASME Boiler Code ASME '55
HM Factory Inspectorate	Flixborough Disaster HMSO '74
Hydrocarbon Processing	Worldwide PV codes Dec. '78
Insurance Monitor	Learning the Lessons of the Flixborough Disaster 11 July '74
Irwing R.R.	Why the ASME Code is in a class by itself, IRON AGE 17 Jan '77
SANDT	CSWIP Certification Scheme for Weldment Inspection Personnel - The Welding Institute
Saunders G.G.	Is Strength a Redundant Engineering Concept? The Metallurgist and Materials Technologist
SEW	Stahl Eisen Werkstoffblaetter (Association of German Steelmakers Material Specifications)
Smith T.A. & Warwick R.G.	A Study of Defects in Pressure Vessels Int. J. PV and Piping 1974
UKEA	Systems Reliability Service Digest Oct. '70
VdTUV Werkstoffblaetter	Vereinigung der Technischen Ueberwachungsvereine e.V. (Association of Inspection Authorities - Material Specifications)
VBG	Verband Berufsgenossenschaften (Federation of Trade Associations)
Wells A.A.	M.S. for Pressure Equipment at sub-zero temps. Brit. Weld. Journal March '64

APPENDIX I

SIMPLIFIED COMPARISON OF CODE RULES

ASME VIII Division 1

Scope - abridged

Included: (a) Containers for containment of pressure different to atmospheric $\geq \pm 103 - 20\,670$ KPa

              (b) Material: Carbon and low alloy steels
                             Non-ferrous material
                             High alloy steels
                             Cast iron
                             Clad material
                             Ductile cast iron
                             Ferritic steel - enhanced properties by heat treatment

              (c) Fabrication: Welding, forging and brazing

              (d) Safety pressure relief devices.

Excluded:
- (a) Fired process heaters
- (b) Pipework systems and related components
- (c) PV within the scope of other sections of the Code
- (d) Pressure containers, forming integral parts of pumps, compressors, engines etc.

Requirements:
- Manufacturers' quality control system
- Authorization by ASME to use code symbols
- Appointment of National Board authorized Inspector who monitors:-
- Manufacturers QC system after audit by ASME and
- Approval of the design and construction of the PV.

Materials:
- Proper identification and certification by Steelmaker
- Material limited to the use permitted in sub-section C of Code, or else proved similar.
- All materials may be used down to temperatures of minus 29°C. for temperatures lower than this figure they shall be subjected to impact testing Charpy Vee Notch $\equiv$ ISO-V.

Design:

Loadings to be considered:

1. Internal and external design pressure
2. Impact loads, including rapidly fluctuating pressures
3. mass of vessel, plus contents, including static head of liquid
4. Superimposed loads - mass of machinery and pipework, etc.
5. Wind loads and earthquake loads
6. Local loads due to attachments or mountings of: internals, vessel supports, lugs, rings, skirts, machinery, etc.
7. The effects of temperature gradients.

Allowable Stresses. "S" (other than bolting)

Minimum of the following values:

Ferritic & Austenitic (1)	Material Type Non-Ferrous	Austenitic (2) and Non-Ferrous
Below Creep Range		
amb or temp UTS/4	amb or temp UTS/4	
amb or temp yield*/1.6	amb or temp yield/1.5	yield /1.11
Creep Range		
1. Average stress - creep rate		0.01%/1000 h
2. Average stress to rupture		100 000 h/1.5
3. Minimum stress to rupture		100 000 h/1.25

* Yield or 0.2% PS.

Weld joint efficiencies

- A relic from the days of rivetted joints.

- Nowadays empirically takes into consideration type of joint and degree of NDT applied to it.

Stress Analysis (Thickness Calculations)

- Basic membrane theory - No defects - $D_o/D_i \leq 1.5$
  $P \leq 0.385SE$

- Ignores secondary stress levels providing standard geometry is maintained.

- Openings reinforced by area replacement technique.

- No rules are provided for local loads or fatigue assessments.

- Rules exist for low stress operation at low temperature without special notch toughness requirements, when $S \ngtr UTS/10$.

- The Code recognised that rules are not provided to cover all details of design and construction, it is intended that the manufacturer, subject to the approval of the Inspector, shall provide details of design and construction commensurate with the safety objectives of the Code.

Inspection openings

- The Code specifies certain mandatory openings for inspection.

Manufacture and Workmanship

- Material identification and control
- Material testing requirements
- Permissible fabrication techniques
- Heat treatment requirements
- Tolerances
- Welding procedures and operations qualified to ASME Section IX.

Inspection, Testing and Certification

- Operators conducting NDE to be qualified to SNT-TC-1A. (Am. Soc. NDT)
- Acceptance criteria for welded joints in two categories

(a) Full Radiography   (b) Spot Radiography

- the following imperfections shall be judged unacceptable

1. any type of crack or zone of incomplete fusion or penetration

1. Ditto

2. elongated slag or cavities (isolated)

   $L \ngtr 6mm$: $t \leq 19\,mm$
   $L \ngtr t/3$: $19 < t \leq 57\,mm$
   $L \ngtr 19mm$: $t > 57\,mm$
   t = weld thickness

2. elongated slag or cavities (isolated)

   $L \ngtr (6mm < 2/3t < 19mm)$
   all t
   t = thickness of thinner plate

3. Slag Group
   $\Sigma L \ngtr t$ in 12t
   except if separation $\geq$ 6LL

3. Slag Group
   $\Sigma L \ngtr t$ in 6t
   except if separation $\geq$ 3LL

   LL = length largest defect

4. Rounded indications
   acceptance charts
   Appendix IV

4. Rounded indications are not a factor in spot radiography.

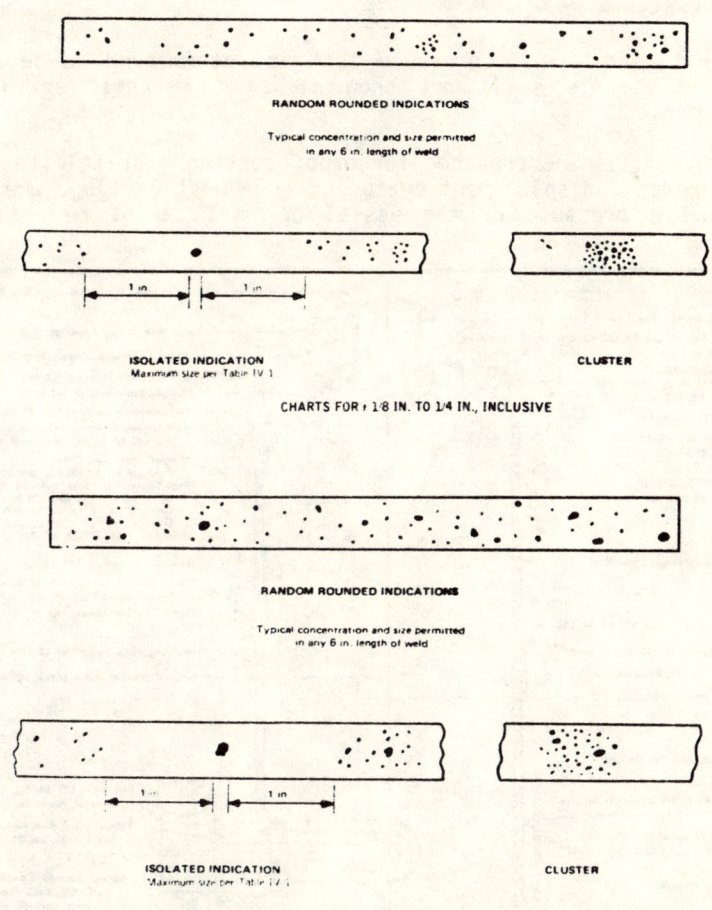

Fig. 16 ASME VIII/1 Porosity Chart

Pressure Testing

- All completed vessels shall undergo a pressure test.

- Hydraulic

Test Pressure Hh = 1.5 P x $\frac{\text{Design stress at test temp.} \quad (S_a)}{\text{Design stress at design temp.} \quad (S_b)}$

- Pneumatic

  Test Pressure $H_p = 1.25\ P \times \dfrac{S_a}{S_b}$

- Metal temperature at pressure test is recommended to be not less than 16°C - vessels shall not show signs of weakness or leakage during this period.

- Special rules are provided for proof testing - brittle coatings, strain measurement, displacement measuring and burst testing, when the maximum allowable pressure of the vessel or part cannot be established with accuracy.

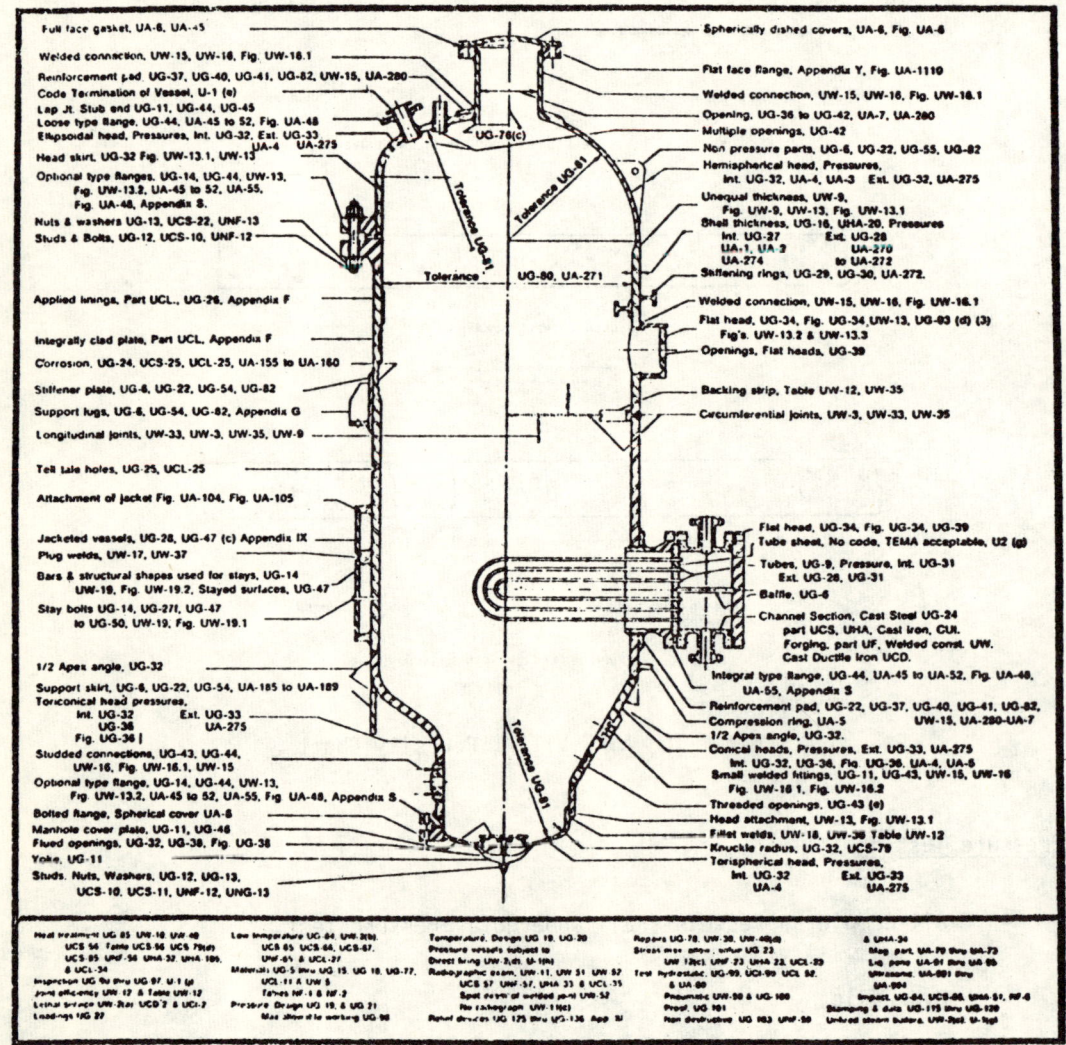

Fig. 17 Guide to ASME VIII/1
(after Hydrocarbon Processing)

Design and the Prevention of Metallurgical Failures   167

                               Certification

- Data report shall be prepared - signed by Manufacturer and Inspector - serves as a certificate.

- Vessel is branded with Code symbol and relevant data. Manufacturer keeps Data Report for not less than 5 years or else registers the vessel with the National Board.

ASME Section VIII Division 2 (Alternative Rules)

Scope - abridged
Included: (a) containers for containment of pressure different to atmospheric $\geq \pm 103$KPa in a fixed location.   without limit
            (b) Material: carbon and low alloy steel
                          non ferrous materials
                          high alloy steels
                          clad materials
                          ferritic steels - enhanced properties by heat treatment
            (c) fabrication: welding, forging
            (d) safety pressure relief devices.

Excluded: (a) transportable cargo tanks
            (b) fired process heaters
            (c) pipework systems and related components
            (d) Pressure vessels within the scope of other Sections of the Code
            (e) Pressure containers forming integral parts of pumps, compressors, engines, etc.

Requirements

Users Design Specification

- Details to constitute adequate basis for selecting materials, designing, fabricating and inspecting the vessel.
- Provide details of whether fatigue analysis is required.
- Provide details of corrosion and/or erosion allowance
- Provide details on whether fluids are lethal.
- Certification of users design specification by a U.S. registered professional engineer, experienced in pressure vessel design and construction.

Manufacturers Design Report.
- Calculations and drawing to show compliance with users design specification and Code Division.
- Certification of manufacturers design report by a registered professional engineer, experienced in pressure vessel design and construction.

Manufacturers quality control system.
- Authorization by ASME to use Code symbols
- Appointment of National Board authorized Inspector who monitors the manufacturers quality control system after audit by ASME and approves the construction of the pressure vessel.

Materials

- Proper identification and certification by the steelmaker.
- Material is limited to those listed in Tables ACS-1, AHA-1, AQT-1, AWF-1.1, 1.2, 1.3 and 1.4, ABM-1, 1.1, 1.2, 1.3, 2, 2.1, 2.2 and 2.3.
- Ultrasonic examinations are required on plates and forgings 102mm thick.
- General toughness requirements for steel products are specified, which are mandatory for lethal contents.
- Certain exemptions from impact testing are made for carbon steels. Also low stress operation ≤ 41.1 MPa but not less than -46°C, for carbon steel.

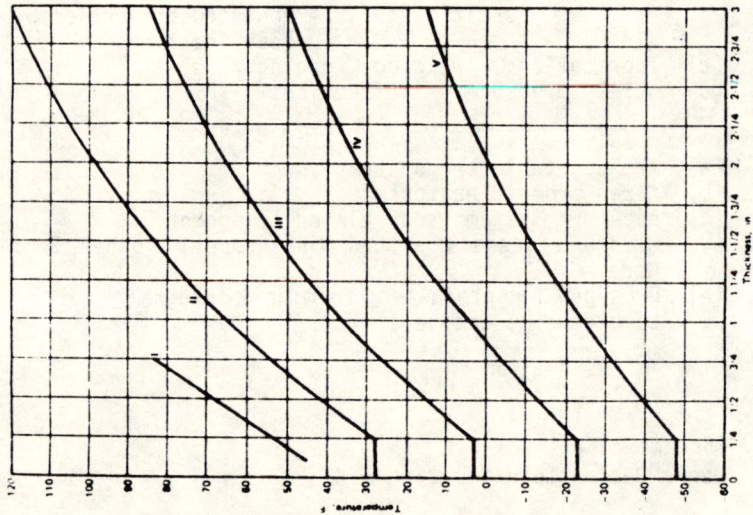

Fig. 18 ASME VIII/2 Impact Test Exemption Curves for Carbon Steels

Notes: (1) Impact tests not required for steel groups above the curves.

(2) Group I SA-36 Plate up to 19mm thick welded to primary pressure component.

(3) Group II Plate Steels SA-36 over 19mm thick when welded to SA-285/515. Included all of the product forms to specifications in Table ACS-1, except as listed below.

(4) Group III Plate Steels SA-442 up to 25mm thick.

(5) Group IV Plate Steels not normalized SA-442 over 25mm thick SA-516/662 Grade B up to 38mm thick.

(6) Group V Pipe Steels SA-524, Plate Steels normalized or heat treated SA-516/662 Grade B over 38mm thick and SA-537 Cl. 1/2-662/Grade A in all thicknesses.

Design

Loadings to be considered;

As for Division 1, but including temperature conditions introducing differential strain loadings and strain induced reactions resulting from expansion or construction of attached piping and other parts.

Design Stress Intensitites "Sm" (other than bolting)

Ferritic, Austenitic and Non Ferrous (1)	Austenitic and Non-Ferrous (2)
Below Creep Range	
amb. or temp. UTS/3	
amb. or temp. yield*/1.5	temp. yield/1.11
Creep Range	
No Data	

* yield or 0.2% P.S.

- Stress intensity   2 X maximum shear stress
- Allowable bearing stress - generally yield stress
- Allowable shear stress   - generally 0.6 X Sm

Stress Analysis

- Basic membrane theory - No defects : $P \leq 0.4 Sm$

- Allowances have to be made for corrosion etc. :

- Theory of failure : maximum shear stress (Tresca) - except certain configurations

- Rules for commonly used Pressure Vessels shapes provided where complete rules not given stress analysis is required

- Analytical stress analysis

- Experimental stress analysis

- Fatigue evaluation

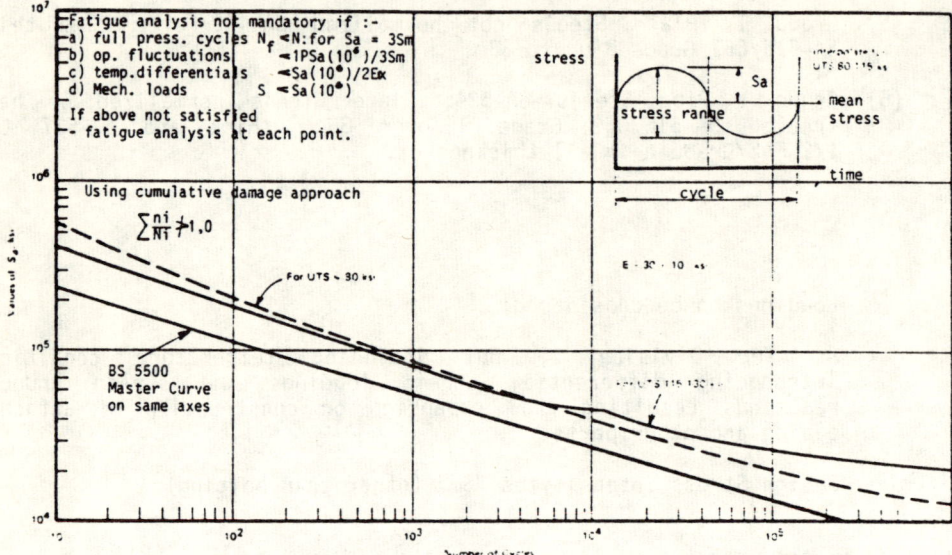

Fig. 19 ASME VIII/2 Design Fatigue Curves

Fig. 20 ASME VIII/2 Stress Categories and Limits of Stress Intensity

Design and the Prevention of Metallurgical Failures

- Opening reinforced by area replacement method - modified stress concentration approach used for specific applications
- No rules for stresses resulting from local loadings
- Special flanges are proportioned in accordance with the rules and stresses Division 1.

Inspection Opening
The code specifies certain mandatory openings for inspection.

Manufacture and Workmanship
- Material identification and control
- Material coupon-testing requirements after heat treatment
- Impact testing of welds and vessel test plates
- Permissible fabrication techniques
- Heat treatment requirements
- Tolerances
- Welding procedures and operators to be qualified to ASME Section IX.

Inspection, Testing and Certification

- Ensure QC procedures maintained, viz:-

    Authoritiy and responsibility
    Organization
    Drawings, design calcs. and specification control
    Material control
    Examination and inspection programme
    Correction of non-conformities
    Welding/NDE/Heat Treatment/Calibration
    Records retention

- Operators conducting NDE to be qualified to SNT-TC-IA.
- Acceptance criterial welded joints equivalent to that for full radiography in Division 1.
- NDE requirements for material groups/weld types defined, including RT, UT, PPT, MT.

Pressure Testing

- All completed vessels shall undergo a pressure test.

    Hydraulic :

$$\text{Test Pressure } P_h = 1.25 \times \frac{\text{Design stress intensity test temp } (S_{ma})}{\text{Design stress intensity design temp } (S_{mb})}$$

Max. primary membrane stress intensity $P_m <$ yield stress test temp. S_y

Max. primary membrane + primary bending,
$P_m + P_b$: $P_m + P_b \leq 1.35 S_y$: $P_m \leq 0.675 S_y$
or
$P_m + P_b \leq 2.15 S_y - 1.2 P_m$: $0.675 S_y < P_m \leq 0.90 S_y$.

Pneumatic :

Test Pressure $H_p = 1.15 P \times \dfrac{S_{ma}}{S_{mb}}$

Max. P_m 0.8 S_y; otherwise as for hydraulic conditions.

- Metal temp. recommended not less than 16°C.

Type of Vessel	Service Temperature	Pressure Test Temperature	Startup Temperature
Base Vessel			
As-welded	Same	30 F higher	(Note 2)
PWHT	Same	Same	(Note 2)
Lethal			
As-welded		Not permitted	
PWHT	(Note 5)	(Note 5)	(Note 2)
Refrigerated Service (Note 4)			
As-welded	Same (Note 3)	30 F higher	(Note 2)
PWHT	Same	Same	(Note 2)

NOTES:
(1) When impact tests are not performed because Fig. AM-218.1 is used, the minimum permissible temperature from Fig. AM-218.1 shall be considered the impact test temperature for applying this table.
(2) If pressure is applied at a metal temperature below the minimum permissible temperature, it shall not exceed 20% of the required test pressure.
(3) Only when nozzle welds and other areas of high localized stress are postweld heat treated, except in the case of 9% nickel steel up to and including 2 in. thickness; otherwise, 30 F higher.
(4) Refrigerated service for purposes of this table is defined as service below 32 F where the temperature is controlled in the process rather than being caused by atmospheric conditions.
(5) 20 F higher for each additional inch of nominal thickness or fraction thereof for thickness over 1 in. but not to exceed 60 F higher.

Fig. 21 Minimum Permissible Vessel Metal Temperatures for Ferrous Metals other than Austenitic in relation to Impact Test Temperature Note (1)

- Vessel shall not show signs of weakness or leakage.
- Special rules for "proof testing" are described in the section on experimental stress analysis.

Certification

A procedure as outlined in Division 1 is used, except the data report shall be kept for a period of 10 years or for the intended life of the vessel whichever is the greater period.

BS 5500

Scope - abridged.

Included (a) containers for containment of pressure where the membrane stresses $\not\geq$ 10% Nominal Design Stress : Section 3.

      (b) Material   : carbon and low alloy steels
                            : high alloy steels
                            : clad materials

      (c) Fabrication : welding

Excluded (a) Storage tanks : near atmospheric pressure - additional pressure $\geq$ -0.6 kPa $\leq$ +14.0 kPa
                                     : low pressure $\leq$ 100 kPa

      (b) Strip wound compound : appropriate for very high pressures

      (c) Transportable cargo tanks.

      (d) Pressure Vessels for specific applications which are covered by other British Standards.

      (f) Pipework systems and related components.

Requirements

- Vessel to be manufactured under the survey of independent Inspection Authority, who checks that the design, materials and construction comply with the requirements of the standard.

- Users design specification - details of normal transient or adverse working conditions - agree a design safe for creep, fatigue and corrosive conditions.
  Regulations for in-service inspection - statutory or other regulations.

- Name of Regulating Authority in country of installation.

Materials

- Proper identification and certification by steelmaker - material limited to those permitted in Section 2.4 of the Standard or else proved similar.

- Materials with hot certified properties.

- Special consideration shall be given to the selection of materials used to operate below 0°C.

Design

- Loading to be considered - as for ASME VIII/1 & 2.
- Constructional categories. 1, 2 or 3.

Construction category	Non-destructive testing (NDT) requirement	Permitted material	Maximum thickness (mm)	Temperature limits	
				Upper	Lower
1	100% (see 5.6.4.1)	All	None, except where NDT method limits	See 3.1.2	See appendix D limitations
2	Limited random (spot) (see 5.6.4.2)	M0 M1 M2 M3 M4	40 30 20 15	See 3.1.2	See appendix D limitations
		Austenitic steel	40		None
3	Visual only (see 5.6.4.3)	C & CMn steel (R_m > 432 N/mm²)	16	> 300 °C	See appendix D limitations
		Austenitic steel	25		None

*For definition of R_m see K.2.

Fig. 22 BS5500 Construction Categories

Nominal Design Strengths (f) - other than bolting.

Minimum of the following values :-

Ferritic   Austentic

Below Creep Range

Hot certified &   50°C*   Hot certified &   50°C*
not certified   not certified
Amb. yield /1.5   Amb. Yield /1.5
Amb. UTS /2.35   Amb. UTS /2.5

Hot certified   150°C*   Hot certified   100°C*
Temp. 0.2PS /1.5   Temp. 1.0PS /1.35
Amb. UTS /2.35   Amb. UTS /2.5

Ferritic   Austentic

Not hot certified   150°C*   Not hot certified   100°C*
Temp. 0.2PS /1.6   Temp. 1.0PS /1.45
Amb. UTS /2.35   Amb. UTS /2.5

  * Intermediate values   * Intermediate values
     50 - 150°C interpolated   50 - 100°C interpolated.

Ferritic and Austentic

Creep Range - Hot certified
$\begin{bmatrix} \text{Mean value of stress** to} \\ \text{produce rupture in time} \\ \text{"T" at temp."t" for grade} \\ \text{of material in question.} \end{bmatrix}$ /1,3

**ISO / TC17 /SC10

- Not certified
$\begin{bmatrix} \text{to be agreed between pur-} \\ \text{chaser and manufacturer.} \end{bmatrix}$ /1,3

- Allowable bearing stress   -   generally 1.5 f
- Allowable shear stress   -   generally 0.5 f

Stress Analysis

- Basic Membrane theory - No defects : $D_o/D_i < 1.3$

- Theory of failure : maximum shear stress (Tresca) - except certain configurations

- Rules for commonly used Pressure Vessels shapes provided - where complete rules not given, stress analysis required unless it can be shown by experience that the analysis need not be performed.

- Analytical stress analysis

- Experimental stress analysis (Proof Testing), (Fatigue Testing)

- Stress intensity limits - as ASME VIII/2

- Fatigue evaluation. Similar to ASME VIII/2 - ISO/DIS2694, except fatigue curve simplified to straight line, see superimposed curve Fig. 17. Safety Factor 15 on cycles : 2.2 on stress level.
Planar features recognised - unwelded lands and defects - LEFM analysis proposed to estimate propagation.

- Openings reinforced by stress concentration approach $\not>$ 2.25.
Account taken of superimposed pipework loads on simplified basis. Alternative area replacement method allowed within certain limits.

- Rules for establishing stresses from local loads, thermal gradients etc. are included.

- Rules for special flange design are similar to those given in ASME VIII/1 except that the higher design stresses are used, but the flange moments are corrected for bolt-spacing to reduce excessive deflection.

- Tentative recommendations are made for vessels constructed from C & CMn steels to operate below 0°C. Minimum temperatures based on experimental results from welded wide plate tests (WWP) (Wells '64) in the specified materials using defects of standard size and a strain of 0.5% ≏ 4 X yield point strain.
  The results are plotted in material thicknesses curves, against minimum design temperatures and ISO-V inpact test temperatures where a corrolation exists with WWP tests.

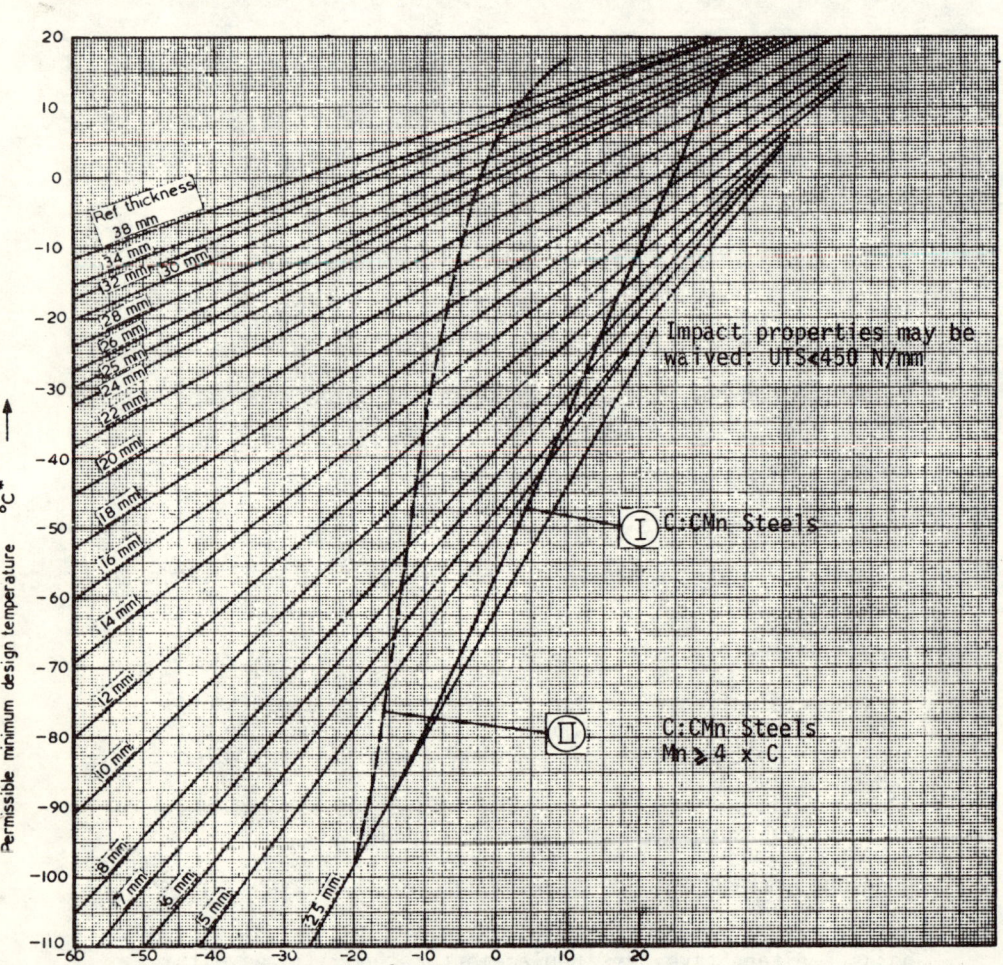

Fig. 23 A BS5500 Limiting Design minimum temperatures for as-welded components

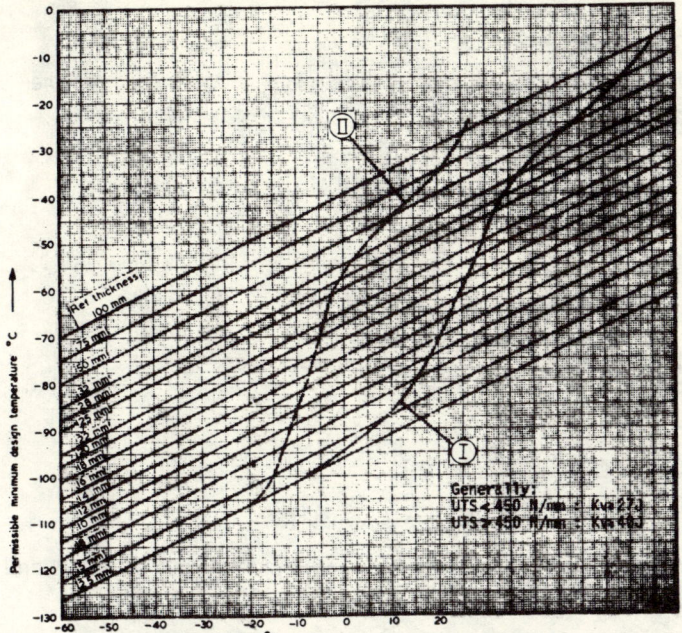

Fig. 23 B BS5500 Limiting Design minimum temperatures for post-weld heat treated components.

- Low stress op. ≤ 50 MPa design temps indicated in Figures 23 may be lowered by 50°C.

  Welds   Yield < 300 MPa = 30J
  Toughness   Yield > 300 MPa = 40J

- Test Temps.

  As welded   -   material reference temp.
  PWHT   -   t > 60   Min. Design temp. t.
                   40 t 60   t + 10°C
                   30 t 40   t + 20°C
                   20 t 30   t + 30°C
                      t 20   t + 50°C

- Fracture Mechanics testing to be agreed between purchaser and manufacturer   - DD 19-1972   BS5447-1977

- Examples of weld preparations given

Inspection openings

- Inspection openings shall comply with BS 470

Manufacture and workmanship

- Fully dimensioned approved drawing required before manufacture - stating all manufacturing parameters

- Material identification and control

- Permissible fabrication techniques

- Tolerances

- Heat treatment requirements

- Welding procedures and operators approved to BS4870 Pt1   BS 4871

Inspection, Testing and Certification

- Principal stages of inspection outlined

- Production control tests - agreement purchaser and manufacturer including toughness testing.

- Operators conducting NDE certified as competent in agreement with Inspection Authority, e.g. CSWIP (SANDT)

- Extent of NDE is specified

- Acceptance criteria   -   parent materials: between purchaser and manufacturer
  - welded joints: as per Table 1.

TABLE 3 A - BS5500 Weld Acceptance Criteria

Abbreviations used:
e is the parent metal thickness. In the case of dissimilar thicknesses e applies to the thinner component;
w is the width of defect;
l is the length of defect;
h is the height of defect;
d is the diameter of defect;
c is the mean nozzle circumference.

Defect type		Permitted maximum
Planar defects	Cracks and lamellar tears	Not permitted
	Lack of root fusion Lack of side fusion Lack of inter-run fusion	Not permitted
	Lack of root penetration	Not permitted
Cavities	(a) Isolated pores (or individual pores in a group)	$d > e/4$ and $d \not> 3.0$ mm for e up to and including 50 mm $d \not> 4.5$ mm for e over 50 mm up to and including 75 mm $d \not> 6.0$ mm for e over 75 mm
	(b) Uniformly distributed localized porosity	2 % by area* (as seen in a radiograph) for $e \not> 50$ mm and pro rata for greater thicknesses
	(c) Linear porosity	Linear porosity parallel to the axis of the weld may indicate lack of fusion or lack of penetration and is therefore not permitted
	(d) Wormholes isolated	$l \not> 6$ mm, $w \not> 1.5$ mm
	(e) Wormholes aligned	As linear porosity
	(f) Crater pipes	As wormholes isolated
	(g) Surface cavities	Not permitted

TABLE 3 B - BS500 Weld Acceptance Criteria

Solid inclusions	Slag inclusions	(a) Individual and parallel to major weld axis NOTE: Inclusions to be separated on the major weld axis by a distance equal to or greater than the length of the longer and aggregate length not to exceed the total length.	Main butt welds	$l \sim e \ge 100$ mm w or $h \sim e; 10 \ge 4$ mm	
			Nozzle and branch attachment welds	Inner half of cross section	Outer quarters of cross section
				w or $h \sim e/4; \ge 4$ mm $l \sim \frac{e}{4} \ge 100$ mm	w or $h \sim e/8 \ge 4$ mm $l \sim \frac{e}{8} \ge 100$ mm
		(b) Individual and randomly oriented (not parallel to weld axis)	As isolated pores		
		(c) Non-linear group	As localized porosity		
Solid inclusions		Tungsten inclusions (a) Isolated (b) Grouped	As isolated pores As uniformly distributed or localized porosity		
		Copper inclusions	Not permitted		

*Area is the product of length and width of an envelope enclosing the affected volume of weld metal measured on a plane substantially parallel to the weld face (i.e. as seen on a radiograph).

- All completed vessels shall undergo a pressure test, either Hydraulic/ pneumatic/ or combined testing.

  Test Pressure $Pt = 1.25p \left[\frac{fa}{ft} \times \frac{t}{t-c} \right]$

  P = design pressure
  fa = design stress test temp.
  ft = design stress design temp.
  t = nominal thickness
  c = corrosion allowance

- Max. general membrane stress at test 0.90 yield stress at test temp.

- Chloride content of test water to be controlled for austentic stainless steel vessels.

- Adequate precautions for pneumatic test - blast protection, degree of confidence in stress analysis, NDE adequacy, resistance to fast fracture, preventing of chilling, etc.

- Metal temps. to be agreed between purchaser and manufacturer but $\ge 7^\circ C$.

- Proof testing details are outlined

- Vessels should not show signs of general plastic yielding or leakage during this period.

Certification

- Manufacturer shall issue certificate of compliance countersigned by Inspection Authority.

- The vessel shall be marked with requirements of the standard - stamping on the vessel shall only be permitted prior to normalising.

A-D Merkblaetter

Scope - abridged

Included (a) containers for containment of pressures different from atmospheric; predominantly under static load.

    (b) Material :   Carbon and low alloy steels
                           high alloy steels
                           cast iron
                           aluminium
                           clad steels
                           ductile cast iron
                           glass fibre reinforced plastics
                           carbon
                           glass

    (c) fabrication : welding, forging and brazing.

    (d) safety pressure relief devices.

Excluded (a) Pressure vessels specific applications which are covered by DIN and other German Regulations
        (b) Pipework systems and their related components.

Requirements

- Vessels in general to be manufactured under the survey of independent Inspection Authority, who approves that the design, materials and construction comply with the requirements of the rules.

- Users design specification - any conditions to be met exceeding the specification in the AD - Merkblaetter, e.g. dynamic load, corrosion allowance, additional tests, use of certain materials, etc.

- Manufacturers to manage production such that the requirements of the rules and good engineering practice is maintained.

- Manufacturers to have available the necessary facilities

- Manufacturers to have their own supervisory staff and trained workers.

- Inspecting Authority entitled to verify in the course of manufacture that the manufacturer maintains his requirements.

Materials

- Proper identification and certification by the steelmaker certification shall be to DIN 50 049, it is required that certain materials be inspected at source by the Inspecting Authority.

TABLE 4 - Summary of permissible materials and type of certificate to DIN 50 049 (A-D W1)

Grades of steel	All contents except of poisonous and liquefied inflammable gases			highly poisonous charge or liquefied inflammable gases		Type of material certificate to DIN 50 049
	Pressure vessels within the limits of the diameter – pressure products $D_i \cdot p$ at atmospheric temperatures and temperatures of the charge of $-10\,°C$ up to the next highest wall temperatures			$l \cdot p \leq 5000$	$l \cdot p > 5000$	
	$D_i \cdot p \leq 7000$ up to 120 °C	$D_i \cdot p \leq 20000$ up to 300 °C	$D_i \cdot p > 20000$ or above 300 °C			
Steels to DIN 17 155	yes	yes	yes²)	yes	yes	H I, H II: 3.1 B other steels 3.1 C
Steels to Stahl-Eisen-Werkstoffblatt 089	yes	yes	yes³)	yes	yes	StE 26, StE 29: 3.1 B other steels 3.1 C
USt 34–1, USt 37–1	yes, s ≤ 6 mm	no	no	no	no	2.1
RSt 34–1, RSt 37–1	yes, s ≤ 12 mm					
USt 34–2, USt 37–2	yes, s ≤ 12 mm	yes, s ≤ 12 mm	no	yes⁴) s ≤ 12 mm	no	2.2
RSt 34–2, RSt 37–2	yes	yes without wall thickness limitation		yes⁴)		
St 37–3, St 52–3	yes	yes without wall thickness limitation	no	yes⁴)	no	3.1 B
Other steels	according to the expert opinion of the Inspecting Authority					

¹) Concerning service pressure < 0.5 bar section 2.5 should be observed
²) Up to the temperatures for which strength characteristics are indicated in DIN 17 155
³) Up to the temperatures for which the strength characteristics are given in 'Stahl-Eisen-Werkstoffblatt 089'
⁴) In addition the application limits for non inflammable and non toxic liquified gases should be observed

- General toughness requirements are specified in the appropriate material specification, DIN, SEW or Vd TUV Werkstoffblaetter.

- Materials are limited to those permitted in Section W of the specifications - other materials are allowed but are subject to approval by the Inspecting Authority.

- Additional requirements for material are specified at design temperatures below 10°C.

- Welding consumables are to be approved by the Inspecting Authority for range of application.

Design

- Loading to be considered - as for ASME VIII/ 1 & 2.

- Design stresses do not exist as such.
  below creep range:

  material guaranteed yield, 0.2/1.0 PS are factored for safety.

  creep range: - lowest of following:-

  average stress to rupture in 100 000 or guaranteed min. yield, 0.2/1.0 PS at temp. are factored for safety.

TABLE 5 A-D/BO Safety Factors on yield or proof stress and creep data

Material	Safety Factor design temp.		Safety Factor test temp.
1. Rolled and forged steel	1.5		1.1
2. Cast steel	2.0		1.5
3. S.G. Iron	Annealed	Not Annealed	
3.1 GGG/70/60	5.0	6.0	2.5
3.2 GGG/50	4.0	5.0	2.0
3.3 GGG/40	3.5	4.5	1.7
3.4 GGG/40.3/35.3	3.0	4.0	1.5
4. Aluminium Alloy	1.5		1.1

Weld joint efficiency - included for certain grades of steel $\geq$ 0.85

Stress Analysis (Thickness calculations)

- Basic membrane theory - No defects: $D_o/D_i \leq 1.2$

- Ignores secondary stress levels providing standard geometry maintained

- openings reinforced by stress concentration approach strains 1% at test pressure

- no rules exist for additional external forces and moments or nozzles

- no rules exist for assessment of local loads

- no rules esist for failure criteria

- rules exist for low stress operation at low temperature without special toughness requirements

- rules exist for ferritic materials operating at low temperature based on special toughness requirements

- simplified rules exist for fatigue loading, giving a safety factor of 10 on cycles

Design and the Prevention of Metallurgical Failures

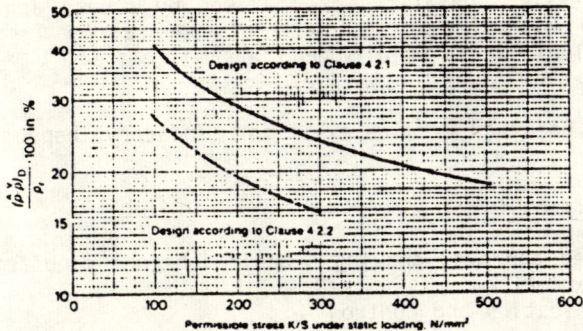

Fig. 24A A-D Continuously tolerable range of pressure variation

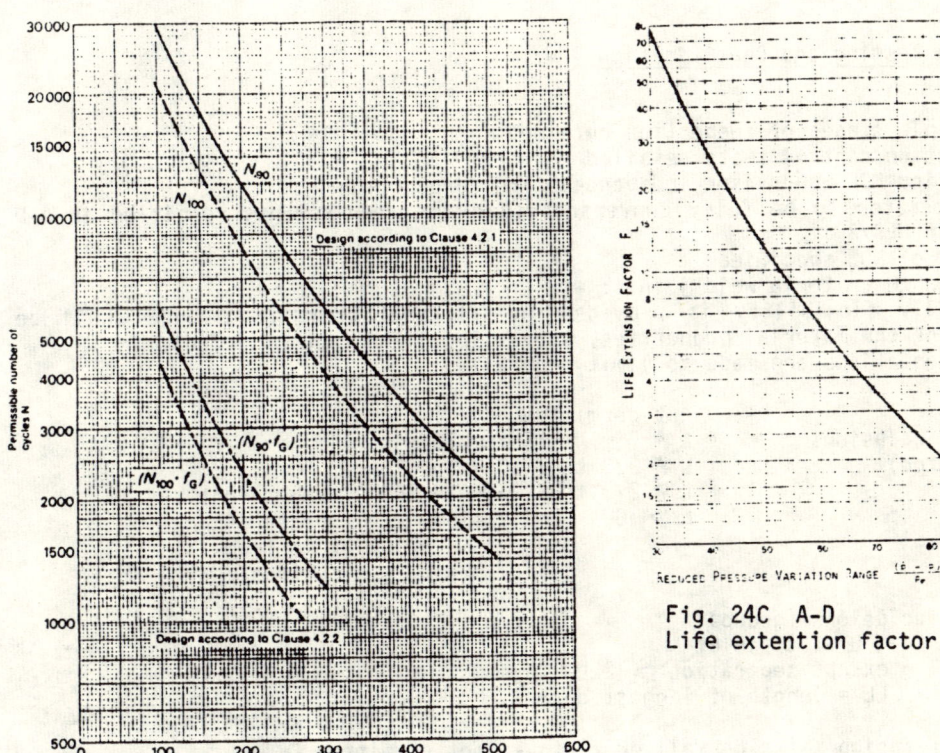

Fig. 24B A-D Permissible number of starts and stops with ranges of pressure variation from 90-100% (Pr-0)

Fig. 24C A-D Life extention factor

- Specification recognises rules are not provided for all design and construction details, deviations are allowed by stress analysis, analytical experimental or experience - subject to approval by Inspection Authority such that technical safety requirements are maintained.

Inspection Openings - Rules exist for mandatory inspection openings.

Manufacture and Workmanship

- Fully dimensioned approved drawings required before manufacture, stating all manufacturing parameters.
- Material identification and control
- Tolerances
- Heat treatment requirements
- Qualified welding supervisors - see also DIN 8563
- Welding procedures qualified to A-D Section HP2/1
- Other joining methods - soldering ,etc. similarly qualified
- Welders qualified to DIN 8560/61, includes a theoretical test.
- Special agreements

Inspection Testing and Certification

- Principle stages of inspection outlined
- Production control tests detailed
- Qualified NDE Supervisor independent from production
- NDE operator to be fully conversant with the requirements specified in A-D Section HP3/3
- Extent of NDE specified
- Acceptance criteria weld joints: -
  Basically flexibility is provided in interpretation, cognisance shall be taken of the material properties, stress level and type of defect. Recommendations are made to limit generally: -

  cracks — not permitted
  solid inclusions — $L \not> 7$: $t \leq 10mm$
  (isolated) $L \not> 2/3\ t$: $10 < t \leq 75mm$
  $L \not> 2/3\ t$: $75 < t \leq 150$ ⎫ when defect
  $L \not> 100$: $t \leq 150$ ⎬ greater than
   ⎭ 10mm from a
   surface.

  Solid inclusions (group)
  $\Sigma L < t$ in $6t$
  except separation $\geq 2LL$
  LL = length of largest defect

  Lack of fusion - side wall or root - not permitted
   - inter run - greater than 3 runs :
     as if solid inclusions

  Rounded
  indications - porosity charts similar to ASME VIII/1

- All completed PV shall undergo a pressure test.
  Hydraulic
  Test pressure p = 1.3 p for steels and aluminium
               p = 2.0 p for cast materials

  Pneumatic
  Test pressure p = 1.1 p
  Maximum temperature test water 40°C

   Certification

- Inspection Authority draws up the Certificate to which are appended material certification and other relevant certificates to be held on file for at least 7 years.

   Testing of Pressure Vessels during Operation

- Rules are included to provide for mandatory acceptance testing of finished fitted pressure vessels in conjunction with overall plant and safety installations together with periodic examinations in service.

FRACTURE MECHANICS AND THE ASSESSMENT OF STRUCTURAL RELIABILITY

G. G. Garrett

Department of Metallurgy, University of the Witwatersrand, Johannesburg, R.S.A.

ABSTRACT

Based on the premise that all structures contain flaws, the aim of fracture mechanics is to describe the ability of a material to resist crack propagation in the presence of these flaws. This may be achieved by providing a quantitative means of relating the failure stress of a structure or component containing a flaw to the size of that flaw through a laboratory-determined value of the material's toughness. Under situations where the avoidance of catastrophic, brittle fracture of high strength materials is of prime importance, it will be shown that stress intensity factors and fracture toughness must be used, rather than yield or ultimate strength. In tougher materials, where extensive plastic deformation may accompany fracture, an alternative method of assessing toughness, based on the critical crack opening displacement (or COD) is described, which has been used very successfully in a wide range of practical applications.

INTRODUCTION

Of the many ways in which a particular structure can fail, the occurrence of fast, brittle fracture can have particularly disastrous consequences, as has been clearly illustrated by a number of spectacular examples over the past few decades. In recent years, requirements for greater structural efficiency and improved operational performance have inevitably led to situations where many components or structures are operating at or near the mechanical limits of the material, with a consequent increased probability of rapid, and sometimes catastrophic, failure. The intensive research effort which has resulted from many costly failures has established the significance of the new scientific discipline of Fracture Mechanics. This has been developed into a powerful tool for design, materials selection and failure analysis for critical components.

DEFECTS, THEIR DISTRIBUTION AND N.D.T. CAPABILITY

All structures contain flaws. These flaws may be metallurgical defects, such as inclusions or porosity, or microscopic crack-like flaws, either intrinsic or initiated in service. Intrinsic flaws are those designed into a structure, such as port holes in a pressure vessel, or windows in aircraft. One should also note the occurrence of stress concentrating features in many components in the form of rapid changes in section, such as keyways, splines or other forms of notches. These are often sites for the initiation of in-service cracks which may arise due to cyclically fluctuating loads (fatigue), corrosion or stress corrosion cracking. Further examples of material defects are given in Table I.

Table 1
Typical Material Defects

Defects existing in mill products	Defects produced by processing (continued)
Chemical contamination	Metal finishing
Inclusions, dirt	Cracks in coating, base metal
Segregation	Pits, blisters
Laminations	Lack of adhesion, insufficient
Internal defects	thickness
Porosity	Hydrogen embrittlement
Pipes	Surface contamination
Cracks	Joining
Surface defects	Weld defects – Cracks
Cracks, tears	Incomplete fusion, residual stress
Laps, pits	Fasteners – Tears, galvanic corrosion
Scratches	
Distortion	*Defects produced in Service*
	Mechanical damage
Defects produced by processing	Particle damage, tool marks
	Improper repair, maintenance
Metal removal	Fatigue cracks
Cracks	Fretting
Tool marks, gouges	Creep
Heat treatment	Environmental damage
Cracks	Corrosion
Distortion	Stress corrosion, corrosion fatigue
Decarburization	Bacterial degradation
Incomplete transformations	Thermal degradation

In addition, many structures today are of an all-welded design which inherently present two major problems in integrity assessment. Firstly, most metallurgical defects associated with welding will occur within the weld metal or heat-affected zone (HAZ) of a joint, and these are the regions most likely to contain high residual stresses. Secondly, once a crack has initiated, the extent of damage is likely to be much greater in welded structures, in comparison with structures fabricated by bolting or riveting, since cracks can readily propagate from one plate member to another. Thus, welded oil pipelines have been known to fail in a brittle manner where cracks have extended to lengths of many kilometres.

Having accepted the basic concept that 'all structures contain flaws', in order to quantitatively assess structural integrity, fracture mechanics analyses require a statistical definition of the largest flaw that can escape detection during inspection, i.e. it should be a 'worst case' situation. Whilst fracture mechanics is often considered primarily as a design tool, it is most effective when interfaced with other disciplines such as materials selection, manufacturing and inspection. A key to the successful implementation of fracture mechanics is often the definition of the starting point for the analysis, i.e. the definition of the largest initial flaw that could reasonably be present, on a given statistical basis. Immediately, however, one is faced with one of the great problems in this field, namely the assessment of a distribution of crack-like defects introduced by the fabrication of a structure. The probabilistic aspects have very recently been extensively reviewed by Johnston (1978) although, as she points out, relatively little experimental data is available. Indeed, most estimates seem to be based on experience and intuition; even in nuclear reactor primary cooling pipework, for example, human judgement based on previous failures has often been the basis adopted to define the probability of finding a crack in a particular location. It is reasonable to expect, though, that there will typically be a large number of small defects with a decreasing number of larger ones, and Fig. 1 schematically summarises one method of presentation (Pettit and Krupp, 1974). However, establishing a statistically reliable flaw size detection limit is normally a very difficult task, influenced as it is by many different parameters (Table II). In particular, an unstated but widely accepted assumption is that the minimum detectable flaw size varies with the experience of each operator. Furthermore, one of the most dangerous pitfalls in the development of an NDT detectability pro-

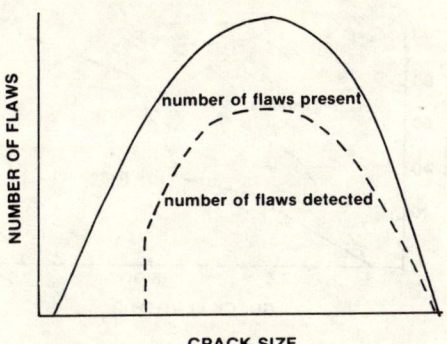

Fig.1: Schematic data presentation for comparison of flaw-size distribution with non-destructive testing detection limits (after Pettit & Krupp, 1974)

gramme is the failure to recognise that the method, personnel and inspection environment must simulate those service conditions to be encountered in the production and/or service situation. Table III, for example, taken from a survey of the U.S. aircraft industry (and therefore subjected to some of the most rigorous standards in the world), indicates that NDT detection limits based on laboratory data are often completely non-representative of actual service inspections. Although Table III provides a reasonable comparative indication of the detection limits which may be achieved using various techniques, it should be emphasised that a given size of crack can only be detected with a certain mathematical confidence. Thus, Fig. 2 illustrates that ultrasonics, for example, may under certain conditions detect a 4 mm defect only half the time, but a 7 mm defect virtually every time. Marriott and Hudson (1977) have recently estimated the probability that defects greater than 25 mm will be undetected in a nuclear reactor pressure vessel to be 1 in ~ 7×10^5. They further estimate that weld defects would account for something in excess of 85% of this figure.

TABLE II

FACTORS AFFECTING THE DEVELOPMENT OF NDI DETECTION LIMITS

(After Pettit & Krupp, 1974)

(a)	Material Characteristics:	grain size, amount and distribution of second phase particles, other metallurgical parameters.
(b)	Type of Flaw:	volume defect, planar defect, surface defect, embedded defect.
(c)	Stage in processing:	surface condition and finish, effect of previous processing steps, residual or applied stresses.
(d)	Part Configuration:	thickness, presence of abrupt geometrical changes, accessability of critical regions.
(e)	Means of defect development:	machining, electrical discharge machining, quenching, welding, fatigue.
(f)	Human Factors:	variation in inspector experience, person-to-person variations in the interpretation of results, use of production v. lab personnel.
(g)	Equipment or Procedure:	variations in calibration of equipment, variations between equipment, variations in inspection procedures and sequencing of operations.
(h)	Inspection Environment:	conditions in the laboratory, in the factory or in the field on an assembled component.
(i)	Detection Limit Criteria:	probability limit and detection limit confidence level required.

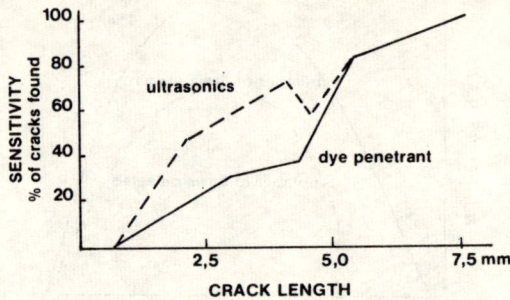

Fig.2: Flaw detection capability (after Packman, 1968)

In summary, therefore, it is evident that defect sizing is of particular significance in the application of fracture mechanics to structural design, and that one of the most important considerations is the acceptable size of defect for a given probability of failure.

TABLE III

ESTIMATED VARIATIONS IN FLAW DETECTION LIMITS BY TYPE OF INSPECTION

(in mm)   (After Pettit & Krupp, 1974)

NDT Technique	Surface Cracks		Internal Flaws	
	Processing	Fatigue	Voids	Cracks
Test Specimens, Laboratory Inspection				
VISUAL †	1,25	0,75	+	+
ULTRASONIC	0,12	0,12	0,35	2,0
MAGNETIC PARTICLE	0,75	0,75	7,5	7,5
PENETRANT	0,25	0,5	+	+
RADIOGRAPHY	0,5	0,5	0,25	0,75
EDDY CURRENT	0,25	0,25	+	+
Production Parts, Production Inspection				
VISUAL	2,5	6,0	+	+
ULTRASONIC	3,0	3,0	5,0	3,0
MAGNETIC PARTICLE	2,5	4,0	+	+
PENETRANT	1,5	1,5	+	+
RADIOGRAPHY	5,0	*	1,25	*
EDDY CURRENT	2,5	5,0	+	+
Cleaned Structures, Service Inspection				
VISUAL	6,0	12,0	+	+
ULTRASONIC	5,0	5,0	4,0	5,0
MAGNETIC PARTICLE	6,0	10,0	+	+
PENETRANT	1,25	1,25	+	+
RADIOGRAPHY	12,0	*	4,0	*
EDDY CURRENT	5,0	6,0	+	+

+ Not Applicable.   † Use with magnifer.   * Not possible for tight cracks.   (Based on 25 mm ferritic steel part, surface 63 RMS)

TOUGHNESS AND FRACTURE TOUGHNESS

A realistic assessment of the risk of fast fracture in structural materials requires an evaluation of both structural and material parameters (e.g. thickness, size and sharpness of defects present, yield strength and microstructure at a crack tip) and operational parameters (temperature, magnitude of stress and strain -- applied and residual, strain rate, etc.). The only way in which all of these factors can be realistically incorporated in any design is to carry out large-scale tests on full service structures containing controlled artificially-introduced defects. Although such tests are carried out from time to time, e.g. on nuclear pressure vessels or submarine hulls, they are usually impractical as well as prohibitively expensive in most cases. However, full-scale tests on representative details from structures are sometimes necessary, and are particularly useful in providing data for correlation with small-scale tests.

In general, however, primarily because of the cost of large-scale testing, much attention has focussed on the use of small-scale laboratory tests which, hopefully, can provide equally reliable information. Of those tests which have been developed over the years, the best known and still the most widely accepted for measurement of 'toughness' -- or resistance to crack propagation -- is the Charpy/Izod V-notch impact test which measures the energy absorbed in fracturing a standard notched test piece under impact loading. Unfortunately, however, there are a number of limitations to the usefulness of the data which are obtained from these tests. Firstly, the restricted specimen size, standardised to 10mm square bar or round diameter, remains constant and is often very small in comparison with the dimensions of the service component. Secondly, the high strain induced by impact loading may be unrealistic to the service loading spectrum experienced by the component or structure. Thirdly, the standard V-notch has a nominal 0,25mm root radius which may well fail to simulate the severity of a pre-existing defect, such as a fatigue crack. Furthermore, not only does the total energy absorbed during the fracture process fail to distinguish between the processes of crack initiation and propagation, but this impact energy measure of material toughness provides no quantitative design parameter which can utilize any non-destructive flaw size determinations. The main use of these impact tests, therefore, is in the material quality control function after suitable correlations have been established with other more quantitative, but still small-scale, laboratory tests measuring 'fracture toughness'. Fracture mechanics analyses have provided such tests.

FOUNDATIONS OF LINEAR ELASTIC FRACTURE MECHANICS

Linear elastic fracture mechanics essentially originated in 1920 when Griffith found from experiments on glass (taken as an ideally elastic, brittle material) that the fracture stress times the square root of the crack length at fracture was a constant. Griffith's explanation for this relationship, given in his crack theory, was based on surface or internal cracks acting as stress concentrations (Fig. 3). Thus the stress at the crack tip may be above the fracture stress and the crack will extend if enough elastic strain energy (stored on loading the specimen prior to fracture: area under the stress-strain curve) can be released to supply the surface energy of the newly-formed crack surfaces (γ_s per unit area). By considering the balance between these energies he showed that the larger the crack, the smaller is the stress necessary to cause fracture, deriving an expression for the fracture stress σ_f in terms of the crack length a:

$$\sigma = \sqrt{\frac{2\gamma_s E}{\pi a}} \qquad (1)$$

where E is the Young's modulus for the material.

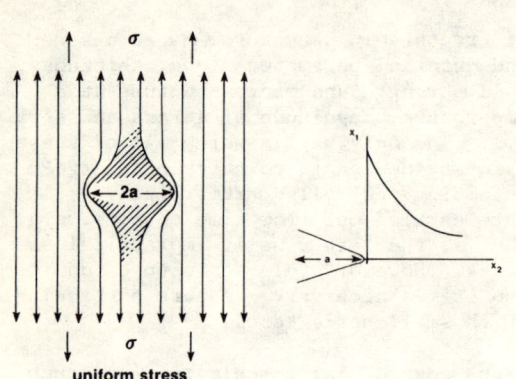

Fig.3: The effect of an elliptical hole on the lines of force in a plate, illustrating the stress concentration at the tip

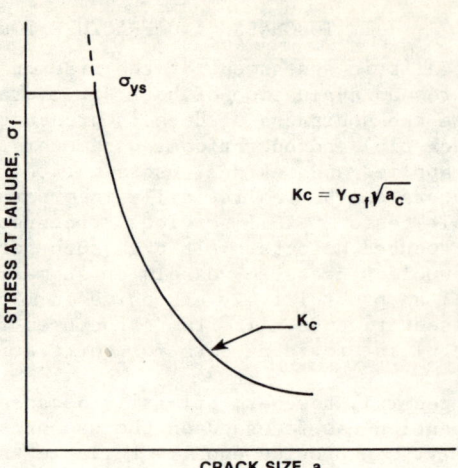

Fig.4. Critical stress intensity factor as a constant in the relationship between the applied stress at failure and corresponding critical flaw size

In this treatment the only effects of the stress which were considered were the elastic deformation of a solid and the local deformation to the extent of breaking the atomic bonds. Metals and their alloys, however, exhibit a more varied behaviour under applied stress, in particular by deforming plastically. Irwin and Orowan independently modified the Griffith equation to take account of this by including a 'plastic work' term, giving:

$$\sigma_f = \sqrt{\frac{2(\gamma_s + \gamma_p)E}{\pi a}} \qquad (2)$$

where γ_p is the work done in plastic deformation per unit area of crack surface. Although γ_p will greatly exceed γ_s in ductile materials, its range varies from material to material and also can increase rapidly with a relatively small increase in temperature.

It is evident from the above equation that one simple criterion for fracture is that the product $\sigma^2 a$ (σ being the nominal applied stress) should exceed a particular value for a given material. Provided a is less than a known value, there is a safe stress level at which crack propagation should not occur. The square root of the parameter $\pi\sigma^2 a$ is known as the 'stress intensity factor' and denoted by K, and having dimensions of stress times length $^{\frac{1}{2}}$ ($MPa m^{\frac{1}{2}}$). When its value exceeds some limit, which is a property of a material at a given temperature and material thickness, the crack becomes unstable. This limiting value is the critical stress intensity, or 'fracture toughness', and is denoted by K_c.

This simple relationship between fracture stress and crack size is shown graphically in Fig. 4 with a yield strength cut-off (since we do not usually design above the yield strength; if fractures do occur in that region they are not of the unexpected, low-ductility type). The meaning of the graphical presentation is simply that so long as the combination of stress and flaw size is below the line, the critical condition would not be expected. However, if the combination is on or above the line, the consequences of sudden, unstable crack growth must be considered.

Factors Affecting Fracture Toughness

(a) **Specimen thickness.** The preceding discussion has related to a sheet of material which is sufficiently thin such that there is negligible stress acting perpendicular to the sheet, i.e. a 'plane stress' condition exists. In a thick plate, material at the crack tip is elastically constrained and through-thickness deformation is prevented (except near the plate surfaces); this is known as 'plane strain' conditions.

Thickness has a marked effect on fracture toughness, as shown in Fig. 5. Also shown is the fracture appearance accompanying the different thicknesses for single edge notch specimens. Fig. 5 shows that thin specimens have a high value of K_c and fracture surfaces are accompanied by the formation of appreciable 'shear lips' or slant fracture. As thickness increases, the proportion of slant fracture surface decreases as does K_c, producing a 'mixed mode' type of fracture. For thick parts almost all the fracture surface is flat and K_c approaches an asymptotic minimum value.

Further increase in thickness does not decrease the fracture toughness nor alter the fracture appearance. This minimum value is called 'the plane strain fracture toughness', K_{I_c}, the subscript I (one) conventionally referring to the most common tensile, crack opening mode by which the specimen is loaded (eg Fig. 3).

This effect of thickness is illustrated by the fact that the minimum K_{I_c} value for a 1 750 MPa yield strength steel can be achieved in a specimen thickness of something less than 20 mm, whereas a 750 MPa steel would require a thickness in excess of four times this to give a so-called 'valid' K_{I_c} result.

The practical implications of the thickness effect are quite significant and will be discussed further in a following section.

(b) **Specimen and crack geometry.** The stress intensity factors discussed above apply to a straight planar crack in an infinite sheet of material. In a specimen or component of finite size the stress pattern around the crack can be evaluated by stress analysis techniques and the relation of the actual failure stress to the appropriate critical stress intensity factor determined. Thus there is a compliance function (γ) which corrects for the geometrical shape of the specimen.

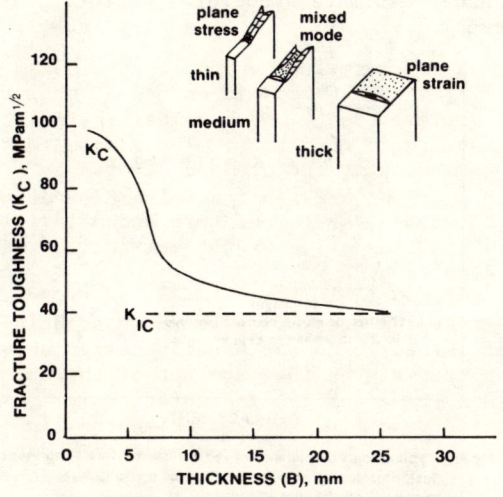

Fig.5. Effect of specimen thickness on fracture toughness

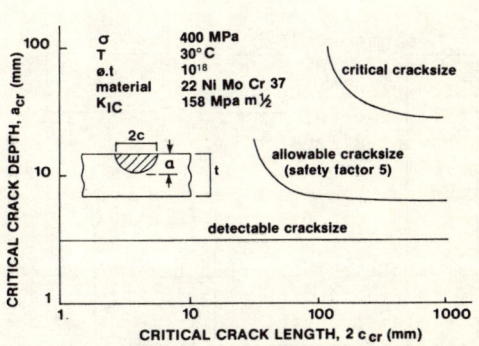

Fig.6: Critical allowable and detectable crack size for a nuclear pressure vessel (after Bartholomé et al, 1974)

There will also be a flaw shape parameter (Q) which allows for the geometry of the flaw. This geometry may be unknown and can only be assumed, eg a surface crack might be semi-elliptical of width $2c$ and depth a, the flaw shape parameter then depending on $a/2c$. For a given set of operating conditions, therefore, the critical crack size will not be a constant but will vary according to its shape, as shown in Fig. 6. However, provided it is possible to detect cracks (by non-destructive testing methods) smaller than this size, allowing for a considerable margin of error, then structural integrity may be assessed.

Fracture Toughness Testing in Practice

The main advantage of the fracture toughness approach over the traditional but more qualitative measures of toughness should now be apparent. Provided that the plastic zone at the crack tip is small relative to the crack length, by establishing the appropriate K_c value from laboratory tests one can calculate either the safe working stress for a component or structure with flaws (of which the maximum size is known) or the maximum flaw size permissible for given working stress.

Methods for determining the stress distribution close to the flaw as well as flaw detecting and sizing techniques are obviously beyond the scope of this paper, but they are currently reasonably well advanced and well documented. It should be noted, however, that both are essential requirements for the successful implementation of a quantitative fracture assessment programme.

The value of K_c for a material can be measured by testing specimens with finite dimensions and the procedure by which this may be carried out is now becoming standardized, at least for K_{I_c} determinations. Typical specimen geometries used are illustrated in Fig. 7.

There are two distinct stages in carrying out a fracture toughness test. The first involves growing a fatigue crack from the machined notch by cyclically loading the specimen, either in tension or bending, until the crack has propagated a pre-determined distance from the bottom of the notch under closely controlled loading conditions. The crack will then be of essentially atomic dimensions at its root thus simulating the worst possible situation that might be encountered in practice. Less sharp cracks of the same length would require higher nominal stresses before propagating in an unstable manner and therefore have higher 'effective' fracture toughness values, depending on their root radius. Unfortunately data so obtained might seriously under-estimate structural performance in the presence of an 'atomically sharp' crack.

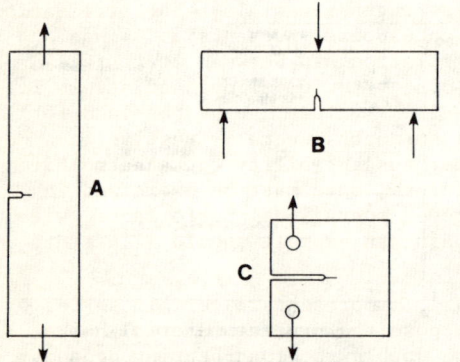

Fig.7: Three common fatigue pre-cracked fracture mechanics specimens: A—single edge notched (SEN), B—three point bending (4—point also used), C—compact tension specimen (CTS)

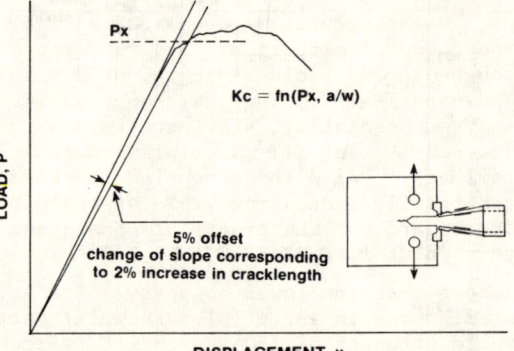

Fig.8: Typical load displacement record for fracture toughness test, illustrating also the location of a clip guage mounted on the test specimen

The second stage in the test involves the application of a steadily increasing tensile load to the fatigue pre-cracked specimen until the crack propagates. This is monitored from an autographic record of applied load versus the displacement across the crack, determined by a clip-gauge mounted across it (Fig. 8). In cases where there is no sharp change in the load-deflection curve corresponding to the onset of fracture, the load corresponding to a 2 per cent increment of crack extension is established by a specified deviation from the linear portion of the autographic record (typically 5 per cent) (Fig. 8). The K_C value is calculated from this load using documented stress intensity values obtained on the basis of linear elastic stress analysis for the specimen geometry used (available from the Standards, 1971, 1972). Whether or not this value corresponds to the minimum or K_{I_C} value for the specimen thickness used may be subsequently determined by simple empirically determined formulae.

It is worthwhile restating that for high yield strength materials, thickness dimensions under 25 mm are usually adequate for determination of valid K_{I_C} values, but with more ductile materials larger specimens have to be used. Thus, in the case of mild steel, specimens 150 mm or more in thickness must be used and the results can naturally only be applied to structures of similar thickness. For thinner cross-sections K_C values must be determined on laboratory specimens corresponding to the thickness of the strucrure or component.

Some Lessons To Learn From Fracture Mechanics

(a) Problems encountered by making parts thicker. Further consideration of Fig. 5 illustrates a most important design philosophy concerning part thickness. Where a component is unsatisfactory, or if higher operating loads are required, a common solution is to make the part bigger. Even allowing for factors of safety (or rather aptly named 'factors of ignorance' as the following example should indicate) this procedure can readily result in an increased susceptibility to catastrophic failure.

Assume that a part 2,5 mm thick, similar to the material and data represented in Fig. 5, is to withstand a load three times the original static design load. If the thickness were tripled and no flaws existed in the material, this would be a satisfactory solution. If, however, a crack of given length existed in both parts, tripling the thickness would in fact decrease the fracture toughness to about one-half the original value ($\sim$ 80 down to 40 MPam$^{\frac{1}{2}}$). Since K_C is the proper fracture design criterion, tripling the thickness would only permit a 1,5 increase in the allowable fracture load, the consequences of which are obvious.

(b) Problems encountered by using high yield strength materials. The utilisation of structural materials, like many human endeavours, presents a 'swings and roundabouts' situation. The very approximate relationship between fracture toughness and yield strength for aluminium alloys, titanium alloys and steels is shown in Fig. 9. Thus, although a wide range of K_{I_C} values can be obtained for a given base alloy, a higher yield or ultimate strength generally results in a decreased K_{I_C} for all materials and thus a greater susceptibility to catastrophic failure. As with increasing thickness, therefore, care must be exercised in arbitrarily increasing yield strength for improved operational performance, with due regard for the practical consequences. Again, an example should illustrate this point more clearly.

Assume that the lower boundary for steel in Fig. 9 represents a certain series of steels used in large (plane strain) parts. Suppose the present material has a yield strength of 750 MPa and it is required to double the load-bearing capacity based on yielding or fracture. If no flaws exist a material with a yield strength of 1,500 MPa would be satisfactory. However, as we have seen, this is an unrealistic assumption for real engineering structures or components. Thus, if a crack of the same size existed in both steels, doubling the yield strength would decrease both K_{I_c} and the fracture load by a factor of three (Fig. 9). For a component of fixed dimensions, therefore, the use of the higher-strength steel would thus *decrease* the load-bearing capacity by a factor of three, instead of increasing it by the expected factor of two!

Under the proper conditions, however, this situation can be improved by the use of higher quality material, as illustrated in Fig. 9. This diagram shows that, for a given yield strength, K_{I_c} can vary greatly depending on the type and quality of the material. Judicious materials selection, therefore, can in many instances pre-empt in-service difficulties.

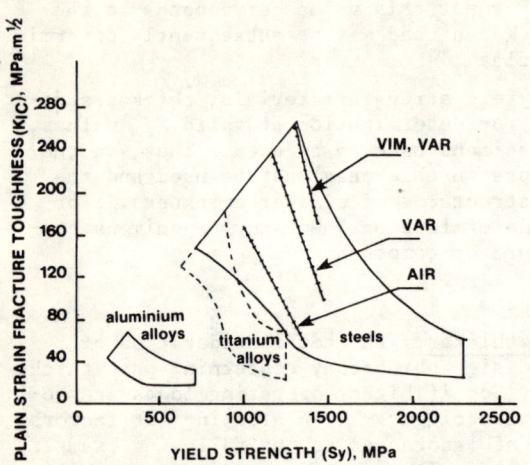

Fig.9: Variation of plane strain fracture toughness with yield strength for a variety of materials. VIM = vacuum-induction melting. VAR = vacuum-arc remelting, AIR = air melting (after W.S. Pellini).

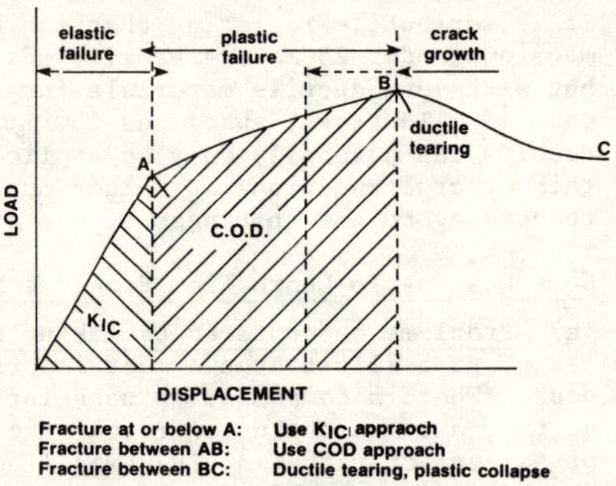

Fig.10: Schematic respresentation of the continuity between the K_{IC} and COD approaches, leading to the adoption of a unified procedure for fracture toughness testing (after Elliott et al, 1971).

YIELDING FRACTURE MECHANICS

Whilst linear elastic fracture mechanics (or LEFM) analyses are reasonably well defined, they can only be rigorously applied when fracture occurs under elastic conditions. In practice, as described above, the analysis can accommodate only small amounts of local plasticity, i.e. the size of the so-called 'plastic zone' formed in the region of a crack tip under load must be small compared with the crack length and the component dimensions. There still remains a problem, however, in quantifying the resistance of a material to crack extension when mechanical conditions at failure cannot accurately be described as linear elastic. Thus, it is often found that the temperature conditions and material thicknesses used for many practical applications of structural steel are such that large amounts of plastic deformation occur prior to fracture. It is therefore only in selected applications (e.g., at large section thicknesses, high yield stress levels, low temperatures or high rates of strain) that LEFM can provide a truly accurate description of the near crack tip region. Furthermore, it is important to realise that the possible modes of failure and thus the controlling fracture criteria can vary from one material or set of circumstances to another.

This point can be illustrated by consideration of Fig. 10, which may be regarded as a load/displacement diagram for any part of an engineering structure. Consider the load to increase until failure occurs; this will depend on the fracture toughness of the material, and the type of fracture can range from the quasi-elastic of the brittle material to the general plastic collapse of the very ductile. At the one end, where little plastic deformation precedes fracture, local conditions around a notch or defect can be adequately calculated using LEFM. At the other end of the ductility range, the conditions for plastic instability may be defined using conventional computational methods. In between, there remains the range of materials in which failure would occur in the region of Fig. 10 labelled AB, where structural failure in the presence of a defect can be by below-yield fast fracture, but with sufficient local plastic deformation to invalidate LEFM. Under these conditions, we require to make alternative measurements of a material's resistance to fast fracture, if possible again using small specimens which are easy to test in the laboratory. Two alternative toughness parameters have been proposed and the Bibliography should be consulted for background detail on both. The first is the critical value of a 'crack tip opening displacement' or COD; the second is the critical value of a quasi-'strain-energy release rate' (J_{I_c}), derived assuming non-linear elastic behaviour. The remainder of this paper will only consider the COD approach, primarily because of the successes achieved using this method; a further paper in this Proceedings deals in detail with some practical applications of the COD method (Harrison, 1978).

CRACK OPENING DISPLACEMENT: CONCEPT, APPROACH AND APPLICATIONS

Concept

A solution to the problem posed in the preceding section follows from the assumption that fracture from a defect occurs under conditions which result in a critical amount of deformation at the tip of a notch or defect. Because of the short and indefinable distances involved, it is preferable to specify this in terms of displacement rather than strain.

The concept of crack opening displacement can be readily appreciated with reference to Fig. 11, which shows the progressive stages during a three-point bend test of both a machined slit and a fatigue crack. As the load increases, plastic yielding occurs at the notch/crack tip resulting in separation of the notch/crack faces without any increase in length of the notch/crack. This separation of the two faces of the notch/crack is then referred to as the COD. In a COD test, a stage is reached at a particular load when the material in the vicinity of the

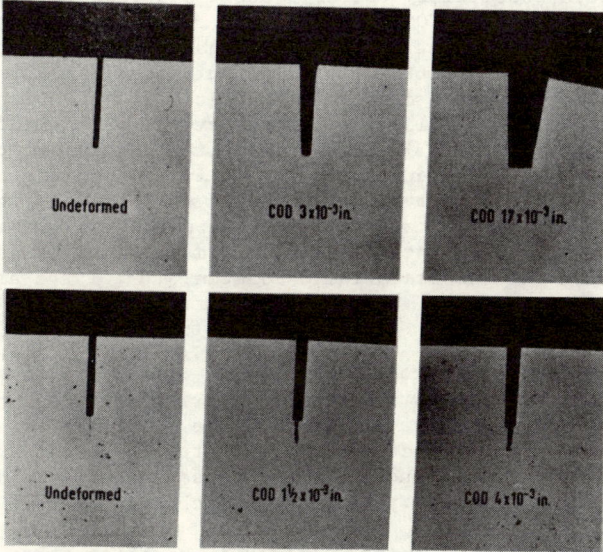

Fig.11: Crack opening displacement (COD) at artificial and natural crack (courtesy The Welding Institute, UK)

discontinuity ruptures. Depending on the extent of plastic deformation, i.e. whether the specimen is generally yielded across its section or not, the ensuing fracture may be of a fibrous (ductile) or cleavage (brittle) nature.

The specific value of COD obtained at the crack tip at the instant of fracture is considered to be a measure of the toughness of the material, and therefore a constant under given conditions of temperature, strain-rate and plate thickness. This critical value of COD can be measured in laboratory tests on conventional specimens identical to those described above for K_c testing. The results obtained can then be used to calculate the failure conditions for real engineering structures containing specified defects, making use either of the appropriate analysis where this exists, or some experimental calibration approach (see following sections).

Basis of the COD Approach

The question must be asked whether results obtained from a small specimen where plastic deformation is extensive, e.g. when the specimen has yielded generally, can be in any way applicable to a large structure or component which may be behaving macroscopically in an elastic manner, even though local crack-tip yield may be quite extensive.

At any instant in the propagation of a fracture, the only material controlling the fracture process is that situated at the crack tip. Once the plastic flow pattern around a stressed crack has developed, however, it seems reasonable to assume that, providing the yield pattern close to the crack tip is equivalent in both the structure and the testpiece, then the fracture phenomenon will be the same (assuming full plate thickness testpieces). Thus a critical value of COD, which is indicative of the specific stress condition at the crack tip, will be the same in both structure and testpiece; providing the thickness of the structure and the testpiece are identical, the conditions for fracture initiation should be similar.

This assumption then, in essence, is the foundation of the COD approach to quantitative fracture assessment using many weldable structural and pressure vessel steels. Based on this assumption, Cottrell and Wells independently proposed the more widespread application of COD as a fracture criterion. Nowadays, a substantial body of experimental data is available, comparing large, notched, wide plate, tension tests with notch bend tests of similar thickness, showing that fracture from a defect indeed occurs at a critical COD value which is the same in large component failure as in the small scale laboratory test. The effect is demonstrated in Fig. 12 (after Knott, 1973) showing three testpieces containing identical stress concentrators all fractured at -80°C. It is obvious that the testpiece with the smallest uncracked ligament has yielded generally across the remaining cross-section before fracture; those of larger size have broken before general yield, but after comparable amounts of local yielding. In each case, the value of COD at fracture (or δ_c) was essentially constant.

Experimental Measurements of COD

The aim of the COD approach to fracture toughness testing is to measure a critical crack opening displacement, δ_c, which is characteristic of the mode in which the material at the tip of a sharp crack will fracture. This is irrespective of whether the crack is to be found in a large structural element subject to an applied stress much less than that necessary to cause general yielding across the structure, or in a small laboratory testpiece which has fully yielded.

As discussed, COD measurements can be obtained in practice from conventional fracture toughness tests, there being obvious advantages to drawing together K_{I_c} and

Fig.12: Sections of mild steel specimans fractured at -80°C, the notch geometry being the same in all specimens. The plastic deformation preceding fracture has been etched using Fry's reagent, × 0.7 (courtesy J F Knott)

δ_c measurements through a single test procedure; the test itself and the corresponding calculations are described in a current British Standard (1979), and the COD measured between the knife-edges (Fig. 8) can be readily related to the crack tip COD, δ.

A point of note is that the recommended procedure for COD testing indicates that δ_c should be determined at the moment of fracture. However, in practice, slow stable crack extension may occur prior to fracture (commencing at some point along the line AB in Fig. 10). In steels, the COD value at crack initiation, δ_i, may be accurately determined using the multiple specimen technique of unloading at various points along the load-COD curve and breaking the specimens in liquid nitrogen: any fibrous crack extension following the fatigue pre-cracking can be detected as an obvious 'thumbnail' showing in the crystalline cleavage facets typical of fracture at -196°C. This method for detecting crack growth prior to fracture can, of course, prove laborious (or indeed impossible in non-ferrous systems): hence a value δ_c is often specified which occurs at maximum load (i.e. $\delta_c = \delta_{max}$).

Clearly, δ_{max} values can be significantly more optimistic than corresponding δ_i values, and recently reservations have been expressed concerning the use of δ_{max} values in either material specification or engineering design (Knott, 1973). Most relevant is the point that arises from the time-dependence of the separation process. Thus, a specimen held under a constant load greater than that required for initiation, shows an increasing COD until the specimen finally breaks. Under constant load conditions, any integrity assessment based on δ_{max} could have serious repercussions.

On the other hand, more recent information has suggested that in a number of situations a criterion for fracture based on initiation may not be entirely suitable for the following reason. Many ductile materials operating above the ductile-brittle transition temperature have relatively low toughness values at the onset of crack extension but on subsequent ductile tearing substantial increases in toughness may result. Thus a crack growth resistance (or R-curve) concept may be employed to describe the apparent increase in resistance to further crack extension as a ductile tear extends in order to avoid over-conservative design. However, it is clear that much care must be taken to evaluate the possible conditions for failure before selecting an appropriate fracture criterion.

Practical Applications of the COD Approach

This section is concerned with a practical method of applying yielding fracture mechanics, based on work carried out at the U.K. Welding Institute (Harrison, 1978), and described in detail in a subsequent paper in this Conference Proceedings). Because of the greatly increased complexity of rigorous elastic-plastic analyses compared with LEFM, the approach is a simplified one. The design information which the approach provides is intended to be conservative and is empirically based on the experimental results of a large number of wide plate tests.

Mathematical analyses and experimental work have shown that under certain conditions COD can be related to the elastic stress intensity factor, K_I, as follows:

$$\delta = \frac{K_I^2}{\sigma_y E} = \frac{\sigma^2 \pi a}{\sigma_y E} \qquad (3)$$

where σ_y, E are respectively the yield stress and Young's modulus, σ the nominal applied stress, and a is the defect length. (If a surface crack, a = half the length of a buried crack.) For criticality, i.e. at the onset of fracture, the symbols δ, K_I, σ and a will be subscripted with a 'c', e.g. δ_c or K_{I_c}, and the same equation applies.

For situations where the effective ratio of the defect size to crack width, a/W, is less than about 0,1 and where the nominal design stress, σ, is less than the yield stress of the base material, Dawes (1974) has proposed that equation (3) can be rewritten as follows:

$$a_{max} = \frac{\delta_c E \sigma_y}{2\pi \sigma_1^2} \quad \text{for} \quad \frac{\sigma_1}{\sigma_y} \leqslant 0.5 \qquad (4a)$$

$$a_{max} = \frac{\delta_c E}{2\pi (\sigma_1 - \frac{\sigma_y}{4})} \quad \text{for} \quad \frac{\sigma_1}{\sigma_y} \geqslant 0.5 \qquad (4b)$$

where a_{max} is the maximum allowable value of 'a' and a factor of safety of 2 on defect size is included. The purpose of these equations is to give conservative predictions of the size of defect which can be allowed to remain in a structure without repair and it is not intended to predict criticality. Thus, a_{max} should be smaller than the critical defect size, a_c.

In equation (3), σ_1 represents a total 'pseudo-elastic' stress in the vicinity of the defect. Whilst σ_1 may in fact be above yield, the structure itself may still behave in a predominantly elastic manner. This is because the yielding in the zone under consideration is contained by the surrounding elastic material. In welded structures such contained yielding can result from residual stresses which may themselves be equal to the yield stress and are additive to any applied system stress. In addition, contained yielding also occurs at stress concentrationing features where pseudo-elastic stresses may well be above yield, and therefore σ_1 must include a further stress concentration factor.

This approach has led the U.K. Welding Institute to the derivation of a 'design curve' (Dawes 1978) which in recent years has been successfully applied to a wide range of structures subjected to a variety of loading conditions. The applications of this approach may be classified into four major groups (Harrison, 1978):

(i) <u>Materials selection</u>. Thus, if the design stress is known and a size of defect which might escape detection by non-destructive testing is assumed, the required level of toughness may be prescribed.

(ii) <u>Acceptance levels for defects</u> (typically weld defects). Based on equation (4), if the material toughness and design stress are predetermined, the NDT acceptance standards may be fixed at an early stage in design or construction. Alternatively, the concept can be used to assess the significance of known defects where repair may be undesirable for reasons either of cost or of the possibility of introducing more harmful defects in the course of the repair.

(iii) <u>Fixing allowable stresses</u>. If the material toughness and inspection level are predetermined, safe operating stresses may be calculated.

(iv) <u>Failure analysis</u>. Comparison of structural failure conditions with the predictions of equation (4).

CONCLUSIONS

In order to avoid catastrophic failures the designer must be aware of the high probability that a component or structure will contain flaws or cracks and that it is necessary to design on this basis.

Fracture mechanics provides quantitative design parameters relating operating stresses to flaw size/NDT detection limits through small-scale laboratory tests measuring a characterisric material 'fracture toughness'. Where conditions are predominantly elastic, the critical stress intensity, K_c, may be appropriate; when fracture is accompanied by significant yielding, as is the case for many structural steel components, crack opening displacement, or COD, can provide the necessary toughness parameter.

REFERENCES

Bartholomé, G. et al (1974). <u>Eng. Fracture Mech</u>. 4 p. 431
British Standards (1972). Methods for crack opening displacement (COD) testing. Document BS5762.
Dawes, M.G. (1974). Fracture control in high strength weldments. <u>Weld. J. Res. Supp</u>. <u>53</u>, p. 369.
Dawes, M.G. and Kamath, M.S. (1978). The COD design curve approach to crack tolerance. Conf. on <u>Tolerance of Flaws in Pressurised Components</u>. Instn. Mech. Engrs. (U.K.)
Elliott D., Walker E.F. and May, M.J. (1971). The determination and applicability of COD test data. <u>Instn. Mech. Engrs.</u>(UK). Paper C 77/71. pp. 217-224
Harrison, J.D., Dawes M.G., Archer G.L. and Kamath M.S. (1978). The COD approach and its application to welded structures. U.K. Welding Inst. Rept. SS/1978/E.
Johnstone, G.O. (1978). What are the chances of failure? A review of probabilistic fracture mechanics literature. <u>Welding Res. Inst. Bull.</u>, Feb. p. 36; March p. 78.
Knott, J.F. (1973). <u>Fundamentals of Fracture Mechanics</u>. Butterworths, London.
Marriott, D.L. and Hudson, J.H. (1977). Prediction of failure risk in pressure vessels due to a manufacturing defect. Proc. Inst. Symp. or <u>Application of Reliability Technology to Nuclear Power Plants</u>, Vienna. Paper SM 218/33.
Pettit, D.E. and Krupp W.E. (1974). The role of nondestructive inspection in fracture mechanics application. In ASM3 <u>Fracture Prevention and Control</u>. American Society of Metals.
Standards: British (1971) DD3; American (1972) ASTM 399-72

BIBLIOGRAPHY

Broek, E. (1974) <u>Elementary Engineering Fracture Mechanics</u>. Noordhoff, Netherlands.
Knott, J.E. (1973). <u>Fundamentals of Fracture Mechanics</u>. Butterworth, London.
Rolfe, S.T. and Barsom, J.M. (1977) <u>Fracture and Fatigue Control in Structures - Applications of Fracture Mechanics</u>. Prentice-Hall, New Jersey.

DEFECT ASSESSMENT BY MEANS OF NON-DESTRUCTIVE TESTING

J. J. Marais

Department of Mechanical Engineering, University of Pretoria, R.S.A.

ABSTRACT

The detection of serious defects and the determination of the defect dimensions are of great importance in fracture control plans based on fracture mechanics principles. In materials with low fracture toughness the size of the maximum allowable defect can be so small as to obviate the detection of such a defect by the normal procedures of non-destructive testing.

It is important not only to detect a serious defect but also to be able to determine its dimensions. Cracks are the most serious defects and are often present in embrittled material. Both in the detection of cracks and in the determination of the crack dimensions, ultrasonic examination has definite advantages over radiographic examination and other non-destructive testing methods generally employed in industry.

INTRODUCTION

Fracture mechanics provides a means by which material properties, design stresses as well as defect orientation and dimensions can be related. The basic equation in this respect is

$$K_c = C \sigma \sqrt{a^*} \tag{1}$$

where K_c = fracture toughness of the material used in the design. K_c is not a true material property but is also a function of the component or structure geometry.

σ = design stress which includes all relevant stresses such as the applied stress and residual stresses if present.

$2a^*$ = critical crack length.

C = defect geometry factor

   = $\sqrt{\pi}$ for a through-thickness crack.

This relationship between the design stress and the critical crack length is shown in Fig. 1. Note the influence of material fracture toughness on the critical defect length.

In the case of linear elastic fracture mechanics, equation 1 can be written as

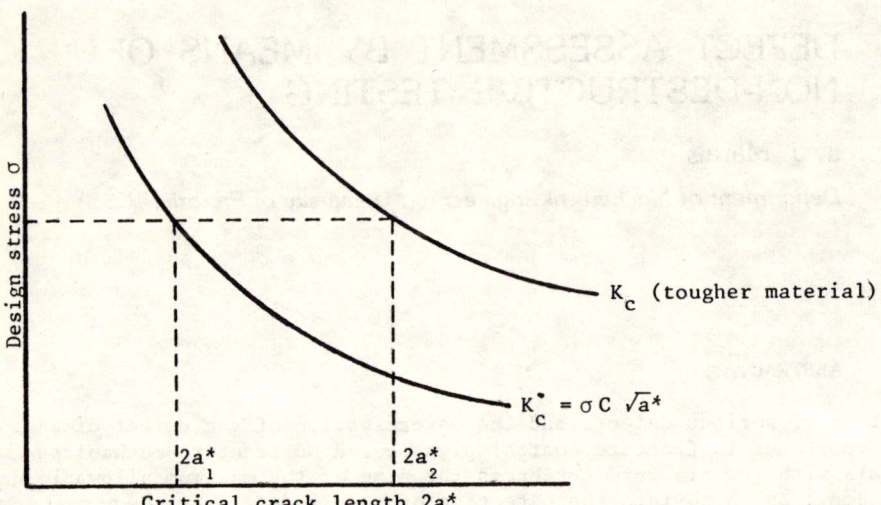

Fig. 1. Relationship between design stress and critical crack length.

$$K_{IC} = C \sigma \sqrt{a^*} \qquad (2)$$

where
K_{IC} = fracture toughness under plane strain conditions and a true material property. The load is also applied perpendicular to the plane of the crack, i.e. mode I loading.

In elasto-plastic fracture mechanics, plastic deformation at the crack tip needs to be taken into account. This condition often prevails in thin sections or in material with a high fracture toughness. There is, however, still a relationship between design stress, material fracture toughness and defect dimensions.

In the crack opening displacement (COD) approach, the design curve introduced by the Welding Institute (UK) is based on a dimensionless parameter ∅ where ∅, in general terms, can be written as

$$\emptyset = f\left(\frac{\delta}{a}\right) \qquad (3)$$

δ = COD and is a measure of fracture toughness
$2a$ = allowable crack length.

The effective stress σ_1 is used in conjunction with the Welding Institute's design curve where

$$\sigma_1 = \sigma_a \times K_t + \sigma_{res} \qquad (4)$$

σ_a = applied stress
K_t = theoretical stress concentration factor

σ_{res} = residual stress.

In both linear elastic fracture mechanics and elasto-plastic fracture mechanics, three major activities are involved, i.e.: material selection (fracture toughness), stress analysis (σ_1 and σ_{res}) and non-destructive testing (C and a).

In this paper the attention will be focussed on the non-destructive testing (NDT) activity. It is important that fracture mechanics principles should already be introduced during the design stage of a component or structure.

NDT IN THE DESIGN STAGE

To illustrate the importance of NDT in the design stage, consider the following case under linear elastic conditions: Two materials have properties as indicated below. In both cases the design stress is 0,5 σ_y where σ_y is the yield strength. There are no residual stresses present. Determine the length of the critical through-thickness crack. From equation 2:

$$K_{IC} = \sigma\sqrt{\pi a^*} \quad \therefore \quad a^* = \frac{1}{\pi}\left(\frac{K_{IC}}{\sigma}\right)^2 \tag{5}$$

Material A	Material B
σ_y = 300 MPa	σ_y = 300 MPa
K_{IC} = 150 MPa $\sqrt{m}$	K_{IC} = 40 MPa $\sqrt{m}$
σ = 0,5 σ_y	σ = 0,5 σ_y
= 150 MPa	= 150 MPa
$2a^* = \frac{2}{\pi}\left(\frac{150}{150}\right)^2$	$2a^* = \frac{2}{\pi}\left(\frac{40}{150}\right)^2$
= 637 mm	= 90,5 mm

From the example it is clear how the material fracture toughness and the design stress can influence the demands made on the sensitivity of defect detection by means of NDT. Consider material B to be welded and not stress-relieved. Take the residual stress to be equal to the yield stress. Under these conditions 2a* decreases from 90,5 mm in the above example to 5 mm.

The combination of fracture toughness and design stress can create a situation where the critical defect has dimensions which are so small that it is virtually impossible to detect by means of NDT. This is an important consideration to be taken into account by the design engineer and he should have a sound knowledge of the capabilities and limitations of the various NDT methods.

NDT METHODS IN GENERAL

Only the methods in general use in industry will be discussed.

Surface Flaw Detection

Visual examination should never be under-estimated as an inspection method on its own or as the first part of any other inspection procedure. With visual inspection geometrical stress concentrations, poorly machined surfaces, some of the more obvious surface flaws such as folds and laps in castings, distortion and undesirable surface irregularities associated with welding can be detected. Certain aids such as magnifying lenses, fibrescopes and tube cameras can be used with visual inspection for detecting fine surface defects or surface defects inside tubes, pipes, reaction vessels or pressure vessels. More sensitive techniques for the detection of fine surface or sub-surface defects are:

Dye penetrant inspection. The defect must come through to the surface. The material must be properly cleaned and must not be porous or absorbent.

Magnetic inspection. The material must be ferro-magnetic and surfaces must be clean. Defects just below the surface can also be detected.

These techniques give qualitative information on surface defects, i.e. location, type of defect and the length of the defect on the surface of the component. If magnetic leakage field measurements are employed, an indication of the depth to which a crack penetrates into the material can be found. This technique can be useful for detecting and finding the dimensions of elliptical surface cracks, but these defects should not be confused with less serious surface folds as shown in Fig. 2.

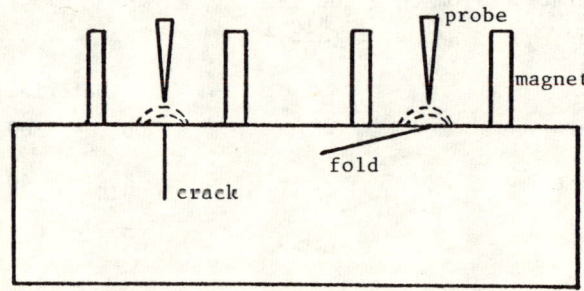

Fig. 2. Surface flaw detection with magnetic field.

In the fracture mechanics approach, the defect orientation and dimensions are important. Of the techniques discussed above, the following combination in the given sequence can be useful: visual examination, magnetic particle inspection or dye penetrant inspection and magnetic leakage field measurements (in ferro-magnetic material only).

Internal Defects

Industrial radiography and ultrasonic testing are frequently referred to as techniques for detecting internal defects. They are, however, of equal importance in revealing defects such as surface cracks which are not visible to the naked eye. Both techniques will detect other surface flaws such as undercut on a weld seam, weld root concavity, over-penetrated weld root, arc marks on welded components and sand inclusions on the surface of a casting. In some of the above cases the defect is associated with the internal surface, to which access for visual inspection might not be possible. Thoroughly conducted radiographic and ultrasonic examinations should thus reveal internal as well as surface flaws.

Which NDT Method to use

All NDT methods have their advantages and shortcomings, therefore, the choice largely depends on factors such as: the type of defects being sought, the type of material, the thickness and geometry of the component or structure, access to the part to be examined, the place of examination, i.e. in the workshop or on site, cost of NDT equipment and the NDT test and the sensitivity of defect detection required.

INDUSTRIAL RADIOGRAPHY AND ULTRASONIC TESTING

The fields of application of these techniques can be summarized as in Fig. 3.

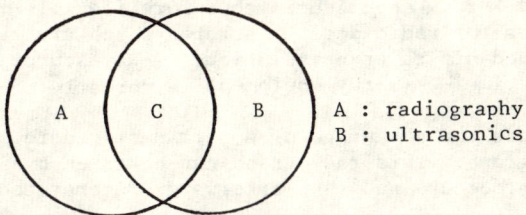

Fig. 3. Fields of application of ultrasonic and radiographic examination.

Part A. Where radiography is specified, welds in austenitic stainless steels, small components on which ultrasonic testing would be difficult, where access for ultrasonic examination is impossible, welds in material thinner than about 3 mm and material where the attenuation of ultrasound is high.

Part B. Material thicker than about 180 mm (steel), detection of fine cracks, testing complicated configurations such as nozzle-to-drum welding and where a component must be tested with other personnel working in the vicinity, detection of laminations in plate material and lack of bonding in cladded material, high speed testing such as the testing of rails at speeds in excess of 50 km/h and where scattered radiation impairs the quality of radiographs.

Part C. Welds in material up to 180 mm thick (steel), castings of limited section thicknesses, thickness measurements and testing for corrosion damage inside pipes, reaction vessels or containers.

In radiography, X-rays are generated in an evacuated glass tube with a cathode and anode plus a heat filament. Electrons are accelerated by means of a high potential difference from the cathode to the anode where they strike a tungsten disc. About 99% of their kinetic energy is converted into heat and less than 1% into useful X-rays.

The component to be radiographed is placed between the X-ray tube and the film. For a high quality radiograph it is important to place the film as close to the component as possible. The greater the distance between the X-ray tube and the component, the better the quality of the radiograph. This is due to the basic laws of optics and by adhering to this procedure, geometrical unsharpness on the radiograph is kept to a minimum. The distance between the X-ray tube and the film also determines the duration of the exposure. It is, therefore, important to use a film to focal-spot distance which will render a high quality radiograph with a reasonable exposure time. Long exposure times have several disadvantages such as overheating of the tube and fogging of the radiograph due to scattered radiation. The maximum steel thickness that can be penetrated with portable industrial X-ray equipment is about 80 mm.

γ-Rays are generated by radio-active materials and the two commonly used in industrial radiography are Ir_{192} and Co_{60}. The maximum steel thickness to be penetrated with Co_{60} is about 180 mm.

Factors that govern the choice between X-rays and γ-rays are: material thickness, access to the component to be radiographed (X-ray equipment is bulky and heavy compared to γ-ray equipment), the availability of electricity for X-ray equipment, adjustment of the radiation is possible with X-ray equipment but not with γ-ray sources (the X-ray technique is thus more flexible), the availability of safe storage facilities for radio-active isotopes which are required by law. X-ray equipment can be switched off to stop radiation. γ-Ray sources on the other hand emit radiation all the time - (γ-ray equipment is normally cheaper than X-ray equipment).

Radiography has the advantage of a permanent record of the inspection in the form of the radiograph. This radiograph can be taken by a semi-skilled operator using an approved procedure and then afterwards interpreted by experienced personnel.

For ultrasonic testing, a highly skilled and trained technician is required for most manual ultrasonic examinations. The technician does the test and the interpretation simultaneously so that there is seldom a direct record of a manual examination other than the technician's report. The element of human error should be borne in mind and only reliable individuals should be recruited to be trained as ultrasonic technicians. Ultrasonic testing has the advantage that its depth of penetration in fine grain steel is up to 10 m and there is of course no radiation hazard.

Practical Aspects of Radiographic and Ultrasonic Inspection

With all NDT methods, the technician must have a reasonable idea of the type of defect most likely to be present in a component or structure.

With radiographic and ultrasonic examination, the geometry of the part to be examined must be known either through visual inspection, thickness measurements or from a drawing. In the case of radiographic examination, this information is required to calculate the correct exposure. The material thickness will determine the most suitable type of radiation source, i.e. X-rays, γ-rays from Ir_{192}, γ-rays from Co_{60}.

With ultrasonic examination the geometry of the component to be tested plays a more important role. The success of such an examination depends to a large extent on whether the technician has all the important dimensions at his disposal. To illustrate this statement, consider a transition piece between two high pressure tubes of different alloys and thicknesses, as shown in Fig. 4. The taper will interfere with testing of the butt weld from the side of the thicker wall tube. This can lead to the incorrect identification of a defect, locating of the defect and determination of the defect dimensions. The testing of a shaft in an assembly is another illustration of the importance of knowing the component dimensions. From Fig. 5 it can be seen that spurious echoes from steps in the shaft could be confused with fatigue cracks. To enable the technician to conduct a sensible test of this shaft, he requires a drawing. Slight changes in geometry are of greater importance with ultrasonic examination than with radiographic examination.

Defect types. In a fracture mechanics approach to a design problem, the worst type of defect, namely a crack, is considered. The conditions at the sharp tip of a crack are favourable for crack propagation. Factors that play a role in this respect are the length of the crack, the fracture toughness of the material and the stresses to which the component or structure is subjected. Depending on these factors, the crack will either not propagate, propagate in a stable manner or propagate in an unstable manner. Defects such as porosity, piping and inclusions are more rounded and are of a less serious nature than cracks. The seriousness of these defects depends on the combination of fracture toughness of the material and the mode

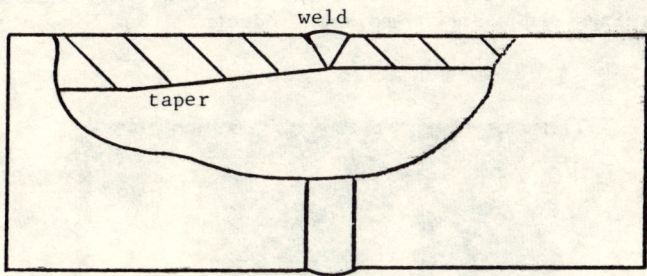

Fig. 4. Transition piece in high pressure tube.

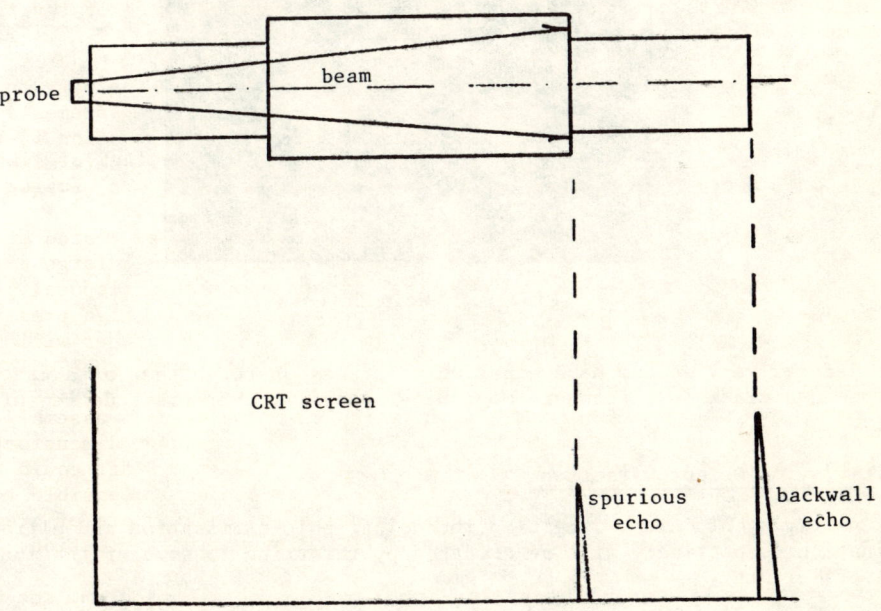

Fig. 5. Ultrasonic testing of shaft (spurious echo at step-down).

of loading of the component. Under cyclic loading, a fatigue crack can be initiated under certain conditions at a less serious defect.

Another category of defects is the "crack like" defects which are of a planar nature. This group consists of defects such as: hot tearing in castings or welds, lack of sidewall fusion in welds, incomplete root penetration in welds, folds in castings and forgings and seams in rolled products. (See Fig. 6.)

heat affected zone crack

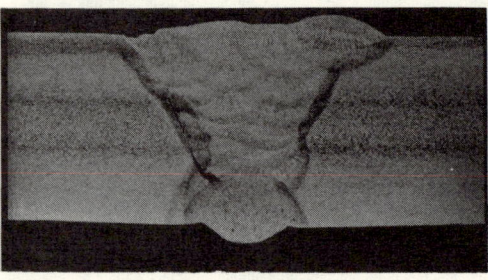
lack of sidewall fusion

Fig. 6. Weld defects

These defects very seldom have edges which are as sharp as that of a crack and if treated as a crack in a fracture mechanics analysis, there is a degree of built-in safety.

Sensitivity of Defect Detection

The sensitivity of defect detection in radiographic examination and ultrasonic examination respectively will be compared by referring to some of the more serious types of defects.

The ultrasonic test technique was basically developed because it was difficult and very often impossible to detect fine cracks in components or structures with the radiographic technique. Three dimensional defects such as gas pores, slag inclusions and slaglines are readily detected with radiographic examination. With ultrasonic examination the reflected echo from these defects often has a relatively small amplitude. It is also difficult to get the accurate dimensions of these defects with ultrasonic examination.

With planar defects on the other hand, the correct ultrasonic test procedure is a more reliable means of detecting the defect and finding the dimensions of the defect than radiographic examination. The orientation of the X- or γ-ray beam is important in this respect. In Fig. 7. this problem is illustrated.

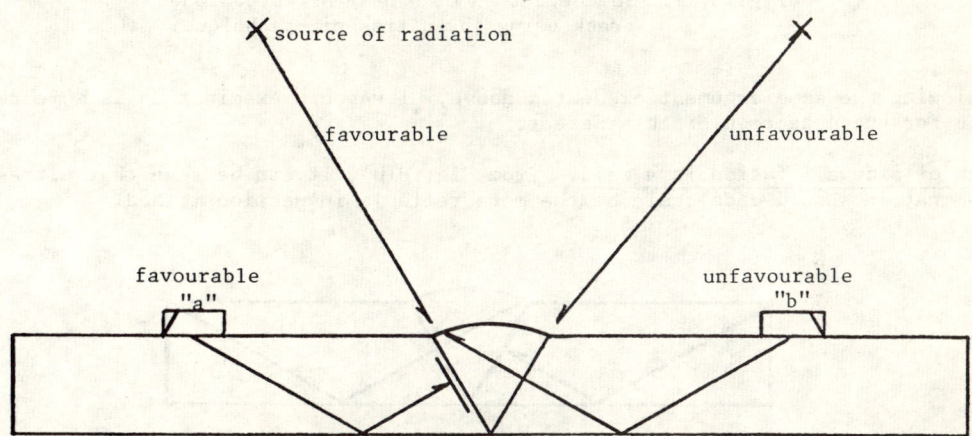

Fig. 7. The influence of defect orientation on its detectability with the radiographic and ultrasonic techniques respectively.

The sensitivity of defect detection with radiography is given by:
$$\text{Sensitivity} = \frac{\text{defect dimension}}{\text{component dimension}} \times 100\% \qquad (6)$$
The dimensions used are the defect dimension in the direction of the beam of rays and the thickness of the component in the same direction. From Fig. 7. it is clear that with the beam striking a planar defect more or less at a right angle, the effective defect dimension is extremely small. The sensitivity required to detect such a defect under the unfavourable conditions indicated, is very often outside the practical limits of radiographic examination. With the beam axis approximately parallel to the defect, the ratio of the effective defect dimension to the material thickness is much more favourable for defect detection.

Consider ultrasonic examination of the same component. Two positions of the ultrasonic test probe are shown in Fig. 7. At position "a" with the correct angle probe, the defect will be detected but not with the probe in position "b". There is, however, a very important difference in the two test techniques: It is not common practice to take a radiograph from both sides of a weld seam but it is common practice to conduct an ultrasonic examination from both sides.

It will be appropriate to compare the sensitivity of defect detection of radiographic and ultrasonic examination with some serious defects as basis:

<u>Heat affected zone crack in a weld</u>. This defect is shown schematically in Fig. 8(a).

Fig. 8(a). The detection of a heat affected zone crack using the ultrasonic technique.

Following the same argument presented above, ultrasonic examination is more reliable for the detection of this defect.

Lack of sidewall fusion in a weld. From Fig. 8(b), it can be seen that ultrasonic examination should once again be the more reliable inspection method.

Fig. 8(b). Detection of lack of sidewall fusion using the ultrasonic test technique.

Incomplete root penetration in a weld. In the fillet weld configuration shown in Fig. 8(c), the only reliable method will be ultrasonic examination. The probe position for defect detection is also indicated. Heat affected zone cracking and its detection is also shown in the same Fig. In Fig. 8(d), incomplete root penetration in a butt weld is shown to be detectable by both radiographic and ultrasonic examination. The orientation of the X- or γ-ray beam is important.

Defect Dimensions

Radiographic examination. The methods available for determining the dimension of a planar defect into the component by means of radiography are mostly of an indirect nature. The difference in photographic density of the defect image and that of the surrounding material can be used to get an indication of the depth of the defect. Taking several radiographs at different angles and/or from different positions of the same defect provides a means for determining this dimension. These methods are, however, time consuming and very often difficult to perform on a site where conditions are less favourable than in a workshop.

Ultrasonic examination. The dimensions of a planar defect such as a crack can be determined more accurately and in a more direct way with ultrasonic testing than with radiographic examination.

Factors of importance in this respect are: the ultrasound beam angle, the beam spread and the beam boundary, the index point of the test probe, the beam path length to the defect as read on the cathode ray tube screen (CRT) of the ultrasonic flaw detector, the geometry and thickness of the component examined, certain distances from reference points and the use of the 20 dB drop system.

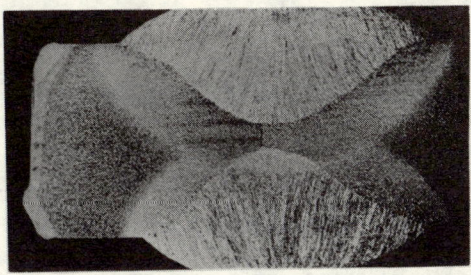

Incomplete root penetration in weld.

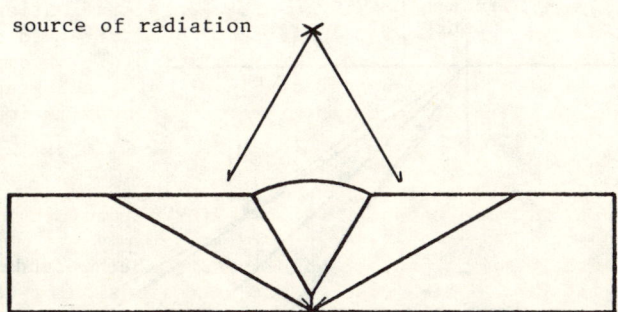

Fig. 8(d) : Detection of incomplete root penetration using ultrasonic and radiographic test techniques.

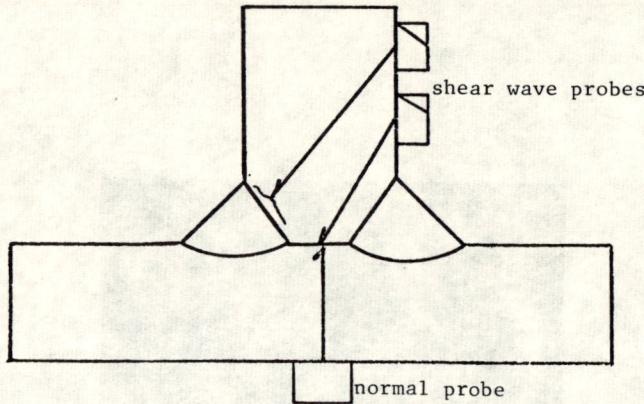

Fig. 8(c). Detection of incomplete penetration and heat affected zone crack in fillet weld using ultrasonic test technique.

A schematic presentation of some of these factors is given in Fig. 9. The basis of the 20dB system is explained in Fig. 10. (See following page).

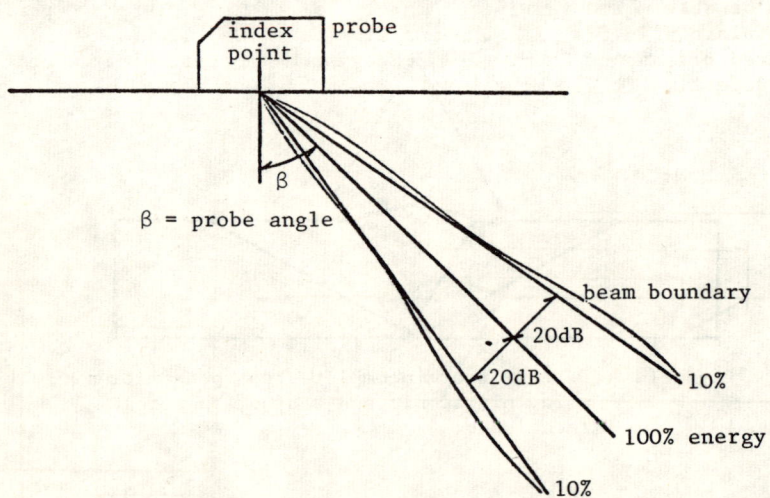

Fig. 9. Ultrasound beam.

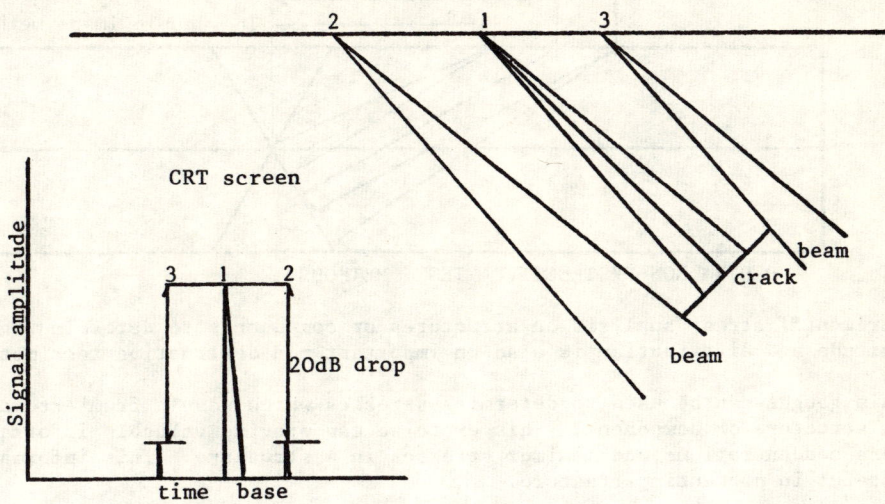

Fig. 10. The 20dB drop principle.

Moving from the central axis of the beam of ultrasound to the outside boundary, represents a 20dB drop in amplitude.

Consider a crack in the heat affected zone of a butt weld. By using the 20dB drop system and a simple double image of the weld, the size, orientation and position of the crack can be determined fairly accurately in a relatively short period. The approach is explained in Fig. 11, where a simple slide rule is used in conjunction with the plotted ultrasound beam. The 20dB system is applied to determine the depth of the defect. Likewise the length of the defect along the weld can be determined.

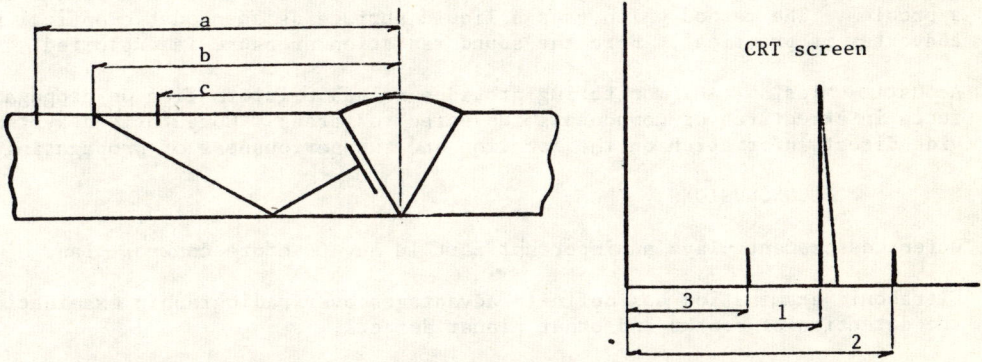

Fig. 11. Determination of defect dimensions and orientation using the 20dB drop principle and the double image method.

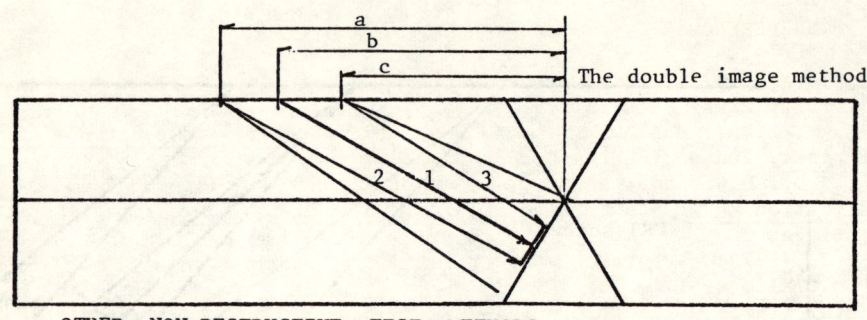
The double image method.

OTHER NON-DESTRUCTIVE TEST METHODS

Experimental stress analysis on structures or components to determine the stress magnitude and distribution is also an important non-destructive test method.

Strain gauges can be used to determine stresses which result from external loads on a structure or component. This exercise can provide valuable information on stress concentrations and maximum stresses in a structure. This information can be useful in preventing fracture.

Residual stresses resulting from cold forming or welding can be measured by using special strain gauges and in one case by drilling a 1,5 mm diameter hole, 1,5 mm deep into the component. Residual stresses are important in the application of fracture mechanics principles. The magnitude of these stresses is often unknown. The use of strain gauges does normally not impair the functioning of the component or structure and can as such be regarded as a NDT method. The use of photo-elastic coatings together with light beams provide a useful method for locating points of high strains in a component or structure subjected to stress.

Holography is a never development in non-destructive testing. Lazer beams are used to measure minute deformations on the surface of a component subjected to a stress. By employing the intensity, wavelength and phase of these beams, a hologram can be produced. This technique is known as holographic interferometry.

Ultrasonic holography employs sound waves to produce a hologram of an object inside an opaque subject. The acoustic hologram is transformed into an optical hologram and a visible image of the object can be constructed. Distortion of the image is a problem. The method which uses a liquid surface as an acoustic-optical hologram converter is practical. Here the sound radiation pressure is exploited.

Acoustic emission (AE) monitoring provides valuable information on propagating defects in structures or components subjected to stress. Computorised systems provide direct information on the location and the seriousness of propagating defects.

CONCLUSIONS

Defect assessment plays an important part in any fracture control plan.

Ultrasonic examination has definite advantages over radiographic examination in the detection of cracks and other planar defects.

REFERENCES

Bijlmer, P.F.A. (1978). Nondestructive testing methods. <u>Fracture Mechanics</u>. The University Press of Virginia, U.S.A., pp. 437 - 453.

Filipczynski, L., Z. Pawlowski and J. Wehr (1966). <u>Ultrasonic Methods of Testing Materials</u>, 2nd ed. Butterworths, London.

Harrison, J.D., M.G. Dawes and M.S. Kamath (Jan. 1978). <u>The COD Approach and its Application to Welded Structures</u>, <u>ref. 55/1978/E</u>. The Welding Institute, Cambridge.

Marais, J.J. (1972). Practical aspects of ultrasonic testing - thin wall tubing. <u>FWP J.</u>, <u>Xll</u>, 4, 17 - 24.

Marais, J.J. (1978). Fracture mechanics and non-destructive testing. <u>FWP J.</u>, XVlll, 1, 47 - 58.

Rolfe, S.T. and J.M. Barsom (1977). <u>Fracture and Fatigue Control in Structures</u>. Prentice Hall, Inc., New Jersey. Chap. 1, pp. 1 - 29.

ESTIMATION OF RISK OF FAILURE OF COMPONENTS DUE TO FAST FRACTURE

D. L. Marriott

Licensing Branch, Atomic Energy Board, Pretoria, R.S.A.

ABSTRACT

Probabilistic fracture mechanics is shown in this paper to provide a basis for rational material selection. A study of available statistics of defect sizes and frequency has led to the proposal of a Design Defect Distribution Curve (DDDC) which is probably conservative compared with the actual incidence of defects. The DDDC combined with an approximate analysis of failure probability enables a choice of material properties to be made to satisfy any required risk level. The effect of non-destructive examination is considered. It is shown that the lower cut off level for defect detection can be related to the probabilistic model.

KEYWORDS

probabilistic fracture mechanics, material selection, NDE reliability, rational design methods.

NOTATION

a	=	crack width
c	=	a (surface defect), a/2 (buried defect)
cutoff	=	min. crack dimension for NDE cutoff
c_o	=	mean crack dimension (a or a/2)
c_{min}	=	minimum crack dimension for negligible contribution to failure
K	=	stress intensity
$\overline{K}$	=	mean fracture toughness
K_f	=	mean fracture toughness of failed components
$K_{f\ min}$	=	minimum fracture toughness of failed components
$p(...)$	=	probability
p_F	=	probability of failure
w	=	load
W_{ult}	=	ultimate load
η	=	load factor (= W/W_{ult})
ρV	=	Average number of defects per component
$(\rho V)_{eff}$	=	Average number of defects after detection and repair

σ = stress
σ^* = effective stress
σ_K = standard deviation of fracture toughness
σ_{Kf} = st. dev of fracture toughness of failed components
σ_y = yield stress

INTRODUCTION

Whether it is recognised or not, risk of failure is an essential parameter in the design of engineering components. While cost considerations consciously dominate design, this is only conditional on the component retaining an adequate margin of failure. Normally the margin is provided by a safety factor defined by a code of practice. Use of safety factors are adequate for most purposes but can be misleading by generating a belief that use of a safety factor is an absolute guarantee against failure. The only rational measure of risk is a statistical one, taking into account the likelihood that variations will accumulate, leading to an unsafe situation. The statistical approach is not independent of the factor of safety, as can be seen by referring to Fig. 1. The diagram refers to a general strength/load relation and is not restricted to fracture. The safety factor is the ratio of the mean strength, $\bar{S}$, to mean load $\bar{L}$. The failure probability is roughly represented by the overlapping area between the two distribution tails. It is easy to see that failure probability increases as the factor of safety decreases. A probabilistic approach is normally too complicated to use in design, but it cannot be avoided in the case of fast fracture because fracture is essentially a statistical problem. The risk of failure by fast fracture depends not only on the fracture toughness, but also on its variation, the probability of defects being present, and the size distribution of these defects.

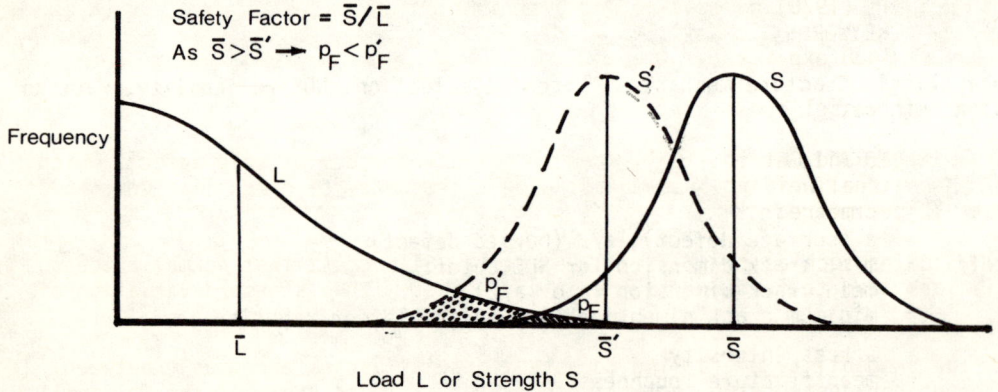

Fig. 1 Relation between safety factor and strength load variability

Two basic forms of fracture assessment exist.

(i) Analysis of component integrity at the design stage in which the presence of defects is still conjectural. This involves choice of material to provide an adequate margin against fracture should a defect be present. The decision depends not only on theoretical fracture calculations but also on the manufacturing and inspection routines. For instance, low failure risk structures can be built in brittle material if the quality of the material is such that defects are very unlikely, or very small, or both.

(ii) Assessment of defects found in service. The only uncertainty over a defect already discovered is a result of the imperfect resolution of non-destructive examination. Even if the size, position, orientation and type of defect are known it is still not really valid to make an assessment by fracture mechanics alone because the risk of propagation depends on the local material properties, and these also may be subject to uncertainty. It is necessary also to consult manufacturing records to establish the probability that the material surrounding the crack may be simultaneously in a degraded state.

This paper will be concerned primarily with the first form of assessment related to design and material choice. The work on which this study is based has been aimed at deriving a failure probability for a nuclear reactor pressure vessel. Some results are of more general interest however, and can form the basis of a rational approach to inspection limits and material selection requirements for pressure vessels in general, as well as other components.

STATISTICS OF FRACTURE PARAMETERS

Information on the variation of fracture toughness, and the frequency and size distribution of defects is difficult to obtain. Nevertheless some data exist which can be used for a preliminary analysis.

(i) Becher and Pederson (1974) record a frequency distribution for crack like defects.

$$p(>c) = 2,56 \exp(-2,56c)$$

where c = defect size in inches

(ii) Marshall (1976) gives a defect distribution of

$$p(>c) = 14,8 \exp(-4,1c)$$

for nuclear vessel construction.

(iii) Salter and Gethin (1971) made an extensive study of weld defects which, for conventional welding procedures, gave a frequency of crack like defects of about 0,2 per metre.

(iv) Studies such as the Marshall Report (1976) suggest that normal process deviations lead to fracture toughness variations with a standard variation of 10%.

(v) The probability of failing to detect defects varies enormously between different non-destructive testing methods. Little is known of the reliability of surface detection methods, but these should be the most reliable. A recent study by Johnson of the Welding Institute (1978), and data quoted by Yang (1978), suggest that surface detection methods are up to 99% reliable for all but very small cracks. Radiography is probably less than 50% effective in detecting crack like defects (Conference on Significance of Defects, 1978). Figures for the failure probability of ultrasonic examination vary from 0,1% for large cracks (Marshall, 1976) to worse than 10% in the studies being carried out in the USA and Europe (Buchanan, 1976).

While the above information is anything but definitive, it is sufficient to construct a rough design case defect population which can be used as a first step in material selection.

Design Defect Distribution Model

This model is shown in Fig. 2.

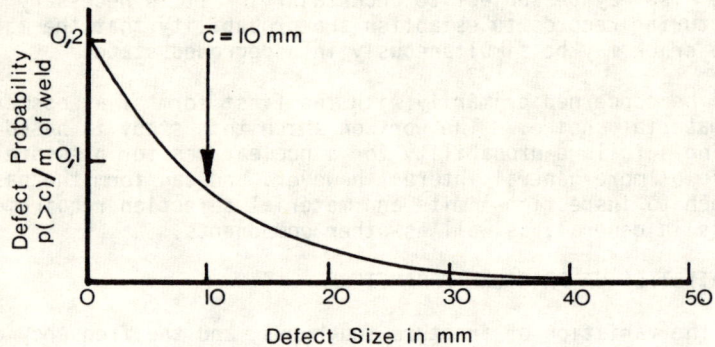

Fig. 2 The Design Defect Distribution Curve

Assumptions made are

(i) Defects in welds and adjacent parent metal or the dominant contributions of total failure probability.

(ii) Exponential distribution of size assumed.

(iii) Average crack size = 10 mm.

(iv) Defect density = 0.2 per metre of weld.

(v) Surface NDE reliability = 99%.

(vi) Ultrasonic reliability for buried defects = 90%.

Hence it is found that the design defect distribution can be described by the following expression.

$$p(>c) = 0.2 \exp(-0.1c) \text{ per metre of weld} \tag{1}$$

where c = crack size in mm.

N.B. The Marshall Study assumed approximately four cracks per vessel. There are about 50 to 100 metres of weld in a nuclear vessel. This gives a defect rate of 0.04 to 0.08 per metre, which is somewhat better than conventional vessel manufacture; as might be expected.

A THEORETICAL MODEL FOR PROBABILITY OF FRACTURE

Assuming defects to be exponentially distributed, fracture toughness to be normally distributed, and defects and material toughness to be independent, it is possible to show (Marriott and Hudson, 1977) that the probability of failure due to initial defects is

$$p_F = \rho V \exp\{-\bar{K}^2/(\sigma^{*2}\pi\bar{a} + 2\sigma_K^2)\} \tag{2}$$

where σ = an effective stress, representing the effect of the varying stress through the component.
ρV = Number of defects per component
$\bar{K}$ = mean fracture toughness
σ = st. dev. of fracture toughness
$\bar{a}$ = mean crack size (surface defect) or ½ mean crack size (buried)

The choice of σ* is a subject of judgement, since residual stresses are usually unknown. It can be shown that a very conservative upper bound on failure probability is given by

$$P_{uF} = \rho V \frac{W}{W_{ult}} \exp\left\{\frac{-\bar{K}^2}{\sigma_y^2 \pi \bar{a} + 2\sigma_K^2}\right\} \tag{3}$$

In other cases, e.g. piping or membrane regions of a pressure vessel shell, is the membrane stress.

Using eqn (2) it is now proposed to show how it is possible to choose a satisfactory level of defects, and assess the effects of non-destructive examination (NDE).

Choice of Material

Using the Design Defect Distribution defined earlier, and a knowledge of the allowable failure probability, it is possible to use eqn (2) to determine the necessary fracture toughness level. This is done most easily by re-arranging eqn (2) as follows

$$\frac{1}{\bar{a}}\left(\frac{\bar{K}}{\sigma_y}\right)^2 = \frac{\pi \ln(\rho V/p_F)}{1 - \ln(\rho V/p_F)/50} \tag{4}$$

In this equation it has been assumed as before, that the standard deviation of fracture toughness, , is 10% of the mean value $\bar{K}$.

Some complex calculations are necessary to determine   in the general case, especially if thermal and residual stresses are involved. Local stress concentrations cause   to increase rapidly from the nominal design stress to near yield. For material selection purposes therefore it is more practical to use the following expression derived from the upper bound of eqn (3).

$$\frac{1}{\bar{a}}\left(\frac{\bar{K}}{\sigma_y}\right)^2 = \frac{\ln(\rho V/\eta p_F)}{1 - \ln(\rho V/\eta p_F)/50} \tag{5}$$

where η = load factor against plastic collapse and for most structures designed to an accepted code of the order of 3.

Example. Thick pressure vessel containing 40m of main seam weld. Allowable failure probability less than 10^{-4}.

ρV = 0.2 x 40 = 8 defects (4 each surface and buried types).

For surface defects, $\bar{a}$ = 10 mm
For buried defects $\bar{a}$ = 5 mm.

From eqn (5), $\frac{1}{\bar{a}}\left(\frac{\bar{K}}{\sigma_y}\right)^2 = 36.8$

For a typical steel with σ_y = 413 MPa, the mean value of fracture toughness required is 177 MPa√m for buried defects, and 250 MPa√m for surface defects. These are extremely high values, and would be unacceptable in practice. However, a number of actions are available to reduce the allowable value to a more practical level.

(i) Invest in a more complex analysis to calculate σ^* exactly, and use the more accurate eqn (2).

(ii) Demonstrate from NDE records that lower values of ρV and $\bar{a}$ are applicable.

(iii) Take advantage of the reliability of NDE to detect defects as a means of reducing the effective defect density to a lower value (ρV)$_{eff}$.

The first course of action is probably only justified in very high integrity structures such as nuclear reactor vessels. Since the Design Defect Distribution Curve has been hopefully chosen as a conservative estimate, NDE records should result in some gains from improved data. NDE information is notoriously difficult to obtain at present, but as more data becomes available, it may be possible to devise a more realistic Design Defect Distribution Curve. Immediate advantage can be taken of the reliability of NDE detection.

Returning to the example it is necessary to consider surface and buried defects separately.

(a) Surface defects ρV = 4
    NDE reliability   = 99%
                (ρV)$_{eff}$ = 0.04
                $\bar{a}$ = 10 mm
                $\bar{K}$ = 170 MPa m
hence

(b) Buried defects ρV = 4
    NDE reliability   = 90%
                (ρV)$_{eff}$ = 0.4
                $\bar{a}$ = 5 mm
hence $\bar{K}$ = 150 MPa m

A mean value of 170 MPa√m with standard deviation of 17 MPa√m is very approximately equivalent to a mean Charpy V Value of 115 Joules, with standard deviation of about 23 Joules (Rolfe and Barsom, 1977). This can be stated in more practical terms as requiring

(a) A mean value of 115 J CVN.

(b) Minimum value of 6 tests of 92 J CVN.

Figure 3 shows the variation of mean fracture toughness for a variety of failure probabilities and defect frequencies.

The figures calculated above are high by current standards. It is possible that this excessive conservatism in the numerical analysis as presented here is due to the handicap of very sparse data. As sources of information are improved it should be possible to reduce allowable levels to more realistic values. The example given should therefore only be taken at this stage as an illustration of how, in principle, material selection can be practically related to a consistent failure risk design criterion.

The gains which can be made from improved basic data can be seen by recalculating

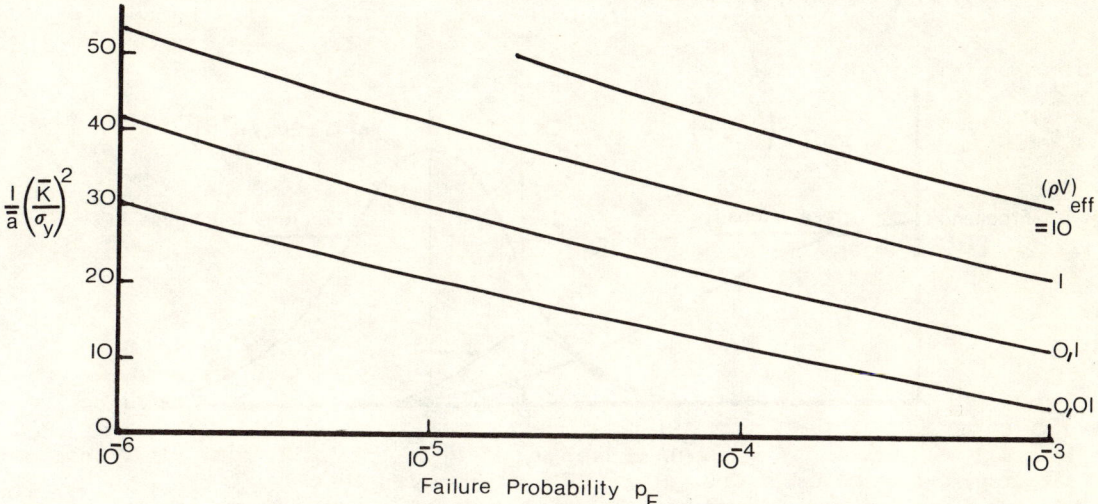

Fig. 3. Material Selection Curve

the example using parameters known to be over-estimates, i.e. conservative for nuclear pressure vessel manufacture standards.

$\rho V < 0{,}04$ crack like defects per metre of weld.
$\bar{a}$ = 6,35 mm

(a) Surface defects, $(\rho V)_{eff}$ = 0,008 : $\bar{K}$ = 108 MPa m

(b) Buried defects, $(\rho V)_{eff}$ = 0,08 : $\bar{K}$ = 103 MPa m

which values correspond approximately to the following Charpy V specification.

Mean value   = 47J CVN
Min of 6 tests = 38J CVN

Determination of NDE Test Threshold

Since most materials have many very small cracks and defects it is normal practice to define a cut-off level below which size defects are considered to be of no importance. This threshhold can be defined from probabilistic fracture mechanics considerations.

Referring to Fig. 4, it is possible to calculate the frequency distribution of stress intensities (or fracture toughnesses) of components which will fail by the presence of a critical defect. This distribution is Gaussian if the fracture toughness is Gaussian, and the defect size distribution exponential, with the following parameters.

$$\bar{K}_f = \left(1 + 2\left(\frac{\bar{K}}{\sigma_y}\right)^2\right)^{-1} \tag{6}$$

$$\sigma_{Kf}^2 = \left(1 + 2\left(\frac{\bar{K}}{\sigma_y}\right)^2\right)^{-1} \tag{7}$$

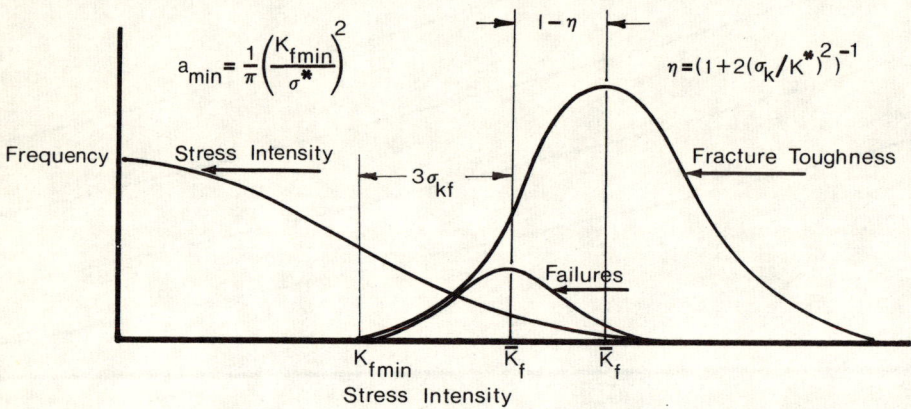

Fig. 4. The probability density function of K for failed components

For practical purposes most failures will occur within three standard deviations of the mean failure stress intensity. A lower bound failure stress intensity can therefore be defined as

$$K_{f\,min} = \bar{K}_f - 3\sigma_{Kf} \tag{8}$$

Knowing the stress level allows the critical defect size to be obtained from

$$a_{min} = \frac{1}{\pi}\left(\frac{K_{fmin}}{\sigma^*}\right)^2 \tag{9}$$

Where there is uncertainty over residual stress, it is advisable to assume that the stress is equal to the yield stress

$$a_{min} = \frac{1}{\pi}\left(\frac{K_{fmin}}{\sigma_y}\right)^2 \tag{10}$$

Example. Using the same parameters as earlier example.

$\bar{K}$ = 170 MPa$\sqrt{m}$
K^* = $\sigma^*\sqrt{\pi \bar{a}}$
where $\bar{a}$ = 10 mm (surface) or 5 mm (buried)

The worst case is obtained when it is assumed that the effective stress σ^* is equal to the yield stress σ_y.

K^* = 413 $\sqrt{\pi \times .01}$
   = 73 MPa$\sqrt{m}$ (surface) or alternatively 51 MPa$\sqrt{m}$ (buried)

Now, $1 + 2(\sigma_K/K^*)^2$ = 1,22 (buried)
or 1,11 (surface)

from eqn (6), $\bar{K}_f$ = 153 MPa√m (surface)
or 139 MPa√m (buried)

From eqn (7), σ_{Kf} = 16,1 MPa√m (surface)
or 15,4 MPa√m (buried)

From eqn (8), K_{fmin} = 105 MPa√m (surface)
or 93 MPa√m (buried)

hence, from eqn (10), c_{min} = 20,5 mm (surface)
or 33 mm (buried)

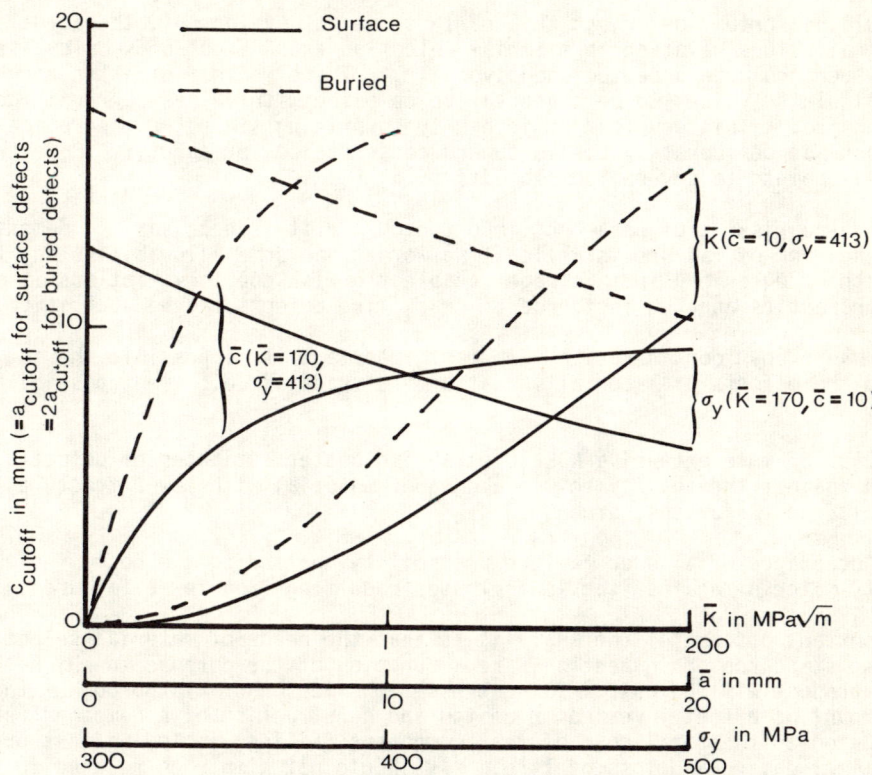

Fig. 5. Typical variations of NDE lower cut-off level with material parameters.

Taking into account the uncertainty in sizing of defects reduces the value of c_{min} still further. Data presented at a recent conference (I. Mech. E 1978) suggests that crack sizes can be often under estimated by a factor of 2. An in-depth study of errors in defect sizing has yet to be done. It would seem, however, that an adequate margin could be achieved by using a factor of 0,4 on the value of c_{min} calculated above. This would give NDE cut-off points for the example considered of

$c_{cut-off}$ = 0,4 c_{min}
= 8,2 mm (surface)
13 mm (buried)

These figures are of the same order-of-magnitude as cut-off levels currently accepted in NDE.

Figure 5 shows the general relationship between cut-off and typical values if $\overline{K}, \sigma_y$ and P_F.

DISCUSSION

It would be fortuitous if, at this early stage in development, the actual numerical values relating to material selection and NDE cut-offs calculated in this paper could be accepted unequivocally. There is in fact a tendency for the calculated values to be conservative compared with values currently accepted in practice. This development is hardly surprising since the theoretical approach was deliberately biased toward conservatism, or safety. However, the analysis points to several possibilities.

(i) Where risk is of paramount importance, and it is necessary to demonstrate that the risk of failure is at least no worse than some allowable value, then the methods described in this paper enable the risk due to variations in material properties and the incidence of crack like defects, to be quantified.

(ii) Where cost considerations are most important it is possible, having quantified the effects of material variation, to trade off different possibilities e.g.

(a) Cost of more extensive NDE to establish better estimates of defect distribution against the ability to use a cheaper material with lower fracture toughness, larger variations, or both.

(b) Acceptance of a lower failure probability, balancing the cost of higher quality material against the lower losses resulting from fewer failures.

An important outcome of the analysis is that the needs of material selection, and the standards of inspection, are a function of the purpose to which the final product will be put. This stresses the lesson of Flixborough - that the amount of effort invested in design and construction of a component should be related to the total cost of that component failing, including lost production time, damage and loss of life. It should not simply be related to the cost of the component itself. As a result of this it is possible that, in time, one may see some movement away from rigid design codes with fixed safety factors and acceptance limits to a more adoptable situation which takes account of product end use in optimising the total cost of material, design effort and losses resulting from failure.

CONCLUSIONS

(i) Using a simple interpretation of probabilistic fracture mechanics it is possible to derive a rational basis for material selection using constant failure risk as a criterion.

(ii) It has been shown that a basis also exists in probabilistic fracture mechanics for decisions on economic cut-off points in NDE.

(iii) The methods developed under (i) and (ii) above are handicapped at present by a deficiency of data. However, the framework of analysis provided here is a guide to what information is required and how it should be used.

(iv) In the absence of adequate information the approximate Design Defect Distribution Curve should provide a conservative estimate when used in calculations for material selection.

REFERENCES

Becher, P.F. and A. Pederson. (1974) Application of statistical linear elastic fracture mechanics to pressure vessel reliability analysis. Nucl. Eng. Des., 27, 12, 413 - 425
Buchanan, R.A. (1976). Analysis of test data on PVRC specification No. 3, Ultrasonic examination of forgings, Revs. 1 and II, Analysis of the non-destructive examination of PVRS plate-weld specimen 51J - Part A. WRC Bull. 221.
Johnson, G.O. (1978). What are the chances of failure? Part 2. Welding Inst. Res. Bull. 19, 3, 78 - 82.
Marriott, D.L. and J.M. Hudson. (1977). Prediction of failure risk in a pressure vessel due to a manufacturing defect. Proc. IAEA sump on Reliability Problems of Reactor Pressure Parts. IAEA, Vienna.
Marshall, W. et al. (1976). An assessment of the integrity of PWR pressure vessels. UKAEA, Harwell, H.M.S.O.
Proc. I. Mech. E. Conf. on Tolerance of Flaws in Pressurised Components. (1978), I. Mech. E., London.
Rolfe, S.T. and J.M. Barsom. (1977). Fracture and Fatigue Control in Structures, Prentice-Hall, New Jersey.
Salter, G.R. and J.W. Gethin. (1971). The significance of defects in welded pressure vessels, Welding Inst. Res. Rep. C322/3/71.
Yang, J.N. Statistical approach to fatigue and fracture including maintenance procedures. (1978), Fracture Mechanics, Proc. 10th Symp. Naval Struct. Mech. Washington D.C., University Press of Virginia, Charlotteville, 559 - 578.

FRACTOGRAPHY: A TOOL FOR FAILURE ANALYSIS

S. B. Luyckx

Boart Basic Research Group, Physics Department, University of the Witwatersrand, Johannesburg, R.S.A.

ABSTRACT

This paper reviews the techniques used in fractography, i.e. in the study of fracture surfaces, and discusses their relative advantages. It also describes the most typical fracture surface features and the information that they can provide. Finally, it illustrates that fractography is useful to both scientists and practical engineers, since it has proved valid in the interpretation of fracture processes as well as in the diagnosis of causes of failure.

KEYWORDS

Fractography; WC-Co alloys; service failures; electron microscopy; fracture mechanisms.

INTRODUCTION

Fractography is the study of the microscopic features of fracture surfaces and fracture profiles, with the aim of determining the causes and the mechanisms of fracture. The first systematic examination of fracture surfaces was done by Zappfe and Clogg (1945), by optical microscopy. However, fractography established itself as a branch of metallography only in the 1960's, with the development of replica techniques, which overcame the problem of examining rough surfaces at high magnification. Replicas of fracture surfaces led to important discoveries on the micromechanisms of fracture, but their lengthy preparation and difficult interpretation made them unsuitable for routine use. Only in the late 1960's, with the introduction of commercial scanning electron microscopes, did fractography become a standard tool for failure analysis. However, the scanning electron microscope has not replaced entirely the earlier techniques: today it is recognized that optical microscopy, transmission electron microscopy of replicas and scanning electron microscopy are complementary techniques, all of which are necessary for a complete interpretation of fracture surfaces.

BASIC TECHNIQUES

The first examination of a fracture surface must be carried out by optical microscopy, preferably by a stereomicroscope, which provides a three dimensional image of the surface, similar to the image provided by the human eye. The stereomicroscope is more suitable than the compound microscope for the study of rough surfaces because of its large depth of field. However, the magnifying power of the stereomicroscope only goes up to about 250χ, and therefore it is only suitable for locating the general areas of interest, like regions of fracture origin, secondary cracks, transition areas between stable and unstable crack growth, etc.

Optical investigations are followed by scanning electron microscopy (SEM), unless the dimensions of the specimen are incompatible with the diameter of the SEM column. Usually, a large specimen can be cut into smaller sections which can fit into the SEM. If cutting is not possible or might contaminate the fracture surface (like spark cutting, in paraffin), the examination can be done by recently developed SEM replication techniques which, for special problems, can be even preferable to viewing the specimen directly (Sarracino, 1973). The advantages of examining replicas by SEM instead of TEM (transmission electrom microscopy) are the following:

(a) The whole replica can be viewed at once, without the holder in the way;
(b) the replica can reproduce more faithfully the topography of the fracture surface, since it can be prevented from collapsing by a rigid backing (McCall, 1972).

In contrast, TEM replicas must be very thin (< 2000 Å); they therefore require a holder on which they flatten to a certain degree and by which they are partially covered. When the specimen has suitable dimensions, the preparation for viewing it in the SEM are minimal: it is mounted on a stub and a good electric contact is established between the specimen and the stub, usually by means of a conducting glue. If the specimen is not conducting or if it is of a low atomic number material, a thin layer of a conducting metal is evaporated on to its surface. Magnetic materials must be demagnetized for best resolution.

The SEM can be considered an extension of the stereomicroscope, because of its large depth of field (some 300 to 500 times greater than the d.o.f. of the compound microscope) which provides a picture of three-dimensional appearance. Its resolution is at least 20 times greater than the resolution of the optical microscope and therefore it is often used preferentially for low magnification work.

In fractography, the SEM is usually operated in the secondary electron mode (Siegel and Beaman, 1975) which provides a picture of the topography of the surface. The back-scattered electron mode (Siegel and Beaman, 1975) is used to locate inclusions or to provide a qualitative picture of the distribution of the phases present, since back-scattering is sensitive to the elemental composition of the surface (Fig. 1a, b).

The resolution of the SEM (200 - 250 Å) limits its use, in fractography, to magnifications of up to about 10,000χ. For viewing details at higher magnifications (for example for examining slip lines or fine fatigue striations) TEM replicas are still extensively used, the resolution of the TEM being considerably better than the resolution of the SEM. Among the many replication techniques which have been developed over the last twenty years, the most commonly used one is the cellulose acetate replica, because the preparation process is simple and the replica is easily removed from most surfaces. A piece of cellulose acetate is softened with acetone and pressed on to the fracture surface. When dry, it is

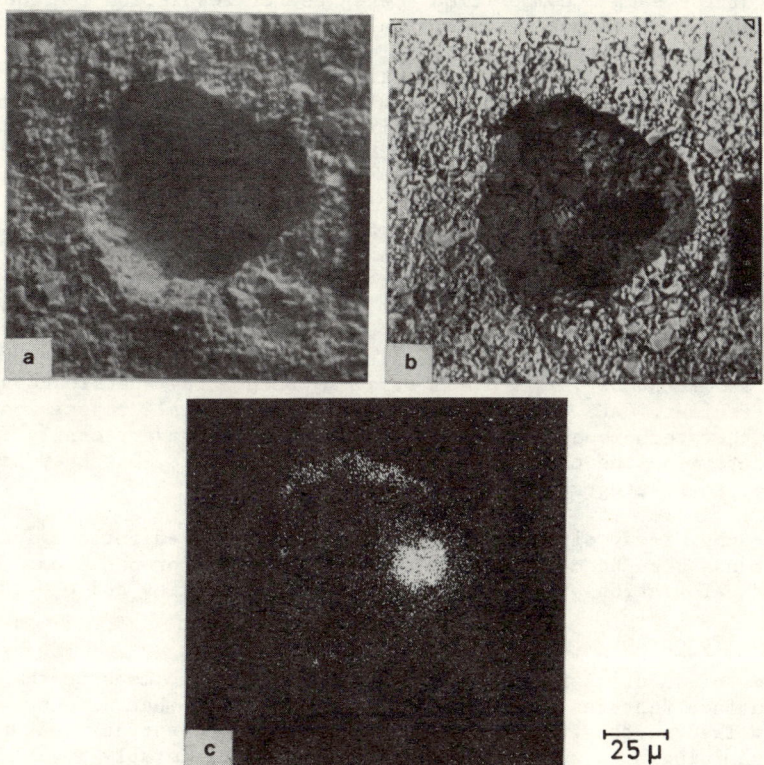

Fig. 1. (a) Secondary electron micrograph of a pore containing an inclusion in a WC-Co alloy. (b) Back-scattered electron micrograph of the same pore. The dark areas (including the inclusion) have a different chemical composition from the rest of the surface. (c) Electron probe micrograph indicating that the dark areas in (b) contain calcium.

removed from the surface, "shadowed" with a heavy metal (Pt, Au, etc.) in an evaporating unit and covered with a thin film of carbon. The acetate is then dissolved in acetone and the shadowed carbon film is viewed in the TEM. The topography of the original surface is deduced from the distribution of the shadowing material, therefore its interpretation is not as easy as in the case of SEM micrographs.

ACCESSORY TECHNIQUES

Chemical Microanalysis.

A common accessory of the SEM is the electron probe, which analyzes the X-rays emitted by the sample when bombarded by electrons. The frequency of the X-rays emitted is a function of the atomic number and the intensity a function of the amount of each element present. Most commonly, electron probes are fitted with energy dispersive X-ray analyzers, which provide instantaneous qualitative representations of the sample composition. When better resolution is required, electron probes are equipped with crystal spectrometers, which detect the elements present in the sample by the wavelength of the X-rays emitted. These instruments are capable of accurate quantitative analysis in the case of flat surfaces, but in the case of fracture surfaces only qualitative determinations can be expected (e.g. Fig. 1c), since different parts of the surfaces are viewed by the detectors under different angles.

An analytical technique often useful to fractographers is Auger electron spectroscopy (AES) (Siegel and Beaman, 1975), which consists of analyzing the energy of the Auger electrons emitted by the atoms of the sample when they de-excite after electron bombardment. The advantages of this technique are that only the first few atomic layers of the surface are analyzed (which makes it particularly suitable, for example, for the study of grain boundaries in intergranular fractures) and that the efficiency of emission of Auger electrons increases with decreasing atomic number, i.e. it is maximum where X-ray spectroscopy is relatively inefficient. AES can only determine the elemental composition of a surface. Other techniques (such as ion spectroscopy) have recently been developed, to determine the chemical composition of surfaces, but they have not yet been applied to fractography.

In fractography, X-ray diffraction is not usually applied for chemical analyses, although it has been used to measure the work of crack propagation, by measuring the plastic deformation on the fracture surfaces (Georgiev and others, 1977).

Quantitative Topography.

The two aims of quantitative topography are, firstly, to measure the size of fracture surface features and secondly, to evaluate to what extent each fracture mode is involved in the fracture process. The size of surface features is measured by examining high magnification stereopairs, preferably by SEM, although TEM replicas can also be used. Stereopairs are pictures of the same area taken from different angles with respect to the electron beam (angle difference $\simeq 7-10°$). By measuring the size of the features of interest in both pictures and by knowing the angle between the stereopairs, it is possible to calculate the real size of the features (McCall, 1972).

As for measuring the amount of a particular fracture mode, attempts are currently being made to extend to fracture surfaces the quantitative techniques used in the examination of polished surfaces (Chermant and Coster, 1979). However, because of the non-planar nature of fracture surfaces, measurements must be repeated after tilting the surface to as many orientations as possible.

Relationship between Fracture Characteristics and Microstructure.

The effect of microstructure on the fracture process is often revealed more clearly by the study of fracture profiles than by the study of fracture surfaces. For laboratory fracture tests, it is possible to prepare the specimen surfaces for microscopic examination before carrying out the test (Luyckx, 1968) so that after fracture, the fracture profiles can be examined with no damage to the fracture surfaces. In contrast, after service failures in order to study the fracture profiles the specimens must be sectioned normal to the fracture surface, after first covering the fracture surface with a protective coating; they can then be polished and etched (Kerr and others, 1976; Van Stone and Cox, 1976). Recently, a technique has been developed which allows the polishing and etching of part of a fracture surface without damage to the remainder and to the viewing of fracture and microstructure simultaneously on the same plane (Chesnutt and Spurling, 1977).

Multi-phase Materials and Inclusions

It is usually important, in multi-phase materials, to determine the role of each phase in the fracture process. Before scanning electron microscopes became commercially available the role of each phase was determined by optical microscopy and by replicas taken before and after etching the surface selectively (Fig. 2) (Hara and others, 1970). The latter being a destructive technique, it is now usually replaced by the back-scattered electron mode of the SEM (Fig. 1b). The back-scattered electron mode is also used for the location of inclusions. Inclusions can then be extracted by suitable replica techniques (Pickwick and Packwood, 1976) and identified in the TEM by electron diffraction.

Fig. 2. Replica of the fracture surface of a WC-Co alloy, after etching it with 10% HCℓ. Only the cobalt has been etched. The features indicated by the arrows are on WC grains while on the unetched surface they were believed to be in the cobalt.

Mating Fracture Surfaces

It is difficult, sometimes, to interpret fracture features by examining only one of the two surfaces produced by the fracture ("mating fracture surfaces"). For example, in the case of WC-Co alloys the examination of matching areas on mating surfaces was necessary to interpret the fracture features in the cobalt phase (Luyckx, 1977) (Fig. 3). In general, in the case of dimples, both surfaces must be examined in order to deduce the direction of the stress and its order of magnitude (Beachem, 1975).

Fig. 3. Matching areas on mating fracture surfaces of a WC-Co alloy. Arrows point to "dimples" in cobalt.

Some other Techniques

In recent years it has become possible to determine the orientation of the grains involved in the fracture process by means of SEM electron channeling techniques (Davidson, 1977). They can be used in the study of brittle fractures, to establish which orientations pose least resistance to fracture propagation. Another new development is ultrasonic fractography (Green and others, 1977). It allows the crack front shape throughout the fracture process to be followed, which can show the effect of inclusions and porosity on fracture propagation.

TYPICAL FRACTURE FEATURES

The study of fracture surfaces at low magnification is usually sufficient to determine, for example, whether a fracture is basically brittle or ductile or whether it is a fatigue fracture. Thus a "fracture mirror" (Fig. 4) around a fracture origin only occurs in brittle materials; "beach markings" (Fig. 5) are crack arrest lines, indicating fatigue; "cups and cones" result from obvious tensile ductile fractures (Fig. 6).

These observations, however, do not say much about the fracture process, since both brittle and ductile fractures can be transgranular or intergranular and both can be due to various causes such as structural defects, stress corrosion, overloading, etc.

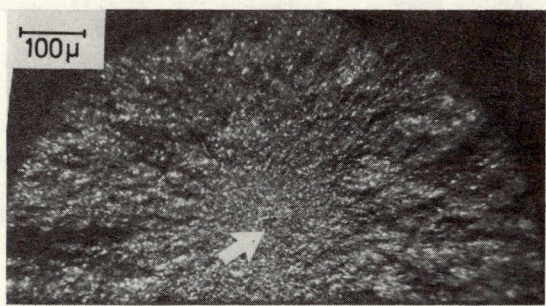

Fig. 4. Fracture "mirror" in a WC-Co alloy. The "mirror" is the circular, smooth area surrounding the fracture origin. The arrow points to the fracture origin.

Fig. 5. "Beach markings" in a WC-Co alloy.

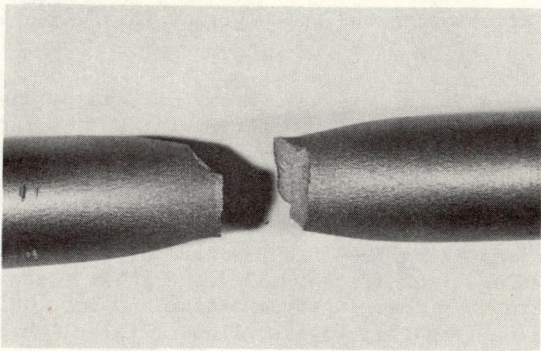

Fig. 6. A "cup and cone" fracture in a low-carbon steel specimen. (Courtesy of Professor F.P.A. Robinson).

Only the microscopic fracture features can provide information on the crack tip mechanisms by which fracture took place and on the detailed causes of fracture. For example, if the fracture surface of a WC-Co alloy is observed at low magnification it is classified as a perfectly brittle fracture; if it is observed at high magnification it shows plastic deformation in the cobalt phase and, occasionally, in the carbide grains (Bartolucci and Schlössin, 1966).

The most common microscopic fracture features can be roughly divided as follows: cleavage patterns, coalesced microvoids, intergranular features and fatigue patterns.

Cleavage Patterns.

Cleavage is generally associated with a low energy brittle fracture. It is a transgranular process occurring in bcc and hcp metals along crystallographic planes and involving little plastic deformation. The propagation of a cleavage crack usually occurs by the simultaneous development of several microcracks, which join and form the typical "cleavage steps" (Fig. 7). Cleavage steps are parallel to the direction of crack propagation. They tend to run together, forming the so-called "river patterns".

Fig. 7. "River patterns" on WC grains in a WC-Co alloy (arrows).

By following the direction of the cleavage steps one can reconstruct the history of the fracture process and trace the fracture origins. As a result, cleavage patterns point to inclusions or other flaws, which are often the fracture nuclei in brittle materials.

Other features often found on cleavage surfaces are "cleavage tongues", which are microscopic ridges resulting from the fracture of deformation twins formed at the tip of the propagating crack (Broek, 1972), "feather features" which are similar to river patterns but emerging from a single point, which is a point of fracture re-initiation (Fig. 8); Wallner lines (Fig. 9) which result from the interaction between the advancing crack front and an elastic wave propagating at the same time.

Fig. 8. Replica of the fracture surface of a WC-Co alloy. The WC grain marked with "A" exhibits a "feather-like" pattern. The WC grain gives an example of cleavage fracture, while the cobalt phase provides examples of microvoid coalescence (arrows).

Fig. 9. Wallner lines (arrow) on a WC grain in a WC-Co alloy.

Coalesced Microvoids.

Microvoid formation and coalescence is a result of plastic deformation, associated with ductile fracture. The microvoids are usually formed at interfaces between the matrix and microscopic inclusions. They grow under stress until they join each other and eventually form the macroscopic fracture. The microvoids appear on fracture surfaces as "dimples" (Fig. 10) whose relative size and shape depend on the microstructure of the material, on the size and distribution of micro-inclusions, on the magnitude and direction of the stress that caused fracture and on the temperature at which fracture occurred (Beachem, 1975). An example of dimples is found in the cobalt phase of WC-Co alloys (Fig. 3). These dimples result from the voids probably nucleated at the interface between cobalt and microscopic carbide precipitates or microscopic inclusions.

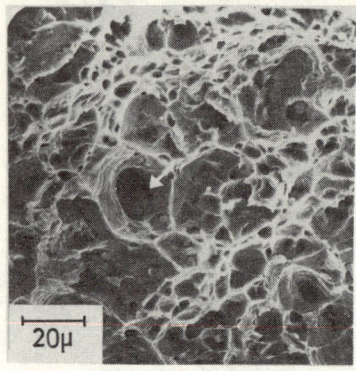 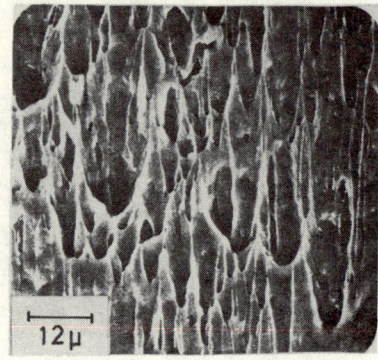

Fig. 10. "Dimples" in a carbon steel. The arrow points to a slag inclusion. (Courtesy of Dr. S. Engell-Nielsen).

Fig. 11. Elongated shear dimples in carbon steel with low carbon content. (Courtesy of Dr. S. Engell-Nielsen).

The dimple shape can assist in the determination of the direction of fracture propagation, since the tip of elongated dimples usually points to that direction (Fig. 11).

Fatigue Patterns.

Fatigue fracture results from a crack progressing under a cyclic stress. A fatigue crack can be transgranular or intergranular. On a macroscopic scale fatigue surfaces are usually smooth and frequently characterized by "beach markings" (Fig. 5); on a microscopic scale they are often characterized by "striations" (Fig. 12). Striations are lines which represent the successive positions of the crack front and which, therefore, are roughly normal to the direction of crack propagation. They reveal the history of the fracture process and show which microstructural elements hinder crack propagation (e.g. grain boundaries) and which foster it (e.g. inclusions fractured ahead of the main crack). One striation is formed during each stress cycle, and therefore one can deduce the crack propagation rate per cycle from the distance between striations. The width of the striations can provide information on the magnitude of the stress. Striations, however, only appear on the fracture surfaces of materials which have many available slip systems (Broek, 1972).

Intergranular Features.

A fracture is called intergranular when it takes place along grain boundaries (Fig. 13). This fracture is observed in materials where the grain boundaries have been embrittled by precipitates or by the environment, or in creep resulting from grain boundary sliding. Intergranular fracture can be brittle or ductile.

Typical intergranular fractures are some hydrogen embrittlement and stress corrosion fractures, where the grain boundaries, having been weakened by the environment, present the least resistance to the propagating crack.

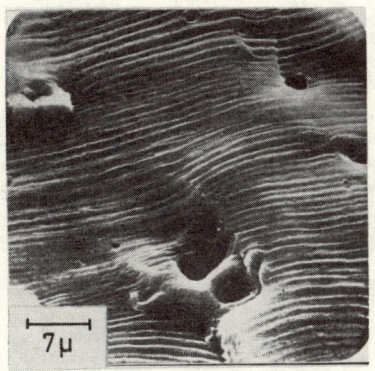 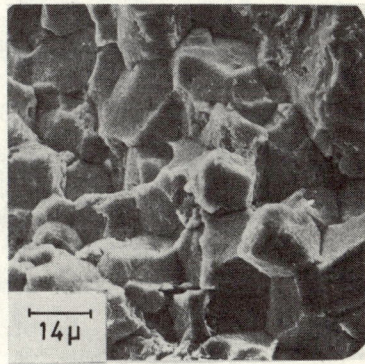

Fig. 12. Fatigue striations in a solution-treated and aged aluminium-copper wrought alloy. (Courtesy of Dr. S. Engell-Nielsen).

Fig. 13. Hydrogen-embrittlement intergranular fracture in quenched-and-tempered high-strength steel. (Courtesy of Dr. S. Engell-Nielsen).

In practice, fracture surfaces often exhibit more than one of the above mechanisms, with one mechanism being predominant. An example can again be given by a WC-Co fracture surface (Fig. 8) where WC grains fail by cleavage or by intergranular fracture and cobalt by microvoid coalescence.

APPLICATIONS

Scientific Applications.

Scientific fractographic investigations have two broad aims: (1) to understand the macroscopic fracture behaviour through the observation of the microscopic mechanisms exposed by fracture surfaces; (2) to find possible correlations between the micromorphology of fracture, macroscopic properties (such as toughness and strength) and loading parameters.

As far as the first aim is concerned, considerable progress has been made, mainly through the use of replicas. For example, it has been understood that ductile fracture always occurs by "glide plane decohesion", i.e. by gliding apart along slip planes. In large single crystals this is a macroscopic phenomenon, while in more complex structures this is only detected at high magnification. This mechanism also operates in the coalescence of microvoids, but only occasionally is it visible as more or less irregular slip traces (Beachem and Meyn, 1964). Fractography has also led to the understanding of the formation of fatigue striations (Broek, 1972). Striations have been explained as the result of slip, both in the crack extension and in the crack closure stages. Fine slip lines have been observed on replicas of striations.

As far as the second aim of fractography is concerned, a correlation between fracture surface features, macroscopic properties and loading conditions is of great importance for the diagnosis of service failures. Progress has been made in developing fractography into a more quantitative tool. For example: in fatigue it is often possible to determine the number of cycles which lead to fracture by counting the microscopic striations on the fracture surfaces; in ductile fractures the amount of deformation can give indications on the load and temperature; elongated dimples can suggest the presence of shear stresses; cleavage crack length can give an idea of the toughness of the material.

Attempts have been made to compile catalogues of the types of fracture morphology associated with various conditions of loading and environment (SEM/TEM Fractography Handbook, 1975). These catalogues, however, cannot include all existing materials, therefore, in most cases, the study of service failures must still be assisted by "model fracture tests" carried out under conditions as close as possible to the actual failure. The "model tests" provide reference fracture features, which are simple compared to the morphology of service failures, resulting from a complexity of factors.

Service Failures.

In the case of service failures the fractographer is often confronted with dirty fracture surfaces. The surfaces must be cleaned with mild organic solvents and in ultrasonic cleaning baths, in order to damage the surfaces as little as possible (Dahlberg, 1976). On large specimens the solvents can be applied by means of a very soft brush. Corrosion or oxidation products can be removed by appropriate etchants (e.g. inhibited acids) but the result may no longer be representative of the fracture surface. Corroded surfaces, therefore, may be of no use in fractography, except as sources of information on the role of the environment in the fracture process. Unfortunately service fracture surfaces are often brought in contact with one another either accidentally or as a consequence of the failure itself, and this can obscure fracture surface features.

The first step of a fractographic examination is to determine the fracture origin. Macroscopically, the origin of a brittle fracture is usually characterized by radial patterns (Fig. 14), while the origin of a ductile fracture is usually the area exhibiting the least plastic deformation. Microscopically, the site of the fracture origin is found by tracing back "river patterns", fatigue striations or dimples.

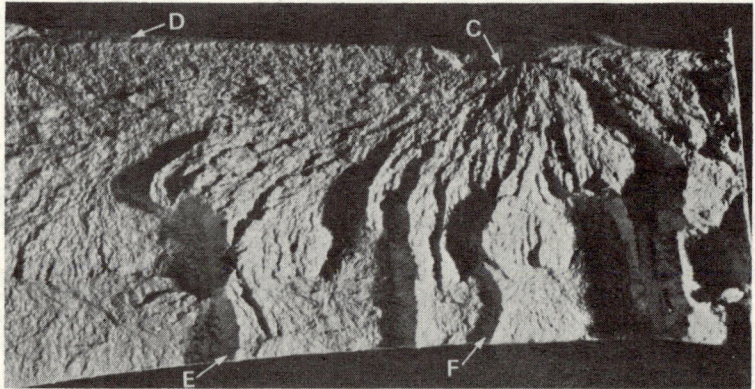

Fig. 14. Brittle fracture surface showing region of fracture initiation (C). Arrows D, E and F point to shear lips. (reproduced from "Fracture Surface Replicas" by The Welding Institute, U.K.).

It is also useful to find secondary cracks which have propagated only partially, since the surfaces of these cracks are usually less contaminated than those of the main fracture. While the fracture origin area exposes the causes of fracture initiation, the rest of the fracture surface exposes the weaknesses of the material, since propagating cracks always follow the path of least resistance. Fractography, therefore, gives direct information on improvements to be introduced: for example, in the elimination of defects, in the microstructure or in the conditions of use.

Fractography and Fracture Mechanics.

For a complete understanding of fracture mechanisms it is necessary to compare the microscopic characteristics shown by fractographic analysis to the energy data provided by the methods of fracture mechanics (FM). Fractography complements fracture mechanics in three ways: (1) it accounts for differences in toughness or fracture energy in terms of microscopic fracture processes; (2) it provides FM with necessary data, such as crack front shape, fracture energy and plastic zone size; (3) it can test FM results.

As far as the first point is concerned, extensive fractographic work is being done on the transition area between stable and unstable crack growth, in order to correlate fracture morphology and fracture toughness values. An example of a successful investigation is given by Chesnutt and others (1976). As far as the second point is concerned, crack front shapes can be deduced from fatigue markings or can be emphasized by the techniques of ultrasonic fractography (Green and others, 1977); fracture energy values can be obtained from measurements of plastic deformation on fracture surfaces (Georgiev and others, 1977); plastic zone sizes can be approximated by shear lip sizes (Pelloux, 1974) or even by the size of the deformed edge along fracture profiles (Fig. 15).

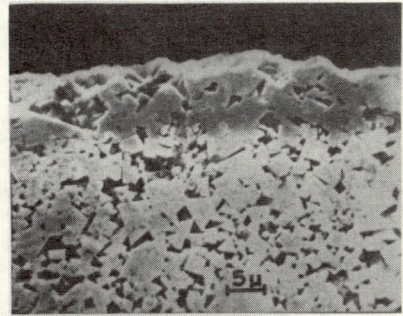

Fig. 15. SE micrograph of the fracture profile of a WC-Co alloy showing the deformed edge, whose size is comparable to the plastic zone size of this material (Chermant and others, 1973).

As far as the third point is concerned, fractographic measurements of fracture-nucleating-pores have been plotted against applied stress and fracture toughness values and the relationships obtained have been in agreement with FM results (Almond and Roebuck, 1977). Another example is given by the spacing of fatigue striations which has been found to relate to variations in the stress intensity factor, again in agreement with FM results.

CONCLUSION

Fractography is a very versatile tool, which is useful to the metallurgist and the materials scientist to better understand fracture mechanisms, to the practical engineer to rapidly diagnose service failures, and to the fracture mechanics scholar to test and interpret his results.

ACKNOWLEDGMENTS

The author wishes to thank Optolabor (Pty.) Ltd. for permission to use the pictures appearing as Figs. 10, 11, 12 and 13, which were taken by Dr. Svend Engell-Nielsen of the Technological Institute, Tastrup, Denmark, and Professor F.P.A. Robinson for providing the specimen shown in Fig. 6. It is also gratefully acknowledged that the work on WC-Co alloys was sponsored by Boart International Ltd., Johannesburg, South Africa.

REFERENCES

Almond, E.A. and Roebuck, B. (1977). Defect initiated fracture and the bend strength of WC-Co hardmetals. Metal Science, 11, 458-461.

Bartolucci, S. and Schlössin, H.H. (1966). Plastic deformation preceding fracture in Tungsten Carbide-Cobalt alloys. Acta Met., 14, 337-339.

Bartolucci Luyckx, S. (1968). Microscopic aspects of fracture in WC-Co alloys. Acta Met., 16, 535-544.

Bartolucci Luyckx, S. (1977). Precision matching of mating fracture surfaces of WC-Co alloys. Proc. 4th Internat.Conf. on Fracture, Waterloo, 2, 223-227.

Beachem, C.D. and Meyn, D.A. (1964). Illustrated Glossary of fractographic terms. NRL Report 1547, U.S. Naval Res.Lab., Washington D.C.

Beachem, C.D. (1975). The effects of crack tip plastic flow direction upon microscopic dimple shapes. Metall. Trans. A, 6A, 377-383.

Broek, D. (1972). Some contributions of electron fractography to the theory of fracture. NLR TR 72029U Report, National Aerospace Lab., The Netherlands.

Chermant, J.L., Deschanvres, A. and Iost, A. (1973). Fracture mechanics, statistical analysis and fractography of carbides and metal carbides composites. Fract. Mech. Ceram., 1, 347-366.

Chermant, J.L. and Coster, M. (1979). Quantitative fractography. J.Mater.Sci., 14, 509-534.

Chesnutt, J.C., Rhodes, C.G. and Williams, J.C. (1976). Relationship between mechanical properties, microstructure and fracture topography in $\alpha + \beta$ Titanium alloys. ASTM STP 600, 99-138.

Chesnutt, J.C. and Spurling, R.A. (1977). Fracture topography - microstructure correlations in the SEM. Metall. Trans.A, 8A, 216-218.

Dahlberg, E.P. (1976). Failure analysis by examination of fracture surfaces. SEM Symposium, Toronto, April 5-9.

Davidson, D.L. (1977). The use of channeling contrast in the study of material deformation. Proc. 10th SEM Symposium, Chicago, 28 March - 1 April, 431-438.

Georgiev, M.N., Danilov, V.N., Mezhova, N.Ya. and Strok, L.P. (1977). Relationship between plastic deformation in the fracture zone and fracture characteristics. Fiz. Met. & Metalloved. (USSR), 43, 403-407.

Green, D.J., Nicholson, P.S. and Embury, J.D. (1977). Crack shape studies in brittle porous materials. J. Mat.Sci., 12, 987-989.

Hara, A., Nishikawa, T. and Yazu, S. (1970). A new replication technique. Plans für Pulver., 18, 28-43.

Kerr, W.R., Eylon, D. and Hall, J.A. (1976). On the correlation of specific fracture surface and metallographic features by precision sectioning in Titanium alloys. Metall. Trans. A, 7A, 1477-1480.

McCall, J.L. (1972). Fracture analysis by scanning electron microscopy. MCIC Report, Battelle Columbus Lab., Ohio.

Pelloux, R.M. (1974). Fracture mechanics and SEM failure analysis. Proc.SEM/1974, I I T Res.Inst., Chicago, 851-858.
Pickwick, K.M. and Packwood, R.H. (1976). An improved extraction replica technique for the microanalysis of inclusions. Metallography, 9, 245-255.
Sarracino, M. (1973). A new technique for the examination of fatigue striations with the SEM. Metallography, 6, 176-182.
SEM/TEM Fractography Handbook (1975). MCIC, Battelle Columbus Lab., Ohio.
Siegel, B.M. and Beaman, D.R. (1975). Physical aspects of electron microscopy and microbeam analysis. John Wiley & Sons.
Van Stone, R.H. and Cox, T.B. (1976). Use of fractography and sectioning techniques to study fracture mechanisms. ASTM STP 600, 5-29.
Zappfe, C.A. and Clogg, M. (1945). Fractography - a new tool for metallurgical research. Trans. A.S.M., 34, 71-75.

SECTION 3B

Solving Fracture Problems— Some Case Studies

THE COD APPROACH AND ITS APPLICATION TO WELDED STRUCTURES

J. D. Harrison, M. G. Dawes, G. L. Archer and M. S. Kamath

The Welding Institute, Abington, Cambridge, England

ABSTRACT

In the United Kingdom, The Welding Institute has pioneered the application of the COD design curve. This paper is a representation of the philosophy underlying design curve applications and illustrates the practical significance of COD by drawing on case studies from various welded structures. Following a brief appraisal of the origins of the design curve, the paper outlines procedures for the use of COD in design. The reliability of a small scale test prediction from the design curve has been investigated on a statistical basis from a survey of more than seventy wide plate tensile test results in which the material had also been categorised by COD. Specific practical examples are then discussed covering the various types of applications, namely, material selection, defect assessment and failure investigation.

INTRODUCTION

The theoretical and experimental basis for crack opening displacement (COD) as a fracture characterising parameter in yielding fracture mechanics is described in a separate report (Dawes, 1977). In order to place the application of the COD approach in context it is convenient to think in terms of the brittle to ductile transitional behaviour encountered with rising temperature in structural steels. If single edge notched bend specimens of thickness, B, equal to that of the structure, width W = 2B and with notch depth a = B (the preferred COD test specimen geometry) are tested over a range of temperatures the following behaviour may be expected. At low temperature failure will occur under elastic conditions, the test will give valid value of K_{Ic} to E399 and the result may be applied in structural analysis using LEFM. With increasing temperature, the toughness will rise until the K_{Ic} measurement capacity of the specimen (the greatest possible capacity for the given thickness) is exceeded and failure will occur only after significant yielding. Crack opening displacement will then be determined from the test record and the result may be applied using yielding fracture mechanics as explained later.

With further increases in temperature the material will behave in a fully ductile manner so that the specimen does not fracture but fails by a simple plastic instability. In such cases structural failure will also be by plastic instability and this will be assessed by limit load analysis.

J. D. Harrison

Proposals for a weld defect assessment method based on a continuous approach covering these three regimes are currently in an advanced stage of drafting by a British Standards Committee. The approach may be summarised as follows:

Specimen behaviour	Structural behaviour	Analysis method
K_{Ic}	Elastic	LEFM
COD	Contained yielding	YFM
Fully plastic	Plastic instability	Limit load

The current paper deals only with the proposed method of application of yielding fracture mechanics. Because of the greatly increased complexity of rigorous elastic-plastic analysis compared to LEFM, the approach is simplified and employs a 'design curve' which is semi-empirical. This curve is considered to be conservative and makes possible swift assessments in practical situations, but more accurate analysis of specific problems are, of course, possible.

DERIVATION OF THE DESIGN CURVE

The evolution of the COD design curve has been described in detail by Dawes and Kamath (Dawes, 1978). The basis was established by Burdekin and Stone (Burdekin, 1966) who studied the extension of the Dugdale strip yield model into the general yielding regime.

The design curve takes the form of a relationship between the non-dimensional COD, Φ, and the ratio of applied strain to yield strain, $\frac{e}{e_Y}$. Φ is defined as $\frac{\delta_c}{2\pi e_Y \bar{a}_{max}}$. The applied strain, e, is taken as the local strain which would exist in the vicinity of the crack, were the crack itself not present. The significance of $\bar{a}_{max}$ should be stressed. The design curve has always been intended, as the name implies, to be one which can be used directly in design. Its purpose is to give conservative predictions of the size of defect which can be allowed to remain in a structure without repair and it is not intended to predict criticality. Thus, $\bar{a}_{max}$ should be smaller than the critical defect size $\bar{a}_{cr}$.

The original curve of Burdekin and Stone was changed to take account of later experimental findings, first by Harrison and colleagues (Harrison, 1968), then by Burdekin and Dawes (Burdekin, 1971), and was finally set out in its current form by Dawes, (Dawes, 1974). This is given by:-

$$\Phi = \left(\frac{e}{e_Y}\right)^2 \quad \text{for} \quad \frac{e}{e_Y} \leq 0.5 \tag{1a}$$

$$\Phi = \left(\frac{e}{e_Y}\right) - 0.25 \quad \text{for} \quad \frac{e}{e_Y} \geq 0.5 \tag{1b}$$

It can be shown that, for small scale yielding δ is related to G_I by:

$$G_I = m\sigma_Y \delta$$

where the plastic stress intensification factor m is equal to 1 for plane stress. Hence:

$$\delta = \frac{G_I}{\sigma_Y} = \frac{K_I^2}{\sigma_Y E} = \frac{\sigma^2 \pi a}{\sigma_Y E} \quad \text{or}$$

$$\Phi = \frac{\delta}{2\pi e_Y \bar{a}} = \frac{1}{2}\left(\frac{\sigma}{\sigma_Y}\right)^2 = \frac{1}{2}\left(\frac{e}{e_Y}\right)^2. \tag{2}$$

Thus equation (1a) has a factor of safety of 2 on defect size based on the plane stress equivalence between δ, K and G at low stresses.

METHOD OF APPLICATION

For situations where the effective ratio of defect size to plate width $\frac{\bar{a}}{W}$ is less than about 0.1 and where the nominal design stress σ is less than the yield stress of the base materials, Dawes (Dawes, 1974) proposed that equation (1) could be re-written in terms of stress as follows:

$$\bar{a}_{max} = \frac{\delta_c E \sigma_Y}{2\pi \sigma_1^2} \quad \text{for} \quad \frac{\sigma_1}{\sigma_Y} \leq 0.5 \tag{3a}$$

$$\bar{a}_{max} = \frac{\delta_c E}{2\pi (\sigma_1 - 0.25\sigma_Y)} \quad \text{for} \quad \frac{\sigma_1}{\sigma_Y} \geq 0.5 \tag{3b}$$

σ_1 is the total pseudo-elastic stress in the vicinity of the defect. Whilst σ_1 may be above yield the structure itself may still behave in a predominantly elastic manner. This is because the yielding in the zone under consideration is contained by the surrounding elastic material. In welded structures contained yield occurs as a result of residual stresses which may themselves be equal to the yield stress and may be additive to the applied stress. Contained yielding also occurs at stress concentrations where pseudo-elastic stresses may be well above yield.

For general applications the following values of σ_1 were suggested:

Crack location	Weld condition	σ_1
Remote from stress concentrations	Stress relieved[1]	σ
	As welded	$\sigma + \sigma_Y$ [2]
Adjacent to stress concentrations	Stress relieved[1]	SCF x σ
	As welded	(SCF x σ) + σ_Y [2]

Part through surface and buried defects

The design curve was originally formulated on the basis of through thickness defects. Dawes (Dawes, 1974) suggested that part through cracks could be dealt with by assuming that, for contained yielding situations, the parameters governing flaw shape effects would be the same as those under linear elastic conditions. It was realised that this approach could not be justified rigorously, but it seems unlikely that elastic plastic solutions for the part through crack will be available for some time to come. The following LEFM expression was used to describe a semi-elliptic surface crack.

$$K_I = \frac{M_t M_s \sigma \sqrt{\pi a}}{\Phi_2} \tag{4}$$

For the equivalent through thickness crack of length $2\bar{a}$, $K_I = \sigma\sqrt{\pi\bar{a}}$.

---

[1] Here it is assumed that postweld heat treatment (PWHT) has eliminated all the residual stresses. Often this will not be so and it is prudent to make some allowance for the residual stress remaining after PWHT.

[2] It has been assumed that residual stress of yield point magnitude will exist in as-welded structures. Whilst this is true for stresses along the weld, transverse residual stresses can often be assumed to be lower than yield in specific cases.

Thus

$$\frac{\bar{a}}{B} = \frac{a}{B}\left(\frac{M_t M_s}{\Phi_2}\right)^2 \tag{5}$$

values of $\left(\frac{M_t M_s}{\Phi_2}\right) = f\left(\frac{a}{B}, \frac{a}{2c}\right)$ were taken from a survey by Maddox (Maddox, 1975) and $\frac{a}{B}$ is plotted against $\frac{\bar{a}}{B}$ in Fig. 1.

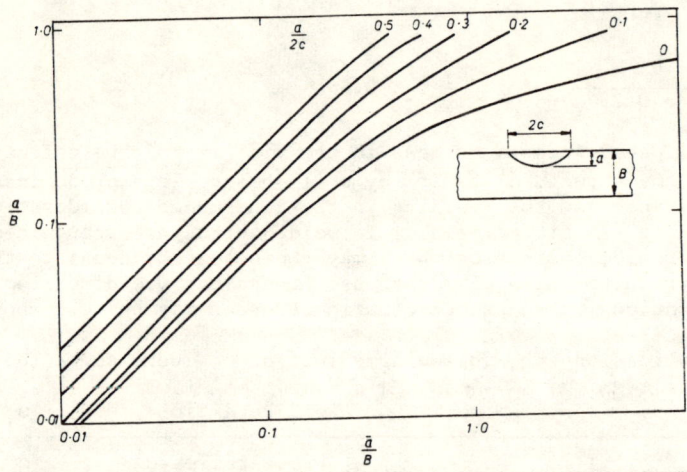

Fig. 1. Relationships between surface crack dimensions and equivalent through-thickness crack dimension $\bar{a}$.

Figure 1 agrees closely with formulae proposed more recently by Newman (Newman, 1977).

For buried elliptical cracks the equivalent equation to (5) is:

$$\frac{\bar{a}}{B} = \frac{a}{2(p+a)}\left(\frac{M}{\Phi_2}\right)^2. \tag{6}$$

M was calculated as the product of the magnification factors M_π and M_o applicable to the stress intensity factor at the end of the minor axis which approaches nearest to a free surface. M_o is the magnification factor at that point due to the presence of the near surface and M_π is that due to the presence of the more remote free surface. M_π and M_o were taken from the work of Shah and Kobayashi (Shah, 1973). For $a/c = 0$, M was derived from Feddersen's relationship (Feddersen, 1967) $M = \left(\sec\frac{\pi a}{B}\right)^{\frac{1}{2}}$.

Figure 2 is a plot of $\frac{a}{p+a}$ against $\frac{\bar{a}}{2(p+a)}$ or $\frac{a}{B}$ for buried defects.

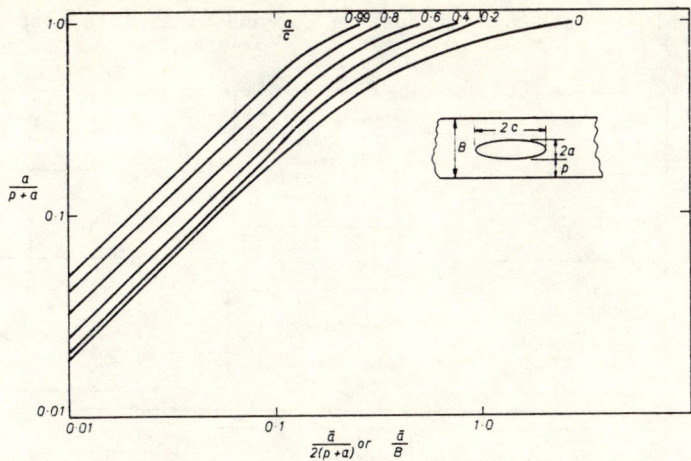

Fig. 2. Relationships between buried crack dimensions and equivalent through-thickness crack dimension $\bar{a}$.

RECENT EXPERIMENTAL JUSTIFICATION FOR THE DESIGN CURVE

The implementation of the COD approach through the simple design curve, which takes into consideration the effects of residual stresses and geometric stress concentrations, has found wide application to welded structures. However, because of its semi-empirical origins and inherent simplicity, the design curve, as already stated, predicts maximum allowable crack sizes and not critical crack sizes, with a margin of safety only <u>vaguely estimated</u> as being approximately 2.0. It was decided therefore, to carry out an assessment of the COD design curve, by making a comparison of the allowable crack sizes predicted by the small scale COD tests and the critical crack sizes at fracture in wide plate tests (Kamath, 1978). From a survey of the published literature and work carried out at The Welding Institute, a total of 73 sets of small and large scale tests were compiled. The results were then analysed on a statistical basis. The main steps and observations from these analyses are summarised below.

Initially the test data was processed, as shown in Table 1, to give safety factors $\bar{s}$ and s, for through thickness and surface cracks, respectively. A comparison between the predicted allowable and the critical crack sizes obtained is shown in Fig. 3, with some of the important probability levels indicated. This shows that, on average, the design curve has a built-in factor of safety of approximately 2.5, and the maximum allowable size derived from the curve implies 95.4% confidence in survival. However, when the scatter in results is taken into consideration, there appears to be little scope for modifying the shape of the design curve.

An examination of the variables in the wide plate tests drew attention to the influence of residual stresses on brittle fracture. For situations where it was reasonable to assume no residual stresses, e.g. plain plate and some stress relieved weldments only, Fig. 4a shows that critical values of Φ, were generally well below the design curve. However, when the results for as-welded plates are added to the plot and if residual stresses are still assumed to be zero (Fig. 4b) it can be seen that a significant proportion of the as-welded plate specimen results fall to the left of the design curve. In other words, the failure stress assumed was lower than that to be expected from the design curve for the known value of Φ. The as-welded wide plate results fall within the same general scatter band as those for plates assumed to be free from residual stress, and were thus safely predicted by

Table 1. Method of processing COD and wide plate test data.

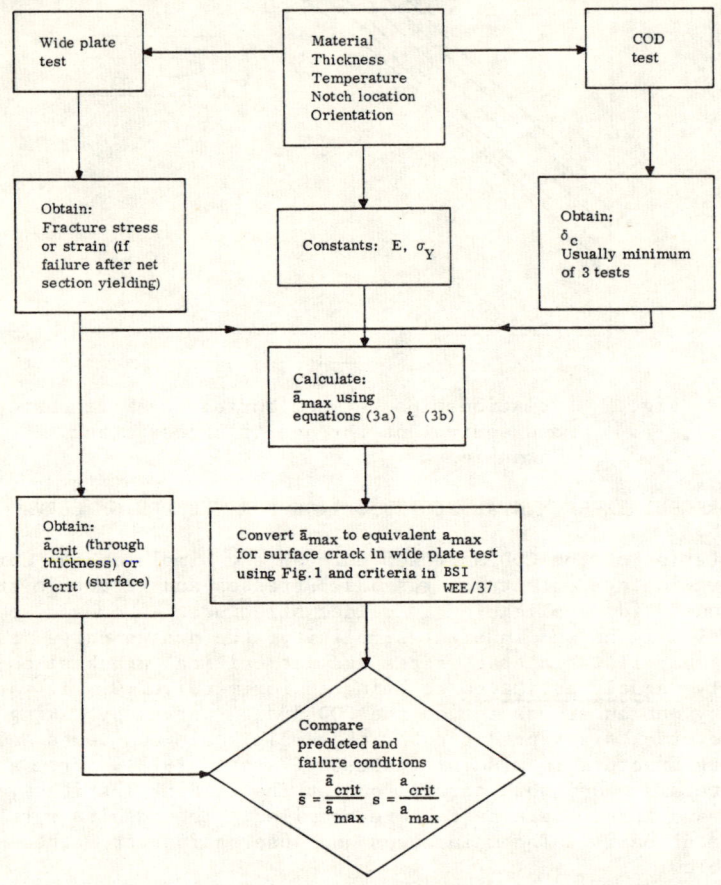

the design curve only when full yield residual stresses were assumed to be active (Fig. 4c).

The assessment also showed that when residual stresses were present there were no significant differences in the average factors of safety for through thickness and surface cracks. However, in the absence of residual stresses, the results suggested that the factors of safety were slightly higher in the case of surface cracks.

NUMERICAL ASSESSMENTS OF THE DESIGN CURVE

Sumpter (Sumpter, 1973) and Sumpter and Turner (Sumpter and Turner, 1973) report the results of elastic-plastic finite element analyses. Some of these were for an elastic perfectly plastic material, but some assume a work hardening law approximately equal to that for A533B pressure vessel steels. The latter were compared by Sumpter with the COD design curve. The following geometries were studied:
Edge cracked plate $a/W = 0.1$.
Crack at the edge of a hole of radius R, $a/R = 0.05$, 0.1 and 1.0
Radial crack at the bore of a pressurised cylinder of thickness T, $a/T = 0.03$.

The COD Approach

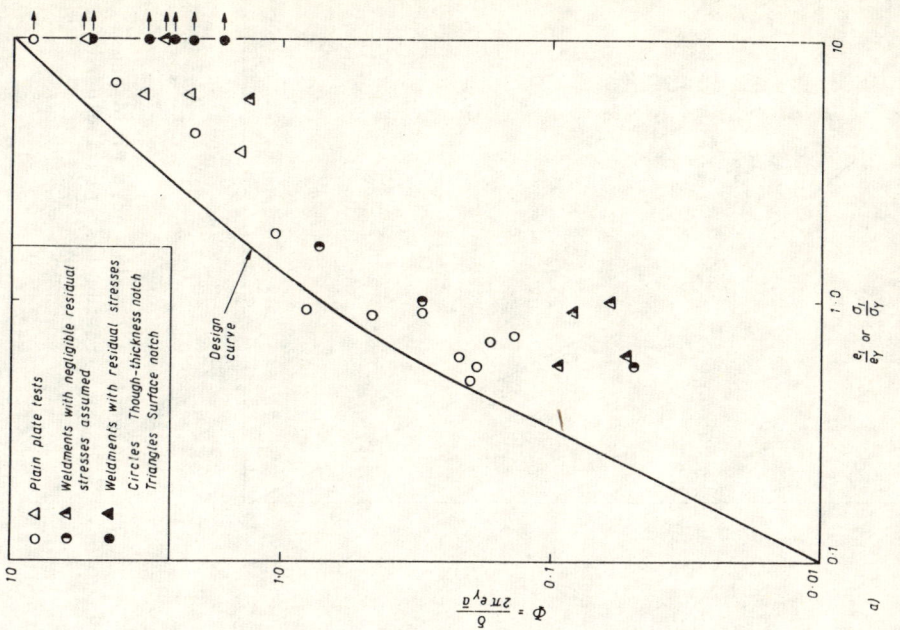

Fig. 4a. Design curve relationships between non-dimensional COD and applied strain or stress (normalised), with experimental results for a) plain plate and weldments with negligible residual stress effects. eg. stress relieved welds

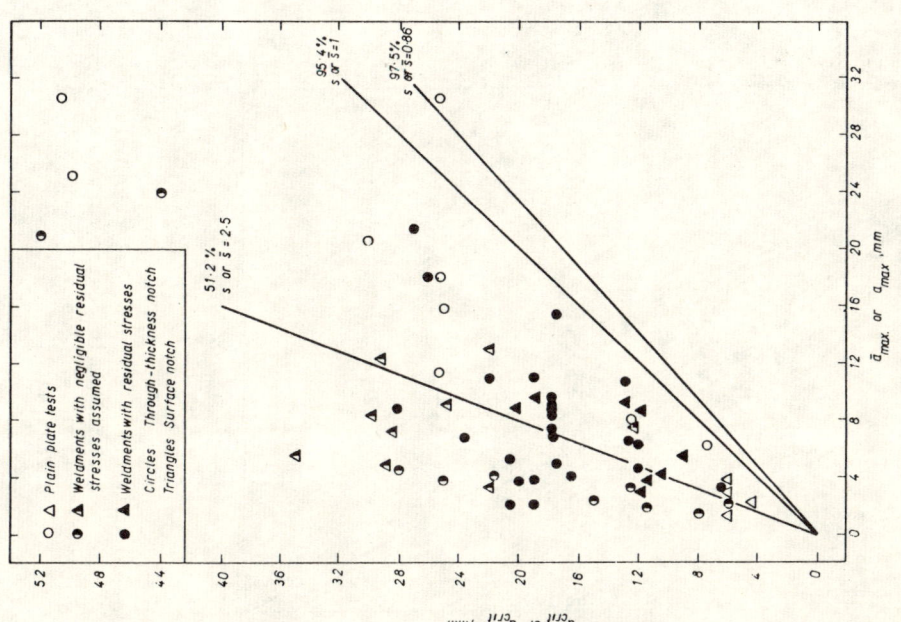

Fig. 3. Comparison of critical and maximum allowable crack sizes showing probability levels and safety ratios.

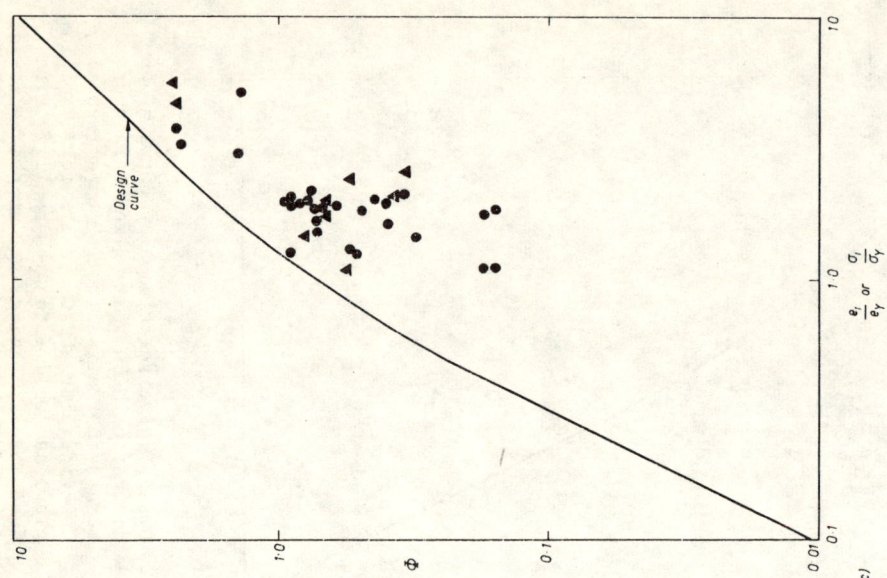

Fig. 4c)cont. All weldments: yield magnitude residual stress included.

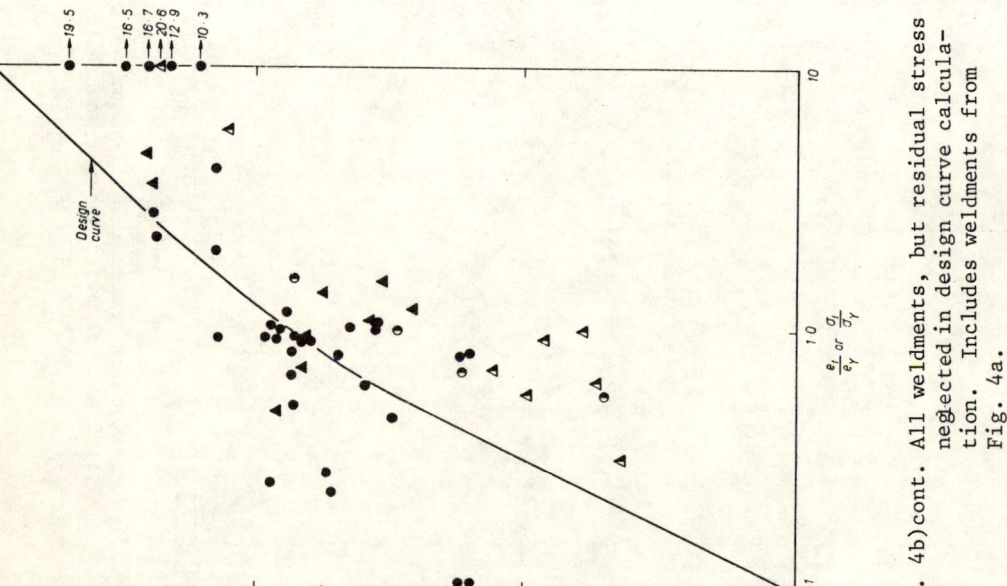

Fig. 4b)cont. All weldments, but residual stress neglected in design curve calculation. Includes weldments from Fig. 4a.

The results are plotted in Fig. 5. For the edge cracked plate there was close agreement between the finite element analysis results and the design curve.

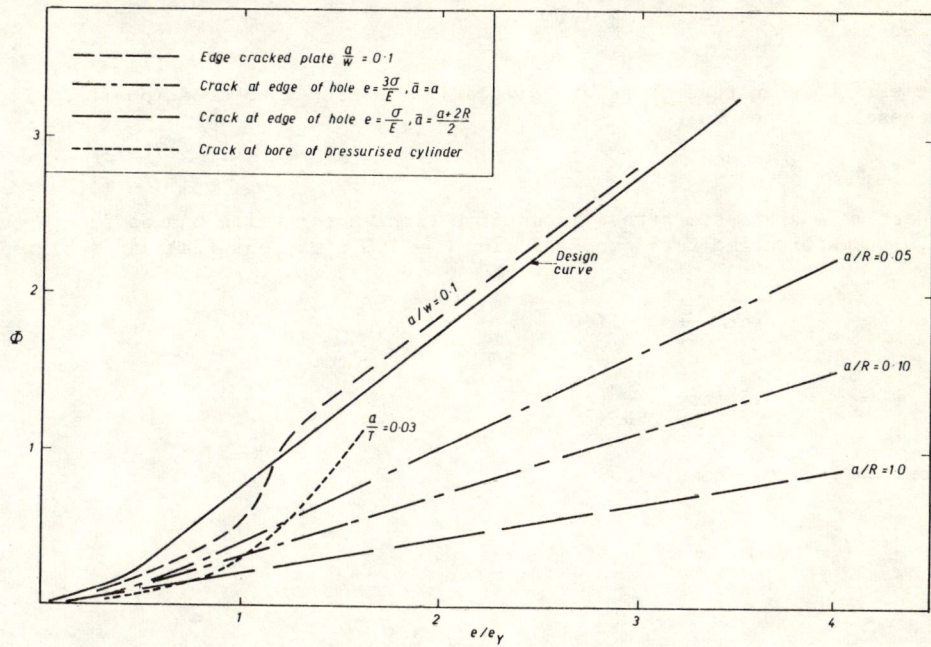

Fig. 5. Comparison between Sumpter's finite element results (Sumpter, 1973) and the COD design curve.

For cracks at the edge of a hole, the design curve was shown to be conservative, but not excessively so for ratios of crack length to hole radius, a/R, of 0.05 and 0.10. As stated in the section dealing with the application of the design curve the recommended procedure is to calculate the value of σ_1 as SCF x σ, Sumpter shows that this procedure becomes very pessimistic for the unusual case of very long cracks at the edge of a hole with a/R = 1.0. However, as Burdekin and Dawes (Burdekin and Dawes, 1971) originally suggested, it is more realistic for a/R > 0.2, to assume that the crack is one of total length a + 2R in a stress field equal to the membrane stress. If this procedure is adopted for the results for a/R = 1.0 the plot of Φ against e/e_Y comes closer to that for a/R = 0.1, but remains very conservative. The mean factor of safety on Φ between the results for a/R = 0.05, and 0.1 and the design curve for a given value of e/e_Y is 2.0.

For the radially cracked cylinder a comparison was made with the design curve for one ratio of crack depth to cylinder wall thickness, a/T = 0.03. The design curve was again found to be conservative with factors of safety on Φ of 4.5 at $e/e_Y = 0.6$, 2.5 at $e/e_Y = 1.0$ and 1.2 at $e/e_Y = 1.6$.

THE J DESIGN CURVE

Begley (Begley, Landes and Wilson, 1974) have proposed a J design curve which is in essence very similar to the COD design curve. The similarity between the two approaches has been discussed by Merkle (Merkle, 1976). The design curve of Begley, 1974, takes the form:

$$\frac{J}{E\pi \bar{a} e_Y^2} = \left(\frac{e}{e_Y}\right)^2 \text{ for } \frac{e}{e_Y} \leq 1.0 \tag{7a}$$

and

$$\frac{J}{E\pi\bar{a}e_Y^2} = \frac{2e}{e_Y} - 1 \text{ for } \frac{e}{e_Y} \geq 1.0. \tag{7b}$$

It will be seen that these are similar in character to equation (1). It is generally stated that

$$J = m \sigma_Y \delta$$

where m is a plastic stress intensification factor which ranges from about 1.0 to 2.0. Substituting for J and assuming m = 1.0 equations (7a) and (7b) reduce to:

$$\Phi = \frac{1}{2}\left(\frac{e}{e_Y}\right)^2 \text{ for } \frac{e}{e_Y} \leq 1.0 \tag{8a}$$

$$\Phi = \frac{e}{e_Y} - 0.5 \text{ for } \frac{e}{e_Y} \geq 1.0. \tag{8b}$$

For m = 2.0 they reduce to

$$\Phi = \frac{1}{4}\left(\frac{e}{e_Y}\right)^2 \text{ for } \frac{e}{e_Y} \leq 1.0 \tag{9a}$$

and

$$\Phi = \frac{1}{2}\frac{e}{e_Y} - 0.25 \text{ for } \frac{e}{e_Y} \geq 1.0 \tag{9b}$$

These are plotted for comparison with the design curve in Fig. 6.

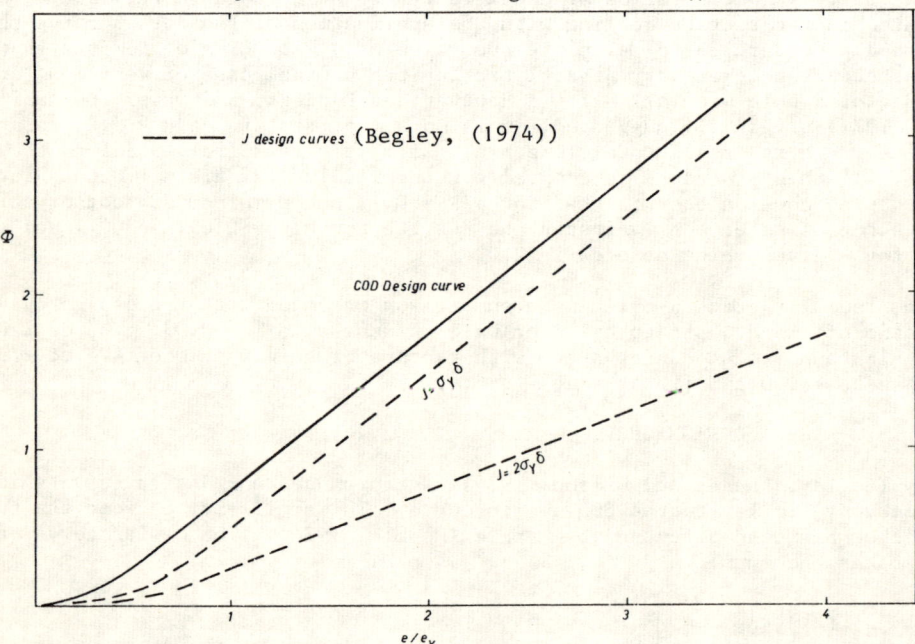

Fig. 6. Comparison between the COD design curve and curves based on (Begley, 1974)

It should be borne in mind that, whilst the COD design curve is intended to be conservative and is empirically based on the results of wide plate tests on specimens where it may be assumed that m varied, the curve of Begley and colleagues is intended to predict critical conditions. Viewed in this light it is felt that the two approaches are in reasonable agreement.

EXPERIENCE IN THE PRACTICAL APPLICATION OF THE COD DESIGN CURVE

The design curve approach has been applied in recent years to a great range of structures in a variety of contexts. The applications may be broadly classified into four major groups:

1. Material selection (design stress and defect levels predetermined).
2. Acceptance levels for weld defects (decisions re: known defects or fixing defect acceptance standards where material and design stress are predetermined).
3. Fixing allowable stress (material and inspection level predetermined).
4. Failure analysis.

Lists of some of the applications with which the authors have been concerned over the last five years are given in Tables 2 - 6. These are by no means exhaustive. Some specific examples are discussed in greater detail below.

TABLE 2   Welding consumable manufacturers for whom COD tests have been carried out with the objective of meeting specific requirements.

Arcos	BOC Murex	ESAB
Big 3 Lincoln UK	GKN Lincoln	Kobe Steel
Lincoln Electric	West Falische Union	Varios Fabrieken
Metrode	Oerlikon	Philips

Material Selection

If the design stress is known and a size of defect which might escape detection by non-destructive testing (NDT) is assumed, the design curve may be used to determine the required level of toughness. Table 2 lists welded consumable manufacturers whose products have been tested by The Welding Institute in order to establish whether toughness levels fixed in this way have been achieved. Table 3 lists a range of structures where this approach has been used. Two specific examples are as follows.

Offshore production platforms. The production platforms for BP's Forties Field in the North Sea involved a quantum jump in size over the great majority of similar structures. Because of the greater depths in the North Sea and the severity of the environment, structural steels of higher strength ($320 N/mm^2$ yield, $500 N/mm^2$ tensile strength) and increased thickness (60-100mm) were employed. Would such structures be safe if the welds in stress concentration regions at the massive intersections (nodes) remained in the as-welded conditions, or should they be PWHT, as would be mandatory for pressure vessels built in similar materials? It was assumed that surface and buried defects up to 12.5mm and 25mm deep respectively might escape detection by NDT in these complex structures. Photo-elastic analysis of a typical node using the frozen stress method indicated a maximum SCF of 8 and the nominal design stress was $75 N/mm^2$ or approximately $1/4\ \sigma_Y$. The required COD value determined from the design curve is given by equation (3b) as:

$$\delta_c = \frac{2\pi \bar{a}_{max}}{E}(\sigma_1 - 0.25\ \sigma_Y).$$

Hence, if the nodes are as-welded:

$$\sigma_1 = (\text{SCF} \times \sigma) + \sigma_R = (8 \times \tfrac{1}{4}\sigma_Y) + \sigma_Y = 3\sigma_Y$$

and substituting for σ_Y and $\bar{a}_{max}$ gives:

$$\delta_c = 0.37\text{mm}$$

Crack opening displacement tests were carried out at the design temperature of $-10°C$ on the parent steel, BS 4360 Grade 50D, which gave a minimum $\delta_m = 0.49$mm but the best weld metal, out of the total of 17 tested, only gave $\delta_c = 0.12$mm.

It was clear that it would be necessary to heat treat the nodes in order to obtain the required defect tolerance. If after PWHT σ_R is assumed to be zero, σ_1 becomes $8 \times 1/4\ \sigma_Y = 2\sigma_Y$. This gives a required COD of:

$$\delta_c = 0.24\text{mm}.$$

Not only did PWHT lower the required COD, but it significantly increased the toughness of the weld metal. Five weld metals were found giving satisfactory toughness and one had a minimum value of δ_m of 0.49mm.

This study (which has been described more fully elsewhere (Harrison, 1977)) showed quite clearly that it was necessary to heat treat the node regions to ensure safety of the complete structures. This decision had a marked effect on the design philosophy adopted.

<u>Specification of toughness of girth welds in large diameter pipeline for service in Arctic regions</u>. Normally girth welds are not highly stressed in the longitudinal direction and hence fracture risks are very low. When the line goes through areas liable to subsidence and/or earthquakes, however, high longitudinal bending stresses can develop. In this particular case, strains up to 0.5% due to the above causes were envisaged which meant that weldments with good fracture resistance were needed, particularly in view of the associated low ambient temperatures (down to $-40°C$). Semi-automatic and manual welding processes were considered in conjunction with pipe to API 5LX70 and 19mm wall thickness. The fracture resistance of the various regions of the heat affected zones (HAZs) and weld metals were thoroughly examined by the COD test at the minimum service and other selected temperatures on samples taken from welds made under field conditions. These tests demonstrated clearly that the fracture toughness of the HAZs was adequate at all temperatures, but that there could be difficulties in the weld metals, both manual and semi-automatic. In order to judge the validity of the maximum allowable flaw sizes predicted from the COD results a series of full size bend tests were carried out in a specially made rig. The girth welds in the pipe lengths contained crack-like defects of suitable size in the centre of the weld deposits which were made under typical field conditions. The weld area was placed at the position of maximum bending moment in the rig and cooled to the required temperature. Load was slowly applied up to failure and the maximum applied strain measured by electric resistance strain gauges. After failure the depth of the actual defect a_{cr} was measured and compared with the predicted maximum allowable depth a_{max}. In calculations, it was assumed that the peak tensile residual stress level transverse to the girth welds would be between 0.5-$0.75 e_Y$. Using these values of residual stress, the minimum COD values from the small scale tests, the measured applied strain values and equations (3b)

and (5), ratios of a_{cr}/a_{max} between 2 and 3 were obtained, which is consistent with the normal experience from wide plate tests as indicated in Fig. 3.

TABLE 3

Cases where the COD design curve has been used as a basis for material selection

Class of structure	Company	Project
Pipelines	Aramco	NGL pipeline
	Aramco	Spiral weld gas pipeline
	Bechtel International	Offshore gas pipeline
	BP	Crude oil pipeline
	BP	Automatic MIG girth welds for Ninian and Forties field lines
	Brown and Root	Offshore flare line
	Brown and Root	Weld overlay on offshore riser pipe
	Canadian Arctic Gas	Natural gas pipeline
	Conoco	Offshore gas pipeline
	Hoesch	Pipeline steel
	Gasunie	Thick walled pipeline
	Shell	Oil pipeline girth welds
Offshore structures	BP	Selection of weld procedures and decision re: postweld heat treatment
	Conoco	Crane pedestals
	Highlands Fabricators	Production platform
	Laing Pipelines Offshore	Production platform
	McDermott	Production platform
	Philips Petroleum	General specification for offshore structures
	Redpath Dorman Long	Production platform
	Shell	General specification for production platforms
	Shell	Jack-up rig
Pressure vessels and boilers	Air Products	Cryogenic vessels (9% Ni steels)
	Clarke Chapman	Boiler drum (decision re: postweld heat treatment)
	International Combustion	Boiler drum
Nuclear	Central Electricity Generating Board	Stainless steel weld for AGR boiler
	Nuclear Power Company	Steam drum SGHWR
	The Nuclear Power Group	Circulator outlet gas duct/liner weld in AGR
Miscellaneous	Allis Chalmers	General purposes
	Aramco	Plates for low temperature service
	Ove Arup	Cast steel weldments for building frame
	Capper Neil	LPG storage tanks
	Capper Neil	Oil storage tanks
	Central Electricity Generating Board	Penstocks for pumped storage scheme
	Chicago Bridge and Iron	LNG storage tanks
	High Duty Alloys	Crash barriers
	High Duty Alloys	Repair procedures for 12 000 tonne extrusion press
	Johns and Waygood	High rise building
	Kockums	High heat input welding for ships
	Lindsey Oil Refineries	Oil storage tanks
	Uddeholms	9% Ni steel for LNG tanks
	Whessoe	5m low speed wind tunnel
	Whessoe	Oil storage tanks

TABLE 4 Cases where the COD design curve has been used to fix acceptance levels for weld defects and inspection sensitivity.

Company	Project
Clarke Chapman	Electroslag welds in boiler drum
Central Electricity Generating Board	Oil Storage tanks
Elliott	Compressor rotor
Gasunie	Gas pipelines
Shell	Oil pipelines
Shell	Offshore pressure vessels

TABLE 5 Assessment of weld defects

Class	Company	Project
Pipelines	BP	Alyeska pipeline
	BP	Refinery pipes
	BP	Ethelyne pipeline
	Danish Welding Institute	Spiral welded oil products line
	Gasunie	Gas pipelines
	Metallurgical Consultants	Spiral welded pipe
	Unit Inspection	Spiral welded pipe
Offshore Structures, etc.	Conoco	Offshore crane pedestals
	Occidental	Deck modules
	Shell	Flash welds in anchor chain links
Pressure vessels and boilers, etc.	British Gas	Gas bullets
	Clarke Chapman	Steam and mud drums
	ICI	Hydro desulphuriser
	Richards and Ross	Nozzle welds
	Whessoe	Pressure vessels
Miscellaneous	Colocotronis	Supertanker
	Dorman Long (South Africa)	Pumped storage scheme penstocks
	Cleveland Potash	Mine shaft lining

TABLE 6 Failure investigations

Company	Structure type
Bechtel International	Pile for offshore platform
BP	Oil tankers
Conoco	Offshore gas pipeline
Conoco	Rolled beams for deck module
Noble Denton	Leg for offshore structure
Shell	Pile for offshore platform
South of Scotland Electricity Board	Steam riser pipe
South of Scotland Electricity Board	Feedwater header forging

Acceptance levels for weld defects

Table 4 lists a number of cases where the design curve has been used to fix NDT requirements at an early stage in design or construction. Table 5 shows cases where the concept has been used to assess the significance of known defects where repair was felt to be undesirable for reasons of cost, delivery or because of the possibility of introducing more harmful defects in the course of the repair.

Defects in the Alyeska pipeline. As a result of disclosures to the press of falsification of radiographs of the girth welds in the Alyeska crude oil pipeline, all the X-ray films for the part of the line completed at that stage were re-examined. This audit indicated that that were defects larger than those permitted by the construction code, API-1104, in some 2955 of the 30 000 welds audited. The defect acceptance levels in API-1104 are set in order to maintain a certain level of workmanship and bear no relationship to the performance of pipelines in service. Nevertheless, the code had been adopted by the Department of Transportation (DOT) as a basis for licensing the pipeline. The pipeline company through BP asked The Welding Institute to help to develop a submission to DOT for waivers to the code. This case was based on the design curve and in terms of the numbers of specimens tested (some 450) represented the largest single case of its application.

The effects of weld procedure (three different procedures), position around pipe circumference and notch orientation were studied. Nine notch positions and orientations with respect to the weld metal and HAZ were investigated. Tests were carried out at $0°$, $-12°$, $-18°$, $-29°$ and $-40°C$. In fact $-12°C$ was chosen as the design basis to allow for the possibility of cold pressurisation during start up. Because of the considerable variety of microstructures sampled there was wide scatter in the COD values, but a lower bound of 0.1mm was used. At $-40°C$ one specimen gave a result as low as 0.025mm. Although these COD values are relatively low, no specimen gave a result which could possibly be interpreted as a valid K_{Ic} value for the specimen thickness of 12.5mm. The stresses considered in the analysis included pressure, thermal, bending due to expansion and self weight, earthquake, residual, etc. It was found as a result of the analysis that none of the defects required repair.

The National Bureau of Standards carried out an independent assessment (Berger and Smith, 1976), in which an analysis is reported based on a different interpretation of the COD test. This suggested that the Alyeska proposals were conservative by factor of about 1.2 to 2.0. However, it is believed that this approach is aimed at predicting critical conditions whilst the design curve already incorporates a factor of safety of 2.5 on the best estimate for criticality. Thus the two approaches seem to be in good agreement.

The DOT accepted the case for waivers in principle, but asked for a further safety margin of 2 to allow for problems of X-ray interpretation. On this basis a small number of repairs were required.

In fact, many of the defects were repaired because the negotiations of the case were time consuming and the pipeline company could not afford to waste this amount of time when the proposal might have been rejected. The important point, however, is that a major US government authority, i.e. the DOT, accepted the principle of using a yielding fracture mechanics analysis to derive defect acceptance levels in a large pipeline project. Furthermore, some defects in a section of pipe crossing the Koyukuk river, which would have cost about five million dollars to repair, were accepted.

Fixing allowable stresses

The use of fracture mechanics to fix design stresses is less common, but one case involving LNG storage tanks can be cited.

A programme of COD and wide plate tests was carried out on weldments in 9% nickel steel 18mm thick mostly at $-164^\circ C$, which was the minimum design temperature. The main objective of the work was to see if the design stress level could be safely raised from the API 620Q value of 196.5 to $290N/mm^2$. Two plate materials, A553 and A353 and two suitable weld metals were examined. The COD tests showed that all the materials were fully ductile at $-164^\circ C$. Values of $\bar{a}_{max}$ between 7 and 19mm were obtained for a design stress of $290N/mm^2$ using COD values at maximum load in conjunction with equations (3b) and (5) and making conservative assumptions about effects of distortion and residual stresses. Through thickness defects 20 - 25mm long were incorporated in various weld regions in wide plate specimens and tested at $-164^\circ C$. The fact that all the plates failed at stresses above the yield stress of the weld metal indicated that the approach was very conservative. The work was carried out about six years ago and in view of the complete ductility of the materials involved it is now thought that a limit load approach, as suggested in the introduction, would be more appropriate and less conservative. Nevertheless, it was clearly demonstrated that an increase in design stress level to $290N/mm^2$ was reasonable in terms of tolerance to severe fabrication defects. This resulted in a significant reduction in the cost of the LNG tanks.

Failure investigations

A number of instances where structural failure conditions have been compared with predictions from the design curve are listed in Table 6. These represent some of the most interesting applications of the design curve, but space permits only one example to be discussed in more detail. The example chosen is the Cockenzie boiler drum. This has already been discussed some time ago by Burdekin and Dawes, (Burdekin and Dawes, 1971) and by Ham (Ham, 1971), but it may now be reassessed in the light of the revised design curve and of the defect shape corrections in Fig. 1.

The failure during hydrostatic test at nominal stress level of $0.55\sigma_Y$ occurred from a large semi-elliptical surface defect 81mm deep x 325mm long at the edge of an attachment. The material, which was a low alloy structural steel had a thickness of 141mm, a yield stress of $376N/mm^2$ and minimum COD value of 0.25mm in the stress relieved condition.

Substitution of the above values into the design curve gives $\bar{a}_{max}$ = 74mm, i.e. $\bar{a}_{max}/B$ = 0.52. From Fig. 1, a_{max}/B = 0.49, giving a maximum depth of 69mm for a surface defect with aspect ratio 0.25.

The factor of safety in this case is 1.17. This is smaller than usual, but it could be influenced by two factors which were ignored in the analysis, both of which tend to increase it. Firstly, residual stress was assumed to be zero. It is probable, however, that some residual stress will have remained since the material was thick and the geometry was complex. Secondly, the defect was close to a nozzle and the local stress may have been elevated because of this. The details available are not sufficient for either of these possible effects to be assessed, however, use of the design curve still gives a reasonable explanation of the failure.

CONCLUSIONS

The COD test is a useful method of studying the fracture toughness of materials in the transition region between linear elastic behaviour, where K_{Ic} should be used and fully ductile behaviour, where a limit load approach is appropriate. It was concluded from the statistical analysis of 73 sets of tests where predictions from the design curve were compared with the results of large scale tests that the average inherent safety factor is approximately 2.5. The design curve represents a confidence level of 95% in survival for the wide plate tests. It was shown that the approach described is comparable with design curves derived from finite element analyses and J analyses. It was concluded from the several practical examples described that the design curve can be successfully used in at least three different ways. These are

i)   selection of materials during initial design stage
ii)   specification of maximum allowable flaw sizes at design or after fabrication to establish the necessity for repairs
iii)   failure analysis

NOMENCLATURE

a	Depth of surface crack or half height of buried crack
a_{cr}	Critical value of 'a' for unstable fracture
a_{max}	Maximum allowable value of 'a'
$\bar{a}$	Half length of through thickness rectilinear crack
$\bar{a}_{cr}$	Critical value of $\bar{a}$ for unstable fracture
$\bar{a}_{max}$	Maximum allowable value of $\bar{a}$
B	Section or specimen thickness
c	Half length of buried or surface crack
E	Young's modulus
e	Strain
e_Y	Yield strain = σ_Y/E
G_I	Mode I crack extension force
J	The J contour integral
K_I	Mode I stress intensity factor
K_{Ic}	Critical plane strain stress intensity factor
L	Plastic constraint factor
m	Plastic stress intensification factor
M_o, M_π	Correction factors for buried cracks at the edge of the crack nearest to a free surface due to that free surface and due to the remote free surface
M	Correction factor for buried cracks = $M_o \times M_\pi$
M_t	Finite thickness correction factor for surface cracks
M_s	Free surface correction factor for surface cracks
p	Depth below surface of buried defects
R	Radius of holes
T	Wall thickness of cylindrical vessels
s	$\dfrac{a_{crit}}{a_{max}}$
$\bar{s}$	$\dfrac{\bar{a}_{crit}}{\bar{a}_{max}}$

W   Half width of centre-cracked and double edge notched specimens
δ Crack tip opening displacement (COD)
δ_c Critical value of δ
δ_m δ at first attainment of maximum load plateau in bend test
π Constant $\simeq 3.142$
σ Applied stress
σ_1 Effective stress = $(\sigma \times SCF) + \sigma_R$.
σ_R Residual stress
σ_Y Uniaxial yield stress

Φ Non-dimensional COD = $\dfrac{\delta E}{2\pi \sigma_Y \bar{a}}$

Φ_2 Complete elliptic integral of second kind

SCF   Elastic stress concentration, or where localised uncontained yielding occurs, the strain concentration factor.

REFERENCES

Begley, J. A., J. D. Landes, and W. K. Wilson (1974). An estimation model for the application of the J integral. ASTM-STP-560.

Berger, J. and J. H. Smith (Eds.), (1976). Consideration of fracture mechanics analysis and defect dimension measurement assessment for the Trans-Alaska oil pipeline girth welds. National Tech. Infor. Service Report PB-260-400.

Burdekin, F. M., and D. E. W. Stone (1966). The crack opening displacement approach to fracture mechanics in yielding materials. J. of Strain Analysis, 1, No.2, 194.

Burdekin, F. M., and M. G. Dawes (1971). Practical use of linear elastic and yielding fracture mechanics with particular reference to pressure vessels. Conference on Application of Fracture Mechanics to Pressure Vessel Technology. Inst. of Mech. Eng. May 1971.

Dawes, M. G. (1974). Fracture control in high yield strength weldments. Weld. J. Res. Suppl. 53, 369s.

Dawes, M. G. (1977). Elastic plastic fracture toughness based on the COD and J integral concepts. ASTM Symposium on Elastic-Plastic Fracture, Atlanta, USA., Nov. '77. The Welding Institute Members Report 54/1978/E.

Dawes, M. G., and M. S. Kamath (1978). The crack opening displacement (COD) design curve approach to crack tolerance. Conference on the Tolerance of Flaws in Pressurised Components. Inst. of Mech. Eng. May 1978.

Feddersen, C. E. (1967). Discussion to plane strain fracture toughness testing. ASTM-STP-410, 77.

Ham, W. M. (1971). Discussion to Conference on Practical Application of Fracture Mechanics to Pressure Vessel Technology, Inst. of Mech. Eng. London, May 1971.

Harrison, J. D., F. M. Burdekin, and J. G. Young (1968). A proposed acceptance standard for weld defects based upon suitability for service. 2nd Conf. on Significance of Defects in Welded Structures, Welding Institute, London.

Harrison, J. D., and W. P. Carter (1973). The use of 9% Ni steel for LNG application. Conference on Welding Low Temperature Containment Plant, The Welding Institute.

Harrison, J. D. (1977). The effect of postweld heat treatment on the toughness of welds for an offshore platform. Performance of Offshore Structures, Pub. Instn. of Metallurgists, London.

Kamath, M. S. (1978). The COD design curve: An assessment of validity using wide plate tests. Welding Institute Members Report.

Maddox, S. J. (1975). An analysis of fatigue cracks in fillet welded joints. Int. J. Fract. 11, No. 2, 221-243.

Merkle, J. G. (1976). Analytical relations between elastic-plastic fracture criteria. Int. J. of Pressure Vessels and Piping, 4. No. 3, 197-206.

Newman, J. C. (1977). ASTM Symposium on Part Through Cracks Life Prediction, San Diego, October 1977. ASTM.

Shah, R. C., and A. S. Kobayashi (1973). Stress intensity factors for an elliptical crack approaching the surface in a semi-infinite solid. Int. J. of Fract. Mech. 9, No. 2, 133.

Sumpter, J. D. G. (1973). Elastic-plastic fracture analysis and design using the finite element method. PhD Thesis, University of London, December 1973.

Sumpter, J. D. G., and C. E. Turner (1973). Fracture analysis in areas of high nominal strain. 2nd International Conference on Pressure Vessel Technology, San Antonio, October 1973.

Draft British Standard rules for the Derivation of Acceptance Levels for Defects in Fusion Welded Joints. BSI WEE/37 Doc. 75/77081 D.C. February 1976. The British Standards Institution.

SIMPLIFIED STRESS INTENSITY EVALUATION OF A NUCLEAR REACTOR PRESSURE VESSEL UNDER A GIVEN ACCIDENT LOADING

W. van der Walt

Licensing Branch, Atomic Energy Board, Pretoria, R.S.A.

ABSTRACT

An introduction is given to the Licensing philosophy applied to nuclear reactors in South Africa. It is followed by a brief description of a typical nuclear pressure vessel and a discussion of three simplified methods for performing a design frac= ture analysis of a thickwalled pressure vessel is given. These methods are suffi= ciently general to be applied in normal industrial applications. Finally, compara= tive results for different loading conditions are given.

INTRODUCTION

The approach adopted by the S A Atomic Energy Board's Licensing Branch in the assessment of the licensability of nuclear power stations to be erected in South Africa is based on a risk approach (Simpson, Winkler and Tattersall, 1974; Tattersall, 1977). This type of approach is based on concepts suggested by Farmer (1967), and probably the most comprehensive nuclear safety assessment based on a risk analysis published to date is the "Rasmussen Study", WASH 1400 (USNRC, 1975).

The application of this method consists of the establishment of a probability based acceptance criterion and the subsequent calculation of the probability of a given failure occurring. In very general terms, the probability of a given release of fission products is then compared with the allowable probability of release as defined in the criterion in order to establish acceptability.

In practical terms this implies that the failure modes for each system and compo= nent that could lead to a release of radioactivity if it should fail, has to be established. This is usually accomplished for systems by establishing a fault tree for the systems concerned. The theory and application of fault tree analyses fall outside the scope of this paper, and are extensively described in literature, e.g. WASH 1400 (USNRC, 1975).

In the case of a component, however, the procedures of fault tree analysis can not be readily applied. In this instance the various failure modes have to be established and investigated individually.

In nuclear plant design a triple system of barriers is employed to protect the population at large from radioactivity. These barriers are, successively, the fuel cladding, the primary circuit and the containment building. The reactor pressure vessel is part of the primary circuit, and is in fact the central component of the primary circuit. No backup systems are available on any present day nuclear steam supply system to mitigate the consequences of massive failure of the pressure vessel.

It is therefore of prime importance that the pressure vessel should be designed and constructed to the highest feasible standards of workmanship.

The pressure vessel is also subjected to extensive analysis in order to establish and demonstrate its integrity under all operational and "emergency" loading conditions. The fracture analyses of various important sections of the pressure vessel form a substantial portion of these analyses, and simplified methods of performing these analyses will be described in this paper.

DESCRIPTION OF A PWR REACTOR PRESSURE VESSEL

As only pressurised water reactor power stations are considered for construction in South Africa at present, only this type of pressure vessel will be considered in this paper.

A typical PWR reactor vessel of the type proposed for use in South Africa is about 10 m in height (excluding the vessel lid), it has an internal diameter of about 4 m and a wall thickness of about 200 mm in the beltline region (core region). The vessel has 6 primary coolant nozzles with inner diameters at the inside of the vessel wall of 900 mm and 735 mm for the inlet and outlet nozzles, respectively. In addition to the primary nozzles a large number of small instrumentation penetrations are situated in the bottom head of the vessel.

The top head of the vessel is bolted on to the vessel shell with 58 studs and is sealed with two metallic O-ring gaskets. The top head also has penetrations and adaptors for the 65 control rod drive mechanisms.

The beltline section of the vessel is constructed from two forged ring sections, each weighing in excess of 50 tons, which are welded together by an automatic submerged arc process. The bottom head assembly is then welded on to the beltline assembly, after which the nozzle ring and flange assembly is welded on. Finally the nozzles are welded in place. The vessel assembly weighs about 260 tons, with the vessel head adding another 54,3 tons and 15,4 tons of bolting materials, giving a total of approximately 330 tons.

The vessel is made from low alloy steel which is clad on the inside with 7,5 mm thick stainless steel.

The vessel is supported by six support pads under the nozzles, which rest on a support ring. It can be appreciated that the stress distribution in the nozzle area will be very complex due to the multiplicity of loadings in this area.

ANALYSIS

Regions of Interest

Five critical regions are generally recognised for the fracture analyses of a PWR reactor pressure vessel (Buchalet and Bamford, 1975). These are the thickness transitions in the upper and lower heads, the nozzles, the flanges and the beltline region. These areas are shown in Fig. 1.

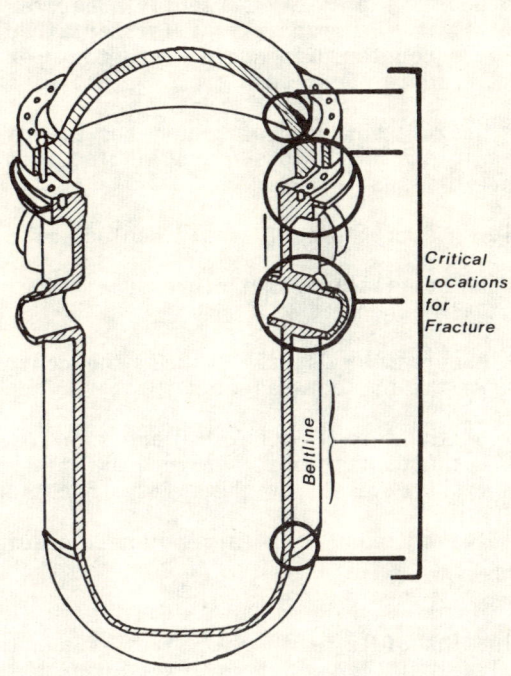

Fig. 1. Typical PWR RPV

The thickness transition areas act as stress concentrators giving rise to high stresses. Since the stress intensity factor is proportional to the stress, crack initiation is more likely to occur in these areas. Furthermore, thermal stresses are likely to be the highest in these areas due to differential expansion. The nozzle areas have large thermal gradients even under normal operating conditions, and additional stresses are induced on the vessel wall by the weight of the vessel which is supported by the nozzles.

The beltline region is the easiest of the five regions of interest to analyse due to its simple geometry. This area is of interest due to irradiation embrittlement of the vessel material.

Loadings

Let us now consider the type of loading that can be expected. Obviously, for each operational condition a different representative loading can be defined. Therefore one would find a shutdown loading condition with a shutdown critical defect size, a range of start-up loadings with corresponding start-up critical defects, etc.

From these the appropriate loading cases for use in the analysis have to be selected. Probably all the normally expected loading cases can be covered by the hydrotest case. After all, the hydrotest pressure is usually 1,25 times the system design pressure (ASME III NB 6211), and the design pressure is the maximum internal overpressure expected under normal operating conditions, including pressure surges. This procedure is probably perfectly adequate for normal conventional applications, but it can be insufficient for certain postulated accident conditions that could be encountered in nuclear reactor operation. The main drawback is that thermal stresses are excluded from the analysis.

In particular two postulated accident conditions are of importance in this respect. The first of these is the large loss-of-coolant accident (large LOCA). Briefly, the accident scenario is as follows:

1) a double-ended rupture of a main coolant pipe occurs,

2) rapid depressurisation of the pressure vessel occurs due to blow-down to the containment atmosphere,

3) the nuclear reactor "scrams", i.e. the control rods are inserted rapidly to shut down the nuclear reaction,

4) to prevent the core from melting and thereby releasing large quantities of radioactivity to the containment, large quantities of cold water are injected into the reactor pressure vessel by the safety injection systems,

5) the cold water leads to a large thermal gradient across the vessel wall, i.e. thermal shock.

Thereby the combination of large thermal stresses and the lower temperature could lead to a critical defect size which is lower than that expected for the hydrotest.

The second postulated accident which is considered to be of importance is the steam line break. This accident will be examined in more detail further on in this report. The scenario is as follows:

1) a break occurs in a main turbine steam line,

2) accelerated flow of the secondary coolant in the steam generator (heat exchanger) causes rapid cooldown of the primary fluid,

3) the reactor scrams,

4) cooldown of the water in the pressure vessel causes a rapid decrease in pressure from 155 bars to 50 bars in 160 sec , followed by an increase in pressure to about 160 bars after 1 500 sec ,

5) thermal gradients in the vessel wall lead to thermal stresses, which, although lower than in the case of a LOCA, must be added to the pressure stresses. The combined thermal and tensile stresses could lead to critical defect sizes which are lower than those associated with a hydrotest.

Calculation of Critical Defect Sizes

The procedure usually given in fracture mechanics textbooks is to use the fracture toughness and applied stress value together with a formula of the following general type for surface cracks:

$$K_{IC} = C\sqrt{\pi}\, M_k\, \sigma\, \sqrt{a/Q},$$

where
- K_{IC} = mode I fracture toughness
- M_k = a magnification factor
- σ = applied stress
- a = crack depth
- Q = crack shape parameter

The value of the magnification factor is usually given in graphical form in literature, depending on the exact form of the equation used; the crack shape parameter, Q, is equal to the square of the elliptical integral ϕ_0,

where
$$\phi_0 = \int_0^{\pi/2} \left[1 - \left(\frac{c^2 - a^2}{c^2}\right) \sin^2\theta \right]^{\frac{1}{2}} d\theta.$$

Fortunately, it is not necessary to evaluate the above integral every time it is encountered in a calculation. Graphs of Q vs aspect ratio ($a/2c$) for values of σ/σ_y are readily available, e.g. ASME XI, appendix A Fig. A 3300-1.

The main problem with this approach is that the analyst is tempted to calculate the critical defect size and then to regard any smaller cracks as acceptable.

A more conservative approach is to apply a safety factor to the fracture toughness value, but this can lead to overconservatism in some cases and thereby adoption of unrealistically small critical defect sizes.

A somewhat more complicated method of analysis was proposed by Buchalet and Bamford (1974). Their original proposal only catered for circumferential and infinitely long longitudinal surface defects, but it was later expanded (Buchalet and Bamford, 1975) to cover semi-elliptical surface flaws as well.

Application of this method comprises the following:

1) Approximate the stress distribution through the vessel wall by a third degree polynomial of the form

$$\sigma_x = A_0 + A_1 x + A_2 x^2 + A_3 x^3,$$

where x is the distance into the vessel wall.

2) Calculate the stress intensity from the relationship

$$K_1 = \sqrt{\pi} \cdot a \left[A_0 F_1 + \frac{2a}{\pi} A_1 F_2 + \frac{a^2}{2} A_2 F_3 + \frac{4}{3\pi} a^3 A_3 F_4 \right],$$

where
- A_n = coefficients from the stress distribution
- F_n = magnification factors as given by Buchalet and Bamford (1975)
- a = crack depth

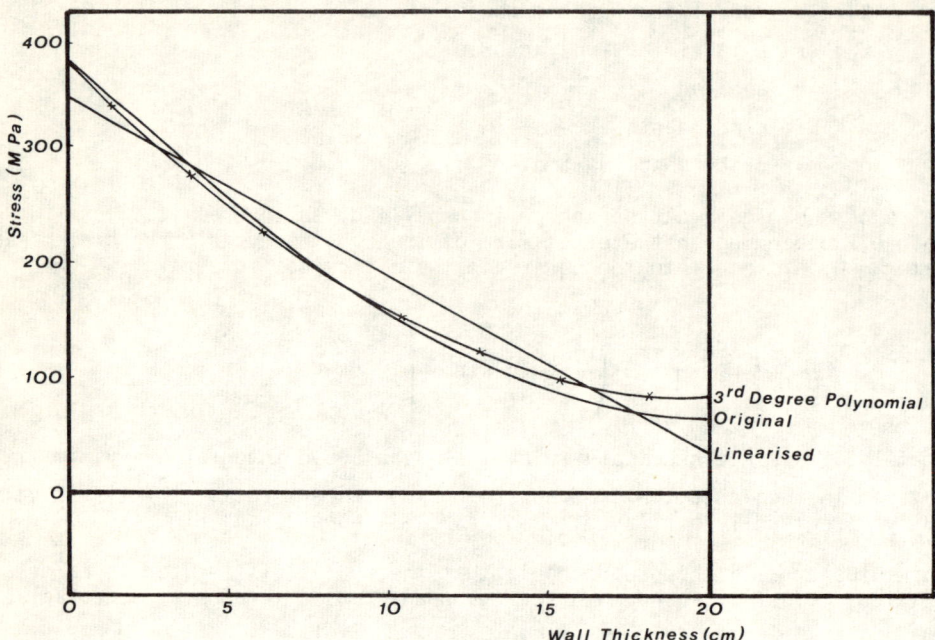

Fig. 2. Stress distributions during a typical accident.

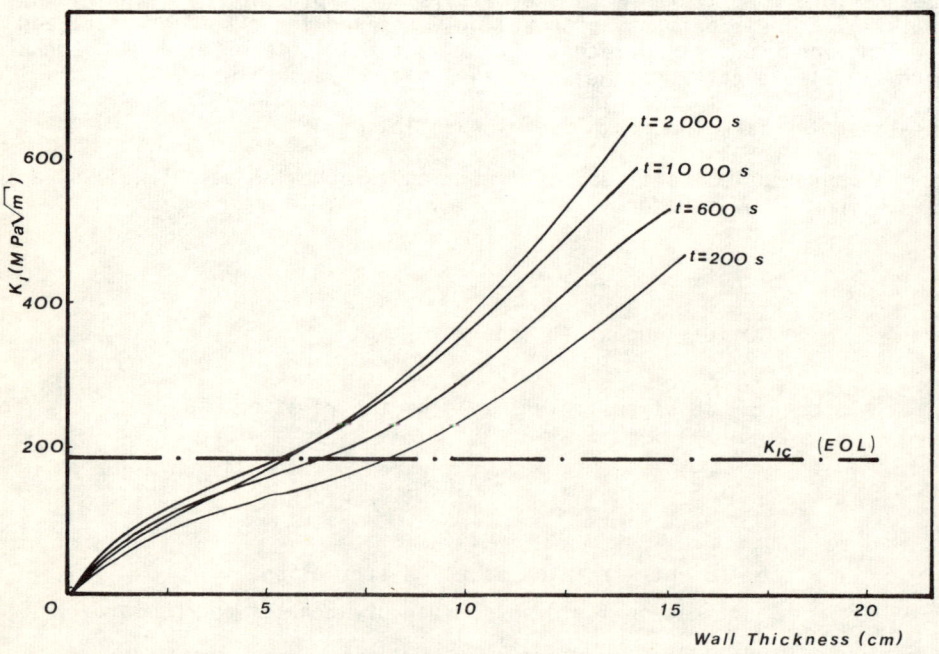

Fig. 3. Polynomial stress intensity distributions.

The magnification factors (Fn) mentioned above are given by Buchalet and Bamford (1975) for various configurations in graphical form, and are used together with the polynomial coefficients and a varying crack depth in the above formula to determine the stress intensity distribution through the vessel wall.

A less complicated method is to use the graphical solutions presented by Rooke and Cartwright (1976). Graphs are given of the relationship between the stress intensity at the deepest point of a surface crack and a reference stress intensity factor against the crack depth. In the case of a nuclear reactor pressure vessel the vessel wall can be approximated by a slab since the radius of curvature is large compared to the wall thickness. The applied stress has to be linearised and separated into the bending and tensile components, after which the stress intensities for the two loading cases can be found. The individual stress intensities for bending and tensile stress as well as the stress intensity are then plotted against crack depth through the vessel wall.

Fracture Toughness Determination

As experimental determination of fracture toughness is time-consuming and requires sophisticated equipment which may not always be available, it is often preferable to use readily available generic data. Alternatively, fracture toughness estimation can be based on Charpy V-notch test results, although great care has to be taken in the selection and application of Charpy V-notch vs fracture toughness correlations. A review of the various correlations with their individual advantages and shortcomings has been published by Pisarsky (1978).

A general expression for the lower bound values of fracture toughness for ferritic steels was suggested by the Pressure Vessel Research Committee (1972). The committee recommended that this K_{IR} curve be used for steels with yield strengths of up to 50 ksi in the absence of data for such steels. This K_{IR} curve has since been adopted by ASME and was included in Appendix G of ASME III (1974). A word of caution should be included at this stage. The K_{IR} formula is given in two places in Appendix G, firstly in the test and then again in the diagram, and through all the Addenda to the 1974 edition of ASME III covering a further three years, it was never given correctly in both places!

Application to a Loading Condition

The first stage in the application of the methods described above is the establishment of the stress distribution approximations. The actual expected stresses are usually available from a stress report, which can then be used to produce the desired approximation. Fig. 2 is a comparison of the calculated stress distribution through the vessel wall at a given time, the polynomial approximation and a linearised distribution. A further stress approximation that can be used is to linearise the stress acting over the crack, which implies that a different stress distribution must be used to calculate the stress intensity at each point through the vessel wall.

The polynomial stress distribution, in the form

$$\sigma_x = A_0 + A_1 x + A_2 x^3 + A_3 x^3$$

can then be used as input to the stress intensity distribution,

$$K_1 = \sqrt{\pi}\ a\left[A_0 F_1 + \frac{2a}{\pi} A_1 F_2 + \frac{a^2}{2} A_2 F_3 + \frac{4}{3\pi} a^3 A_3 F_4\right],$$

which will give the stress intensity distribution for an infinitely long surface crack. The results of this calculation are given in Fig. 3. The calculation is not very involved and can be performed on currently available programmable pocket calculators.

Any linear or linearised stress distribution can be used as an input to the Rooke and Cartwright method (after Shah and Kobayashi).

Results for linearised stress distributions at various stages of the situation under consideration (SLB) are given in Fig. 4.

From the above-mentioned stress intensity results it can be seen that the crack shape as characterised by the aspect ratio has a prominent effect on the stress intensity factor. Generally, a long shallow crack has a higher stress intensity under a given loading than a shorter and deeper crack. However, it is more likely that a long shallow crack will be found by surface crack detention methods than a shorter deep crack.

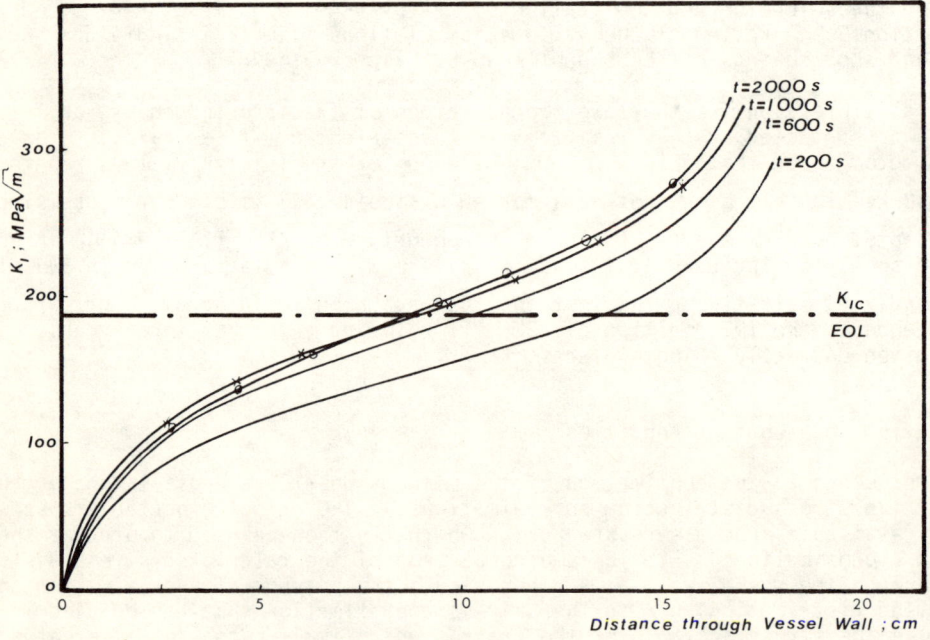

Fig. 4. Stress intensity distributions calculated from linearised stress distributions.

Figure 5 is a comparison of the polynomial method stress intensity distributions with similar distributions as found after Rooke and Cartwright. The polynomial distributions are for infinitely long cracks and cracks with an aspect ratio of

0,1. The distributions using the Rooke and Cartwright data were calculated using stress distributions which were linearised over the section and over the crack length respectively.

It can be seen from the comparison that the polynomial method provides the more conservative results, except for shallow cracks, and the linear stress distri= bution provides the least conservative results. However, as Buchalet and Bam= ford (1975) stated that their method (the polynomial method) is probably over= conservative, it is recommended that the linear stress over the crack tip should be used, for ease of calculation and for conservatism. On the other hand, if calculations are done only for purposes of comparison (i.e. to find at which time during an accident the worst conditions for fracture exist), then it is probably adequate to do the analysis by the method which is easiest to apply.

The second stage of the calculation is to establish the fracture toughness profiles through the vessel wall if the toughness should vary. This will be the case where the temperature varies through the vessel wall, or if the vessel is subject to irradiation. This can easily be done if fracture toughness data is available in the form of the K_{IR} curve. For a temperature distribution, the fracture toughness is simply taken from the K_{IR} curve at the temperature for each point through the vessel wall, and plotted against the distance into the vessel wall.

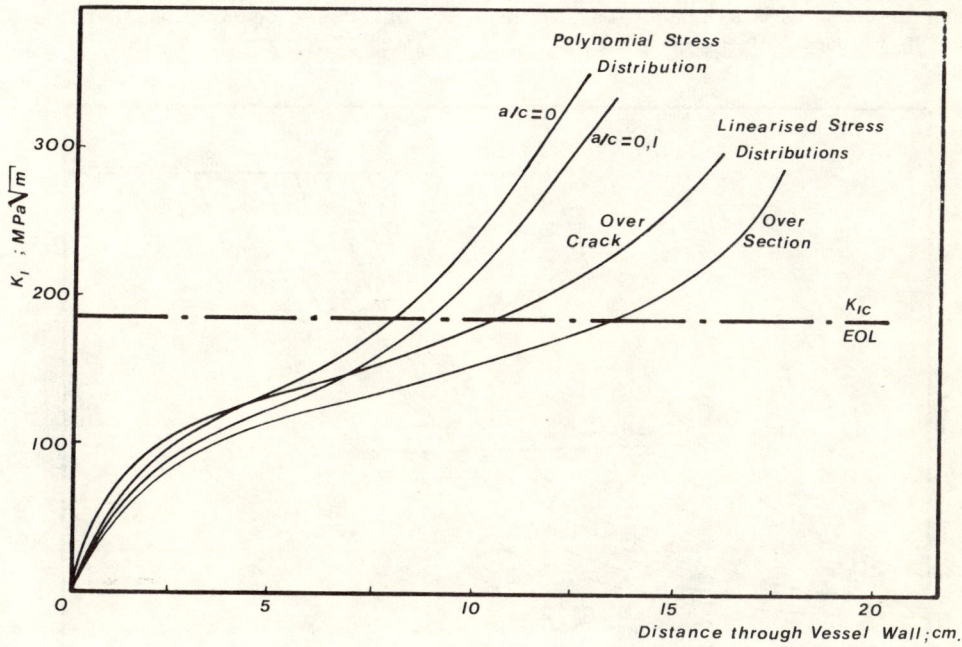

Fig. 5. Comparison of stress intensity distributions.

In the case of irradiation, however, the situation is somewhat more complex. Firstly, the amount of irradiation at each point in the vessel wass has to be determined from a neutron fluence vs fractional distance through vessel wall relationship. For each point, then, the shift in nil-ductility transition temperature is found for the fluence at that point. Using this RT NDT shift, a fracture toughness vs temperature relationship (as in Fig. 6) is established for each point through the vessel wall by shifting the K_{IR} curve through the distance given by the RT NDT shift. Then the fracture toughness for each point through the vessel wall can be found from the K_{IR} curve appropriate to that depth and temperature.

Critical Defect Size

By plotting the stress intensity distribution and the fracture toughness profile on the same axes, the critical defect size can be found at the intersection of the fracture toughness and stress intensity curves.

Figure 7 shows the stress intensity and fracture toughness distributions for the worst time during the SLB accident and also for a typical RPV hydrotest. It can be seen that even though the stress intensity distributions are very similar, there is a large difference in critical defect size due to the different fracture tough= ness levels.

Effect of Using K_{IR} or Actual Toughness Data

When using the ASME or PVRC K_{IR} fracture toughness curve, it must be realised that this is a lower-bound curve based on the lowest K_{IC} data recorded. This approach is conservative when used for design, but could be non-conservative and misleading if used for analysis.

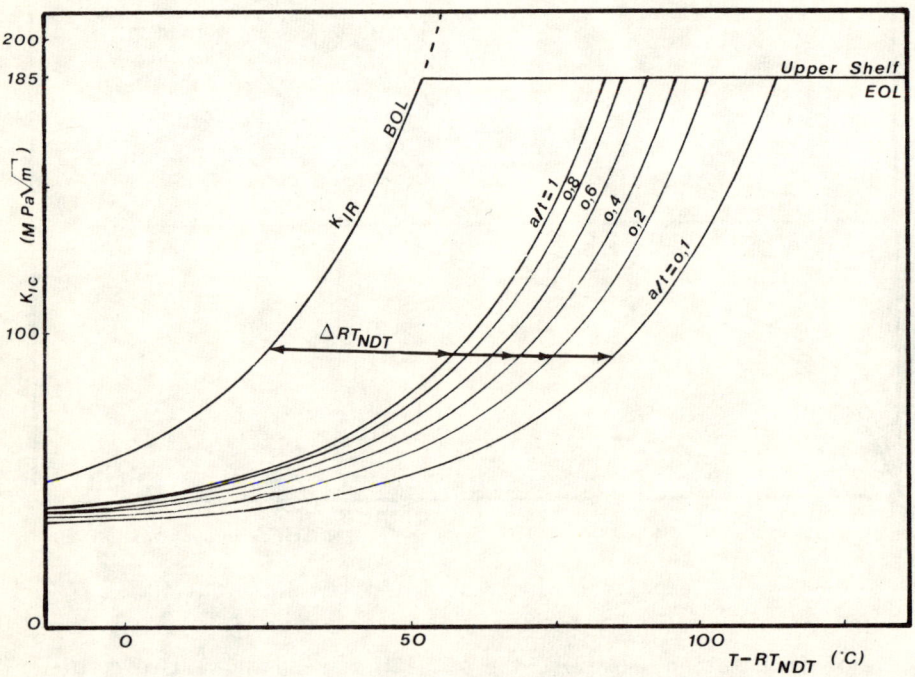

Fig. 6. Reference fracture toughness curves.

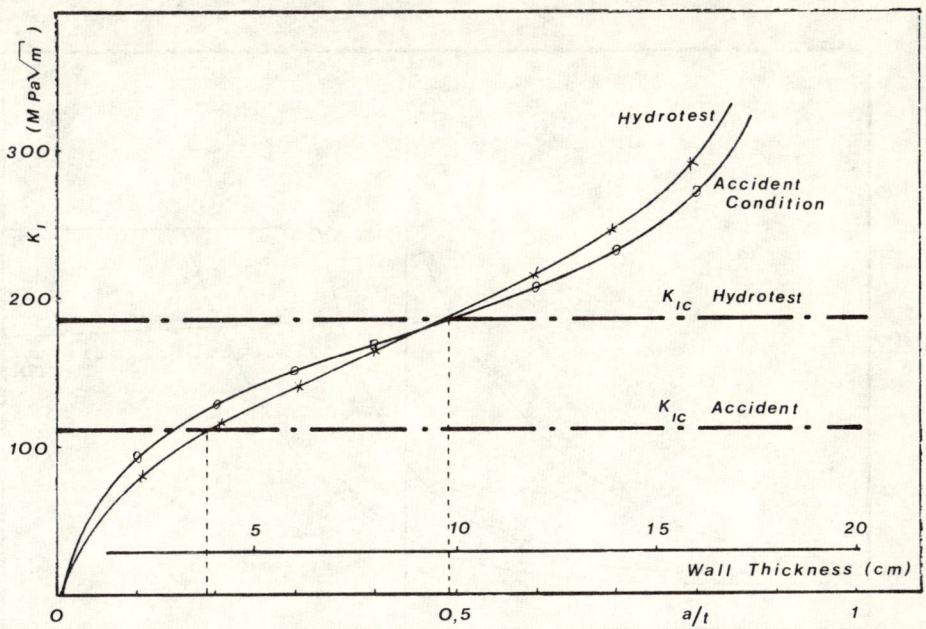

Fig. 7. Stress intensity distributions for hydrotest and accident conditions.

As an example of a non-conservative result, consider Fig. 7 again. The immediate conclusion to be drawn is that the hydrotest is a very severe test from a fracture point of view as it leads to a smaller critical defect size than the accident condition under consideration. However, Fig. 8 is a plot of the K_{IR} and actual fracture toughness data as measured (Ingham and Sumpter). It can be seen from this figure that the low critical defect size for the hydrotest using the K_{IR} data is due to the fact that the lower bound K_{IR} line is in the transition region at hydrotest temperatures. On the other hand, the actual data have already reached upper shelf conditions at the hydrotest temperature. Therefore, use of the K_{IR} line in this instance gives the false impression that a vessel which has passed the hydrotest is safe for operation under accident conditions, whereas in reality the critical defect size for the two cases is of the same order.

Crack Arrest Calculations

A further development of this technique is in predicting crack arrest behaviour. Available crack arrest toughness data indicate that a similar relationship exists between K_{1a} and temperature as that between K_{IC} and temperature. K_{1a} values are generally lower than K_{IC} values. From the K_{1a} data, a crack arrest toughness distribution can be constructed through the vessel wall similar to that for fracture toughness. Using the same procedure for calculating the stress intensity distribution, this is again plotted on the same axes as the toughness distributions. Since the stress intensity distribution has the general form of a third degree polynomial, a situation can exist where the stress intensity distribution increases initially through the vessel wall but starts to decrease at a certain

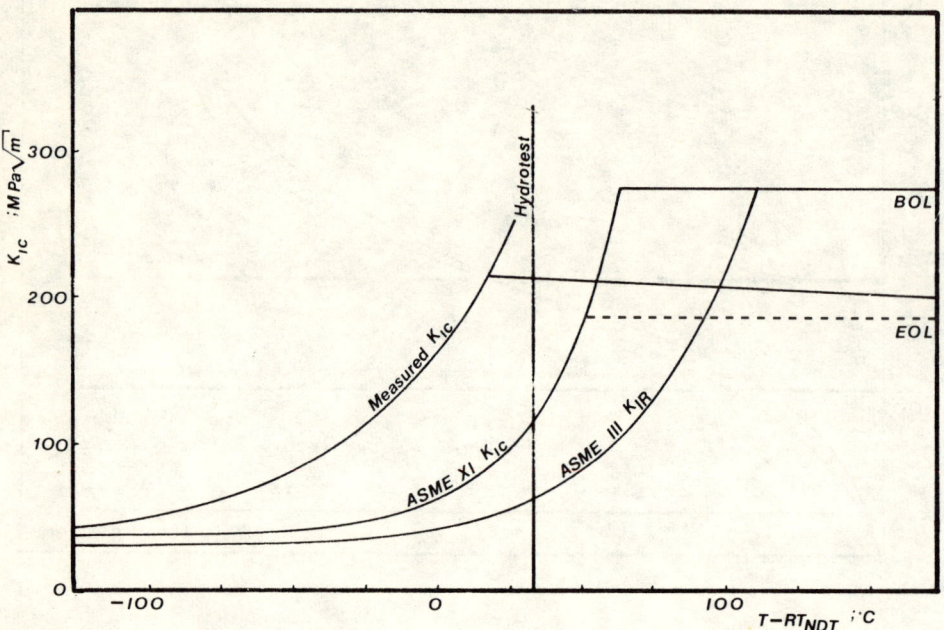

Fig. 8. Reference and measured fracture toughness curves.

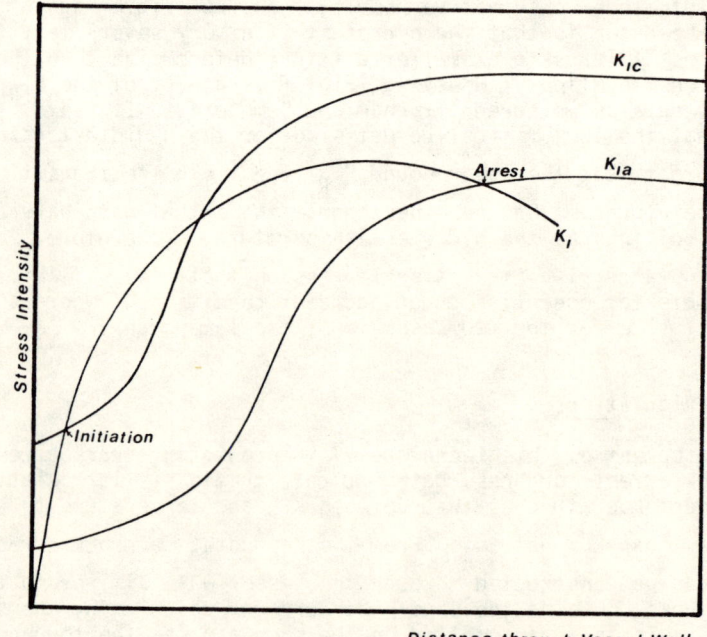

Fig. 9. Crack arrest methodology.

distance. If this decrease is such that the stress intensity line intersects both the K_{IC} and K_{1a} lines, then it can be postulated that a crack will initiate at a size given by the intersection of the K_1 line with the K_{IC} line, and propa= gate until it reaches the size given by the intersection of the K_1 line with the K_{1a} line, at which stage it will arrest.

CONCLUSION

Of the methods investigated, the polynomial method is the most conservative, whereas the average linear stress distribution with the stress intensity factor solutions given by Rooke and Cartwright is the least conservative. For design purposes, a stress distribution which is linearised over the crack length, to= gether with Rooke and Cartwright's solutions, is recommended, but for compara= tive purposes the general linear stress distribution is recommended due to ease of calculation.

For design purposes, the K_{IR} toughness curve (or similar lower bound curves) can be used to obtain conservative results, but care must be taken in the analysis of design and operational conditions.

Finally, a fracture analysis can be performed with sufficient accuracy and con= servatism for most applications, using simplified techniques without the need to resort to expensive computer techniques.

REFERENCES

American Society of Mechanical Engineers (1974). Rules for construction of power plant components. Boiler and pressure vessel code, section III Division 1. ASME, New York. Subsection NA, pp 491 ' 495.
American Society of Mechanical Engineers (1974). Rules for inservice inspection of nuclear power plant components. Boiler and pressure vessel code, section XI. ASME, New York. pp 124 - 126.
Buchalet, C.B. and W.H. Bamford (1974). Stress intensity factor solutions for continuous surface flaws in reactor pressure vessels, WCAP-5292. Westinghouse Electric Corporation, Pittsburgh.
Buchalet, C.B. and W.H. Bamford (1975). Method for fracture mechanics assessment of nuclear reactor vessels under severe thermal transients, 75WA/pvp-3. ASME, New York.
Farmer, F.R. (1967). Containment and siting of nuclear power plants. Proc. Symp. Vienna. IAEA, Vienna.
Ingham, T. and J.D.C. Sumpter (1978). Design against fast fracture in thick walled pressure vessels. I. Mech. E., London.
Pisarsky, H.G. (1978). A review of correlations relating Charpy energy to K_{IC}. Welding Institute Research Bulletin, 19. pp 362 - 367.
PVRC Ad Hoc Task Group on Toughness Requirements (1972). PVRC recommendations on toughness requirements for ferritic materials. In R. Roberts (Ed.) (1976), Pressure Vessels and Piping, Vol. 3. ASME, New York. pp 140 - 163.
Rooke, D.P. and D. J. Cartwright (1976). Compendium of Stress Intensity Factors. HMSO, London. pp 247 - 299.
Simpson, D.M., B.C. Winkler and J. O. Tattersall (1974). On the use of the risk concept and cost-benefit analysis in the safety assessment of nuclear instal- lations. IAEA-SM-184/31. Proceedings of the Seminar on the Radiological Safe- ty Evaluation of Population Doses and the Application of Radiological Safety Standards to Man and the Environment, Portoroz, Yugoslavia. IAEA, Vienna.
Tattersall, J.O. (1977). Nuclear plant safety. The Professional Engineer, Vol 6, No. 4.
USNRC (1975). Reactor Safety Study (WASH-1400). USNRC, Washington.

AVOIDING FRACTURE IN PRESSURE VESSELS

J. R. Campbell

British Engine Insurance Company, Johannesburg, R.S.A.

ABSTRACT

This paper attempts to highlight the problems facing the practicing Engineer in assessing the acceptability of known defects in service. Specific examples are briefly considered and the current code philosophy in various areas is examined. Some of the problems involved in making a general extension to the present Pressure Vessel Code allowable defects are suggested and specific recommendation made on improving concerning fracture resistance. The paper concludes that there is obvious economic benefit, but many technical difficulties in being able to tolerate a higher level of defects in real structures.

INTRODUCTION

The failure of vessels due to brittle fracture is, thankfully, not a common occurrence when one considers the large number of vessels in service. Nevertheless, in the past few years, there has been an increasing number of failures of large Pressure Vessels loaded within their design limits and these have naturally caused concern. Engineering Design relies heavily on past experience and it is only when faced with a series of similar failures that a fresh attempt is made to re-appraise the validity of the accepted design criteria and the effectiveness of the existing inspection procedures. Although few, if any, accurate statistics are available, we in South Africa would appear to have been extremely fortunate in that the number of serious fractures causing plant outages and loss of life have been relatively few and usually attributable to over pressure due to operator error, i.e. gags left in safety valves following hydraulic test, etc. The problem facing the practicing Engineer in avoiding brittle fractures and assessing the significance of actual defects in service can best be illustrated by considering a variety of examples of actual defects in Pressure Vessels.

EXAMPLES OF DEFECTS IN SERVICE

The Potchefstroom Disaster

A classical example of brittle failure in South Africa was the Potchefstroom Disaster of 1973 (Fig. 1) and I will consider this in some depth as it is an extremely valuable illustration of the problem of assessing the significance of real defects.

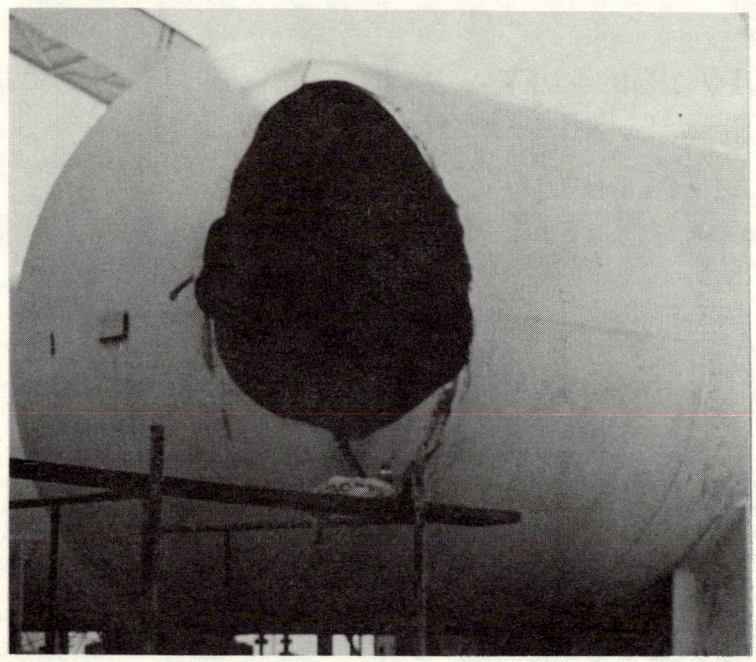

Fig. 1. General view of failed Anhydrous Ammonia Vessel

History

The vessel was constructed in 1967 to B.S.1515, in semi-killed carbon steel (B.S. 1501-151-28A) for a design pressure of 1.84 MPa (267 p.s.i.g.) and a stated design temperature of 49°C (120°F). The normal working temperature was more in the region - 6.7°C to 12.8°C (20°F to 55°F) at a pressure of approximately .62 MPa (90 p.s.i.g.) and this, in itself, illustrates a serious and not uncommon problem in fracture control. Frequently, users specify the maximum conditions rather than an operating range of pressures and temperatures and a vessel is thus designed and material selected with only the upper temperature limit in mind.

The vessel, which stored anhydrous ammonia for fertiliser production, was due for a statutory hydraulic test in November 1971. Since the user considered this extremely inconvenient from a production view-point, an application was made to the Department of Labour for an exemption which was granted providing that N.D.T. checks were performed as soon as practical and a hydraulic test made when the vessel was emptied for cleaning or repairs. This N.D.T. test revealed the defects shown on Fig. 2, the shaded areas being described as laminar defects, but the following quote is taken from the report:

"Although these laminar defects cover large areas they could not be described as laminations due to the fact that each area consists of numerous laminar defects all at different levels and having the appearance of laminar tears which could have developed during the forming of the dished end".

Two areas of lack of sidewall fusion of the seam-weld were also reported, the longest being approximately 37mm (1.5"), 12mm (0.5") under the inner surface and 5mm (0.2") in cross section.

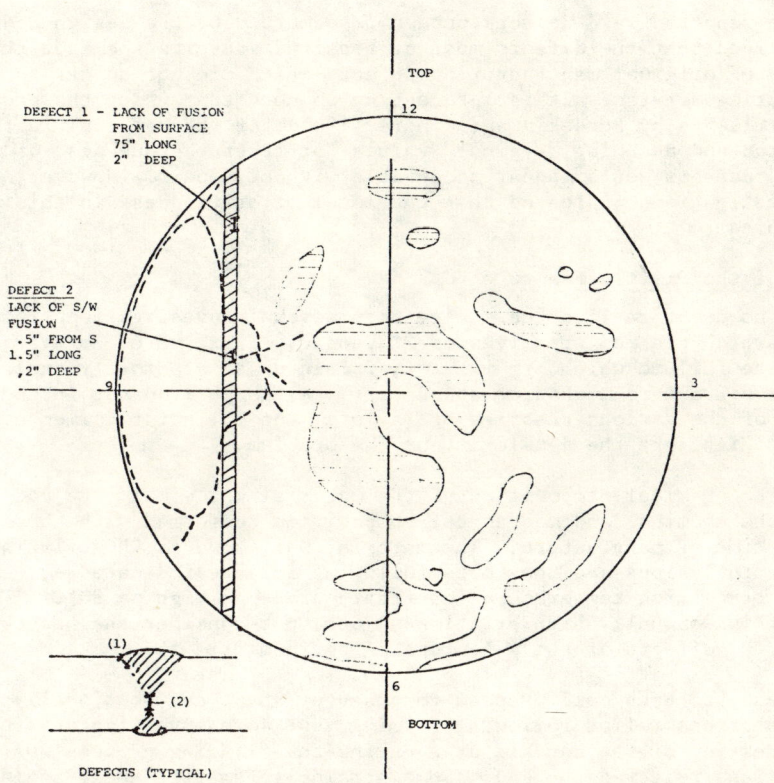

Fig. 2. Defects in Dished End - Viewed from Inside

Acceptability of the Defects

A - Laminations. The original construction of the vessels would not have required lamination checking of the plate materials other than visually at the plate edges and thus the manufacturer and inspection authority would have been completely unaware of their presence. B.S.1515 gives no basis for the acceptance of such defects, but the latest draft for development DD 21 : 1972 which does provide a basis for acceptance, would not permit laminations of the size shown.

The Engineer responsible for the assessment of the defects made, in the circumstances, an extremely brave decision in accepting these laminations for future service without repairs, providing that they were regularly monitored by ultra-sonics for possible increase during service.

The subsequent extensive metallurgical investigations into the cause of failure and the fracture surface (shown dotted on Fig. 2) indicate that these reported laminations played no part in the eventual failure.

B - Weld defects. On the acceptability of weld defects B.S.1515 made the following comments:

The following imperfections shall be judged unacceptable:

1. Any type of crack or zone of incomplete fusion or penetration.

2. Any elongated slag inclusion which have a length greater than 1/3 T, i.e. approximately 8mm in this case.

Clearly, the reported weld defects were not permitted by the design code and the Engineer decided that the defects must be repaired, despite the fact that they too were considered original manufacturing defects which did not appear to have been extending during service. It is interesting to note that using the proposed defect acceptance criteria by Burdekin and others (1967) for the material in the "as welded" condition and assuming "typical" values for fracture toughness of the weld metal, these defects would appear acceptable without repair. However, the post failure investigations indicated that the level of brittleness in this end was by no means typical.

Discussion of the Brittle Fracture

As you will no doubt realise there have been several investigations into this tragic failure which claimed the lives of 22 people. There were, however, several aspects of the failure which, in my view, remain unsatisfactorily explained, particularly from a fracture mechanics standpoint. Although I do not intend to present an analysis of the various theories of failure, the following comments will, I hope, serve to highlight the complexity of the problem.

Material. The chemical composition of the material was within the code permitted limits and the chemical and mechanical composition conformed with the specification at the time of manufacture. The material was, however, heavily laminated with non-metallic inclusions and the tests following failure indicated extreme brittleness with a transition temperature in certain areas as high as 80°C (176°F). To understand this unusually high brittleness one must consider the process of manufacture and the effects of the weld repairs performed in 1971.

The dished end had been cold pressed to a saucer shape and sectionally hot flanged. There was no stress relief following forming, but some investigators felt that the successive heating of the knuckle area during the flanging process would have effectively stress relieved the end. Some strain-age embrittlement could be expected in the cold formed areas and it has been suggested that this effect could only have been adequately removed by heating the complete end to normalising temperature. It is thus possible that some of the material in the area of interest was in a fully strain-aged condition prior to the repairs of 1971, but the metallurgical authorities on this aspect are by no means unanimous.

During the first weld repairs (see Fig. 2) the upper and lower defects were ground out and welded using low hydrogen electrodes. The upper repair was accepted, but the lower repair showed excess porosity and was rejected. The lower defect was then re-repaired by grinding out and re-welding. The radiograph of this repair showed extensive cracking both within the weld and in the parent plate.

This cracking was believed to be due to the adverse weather conditions prevailing during the repair and the cracks were re-ground and a further weld repair effected. The radiograph of this repair also showed porosity in excess of that permitted by B.S.1515, but possibly to avoid a further repair, the Engineer decided to accept this repair. There was no stress relief performed following these repairs, which is still normal practice with vessels not subject to stress relief during manufacture. The vessel was then hydraulically tested and returned to service.

Having examined in detail the history of this dished end it is clear that the material composition varied from normalised relatively fine grain material in certain areas to courser grain, laminated and embrittled material in other areas. Superimposed on this problem is the assessment of the nett effect of successive weld repairs on the surrounding material since some investigators felt that the stresses induced by the repairs would encourage further strain-age embrittlement.

Stress levels. The vessel was repaired in November 1971 and hydraulically tested to 2.4 MPa (347 p.s.i.g.) for a period of 30 minutes. The test temperature was not recorded but it can be assumed in the region of 20°C (68°F). The vessel operated satisfactorily at a normal operating pressure of .62 MPa (90 p.s.i.g.) until May 1973 when the vessel was shut-down for repairs to leaking valves. Following these repairs the vessel was again hydraulically tested to 2.4 MPa without any sign of weakness. This test was approximately six weeks prior to failure.

The vessel failed at mid-day on Friday 13h July 1973 at a pressure of approximately .62 MPa (90 p.s.i.g.) while it was being filled from a mobile tanker and at an approximate temperature, allowing for possible chilling during the decanting process of 5°C (41°F) (not recorded). The vessel records produced at the enquiry indicated that temperatures as low as -8°C had been recorded several weeks before the failure with the vessel operating at .62 MPa (90 p.s.i.g.)

Defect. The exact defect which caused the failure was not precisely identified - some investigators commenting that the material was so brittle that a minute defect would be sufficient to create a fracture. At least two theories of failure were suggested, one based on the area a b (shown on Fig. 3) and in the photograph shown in Fig. 4, and the other on the area g (shown on Fig. 3) and shown in detail in Fig. 5. No positive evidence of stress corrosion cracking was detected.

As mentioned, I do not intend to consider extensively the various theories of failure, but I have detailed the circumstances surrounding this failure in depth in order to indicate the danger in assuming a nominal fracture toughness for a material and calculating an acceptable defect size. The exact history and true metallurgy of a specific area are the major factors in determining defect sensitivity. Even allowing for the differences in temperature between the temperature at failure and the probable temperature during pressure testing, it is difficult to explain the mechanism of failure in fracture mechanics terms particularly since most of the investigators considered that the material even at 20°C was in many areas below its transition temperature.

Defects in E.R.W. Boiler Tubing

My second example of actual defects in service is one which highlights the economic considerations and the problem of assessing the true service environment. The material concerned was boiler tubing for a water-tube boiler (see Fig. 6), specified to B.S.3059 : Part 1 : ERW 33. This electrically resistance welded tubing was 50.8mm outside diameter and 3.25mm wall thickness. In accordance with normal practice the tubing was being cold expanded into the boiler drum as shown in Fig. 7 & 8, when numerous cracks appeared in the heat affected zone of the tube weld.

A micro examination of a typical tube is shown in Fig. 9 and indicates oxide or sulphide stringers (inclusions) running longitudinally in the tubes and distributed throughout the circumference of the tube. In the heat affected zone the shrinkage stresses following welding opened up these stringers in the micro cracks as shown on Fig. 10 and during the expansion process they frequently opened up into major cracks, as shown on Fig. 11.

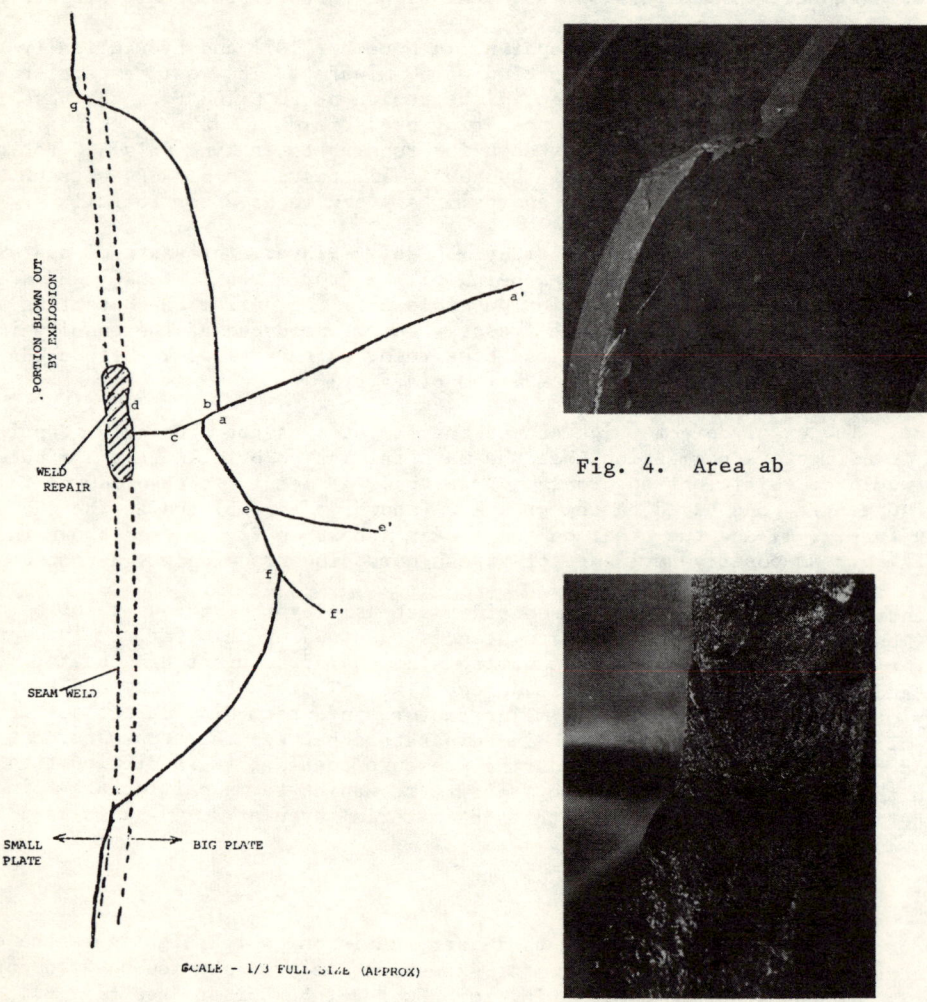

Fig. 3. Fracture Pattern.

Fig. 4. Area ab

Fig. 5. Area g.

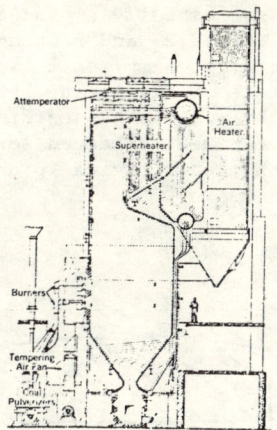

Fig. 6. Medium sized water-tube Boiler (typical).

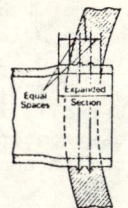

Fig. 7. Tube expansion (typical).

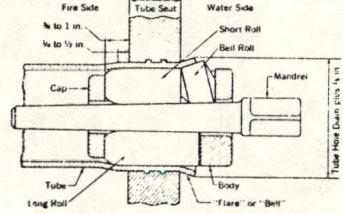

Fig. 8. Position of Expander after Expansion/Flaring.

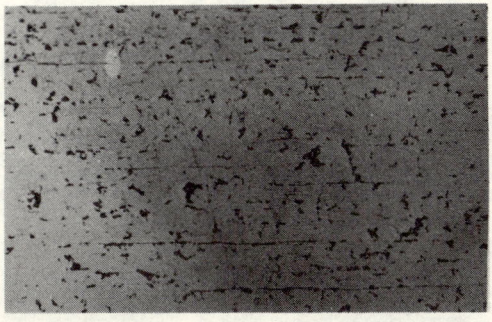

Fig. 9. Oxide or Sulphide Stringers.

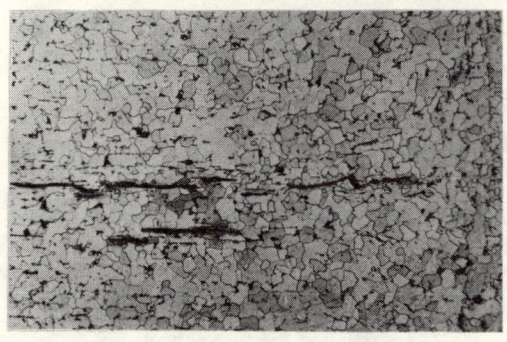

Fig. 10. Micro Cracks in H.A.Z.

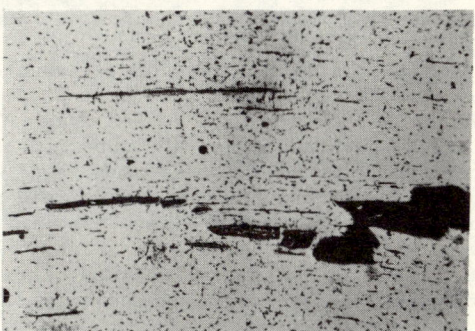

Fig. 11. Cracking during Expanding.

The extent of the stringers and micro cracks varied throughout the length of the tubing and a large number of tubes had been successfully expanded into the boiler. The tubing was re-tested by Drift Expansion (Fig. 12) and Flattening Tests (Fig. 13) to the original specification, and the majority of samples (22 were tested) failed in the Flattening Test. During the manufacture of ERW tubing, the excess weld bead both externally and internally is mechanically removed by scarfing and this frequently results in stress concentrations in the heat affected zone due to surface imperfections. This problem combined with the metallurgical problem to cause the extensive cracking.

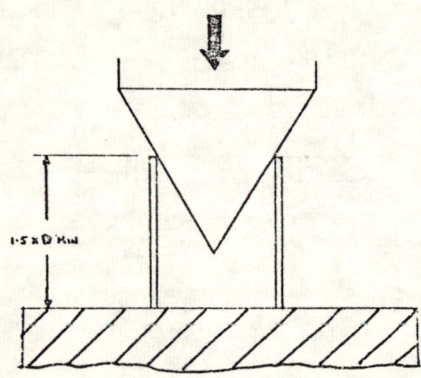

Fig. 12. Drift Expansion Test. Increase in diameter without failure. Typically 17%.

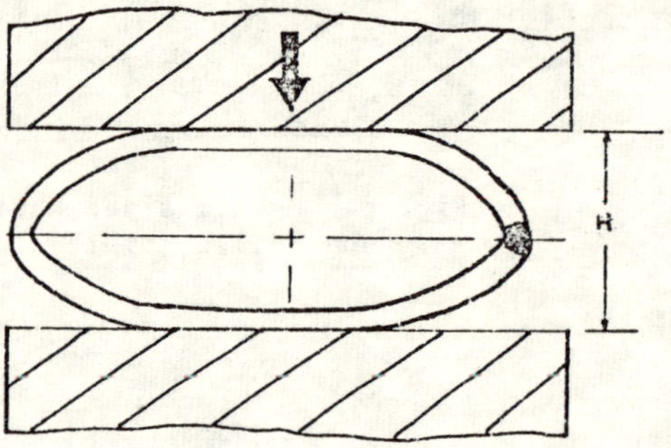

Fig. 13. Flattening Test. Must flatten to H without failure. H (typically) 40% D.

The boiler was intended for the processing of sugar cane and the user was naturally anxious for a decision on the acceptability of the defects for future service. The following factors had to be considered.

<u>Stress level</u>. It was commented that the stress level during the expansion process was considerably higher than that likely to be experienced during the normal operation of the boiler. The design code (B.S.1113) does not actually specify the required clearance between the tube and the tubehole, but typically on this size of tubing the clearance can be as high as 2 to 3mm and thus the stress during expansion would clearly have exceeded yield magnitude. However, in service the tubing would be subject to stresses due to internal pressure, fluctuating thermal stresses due to working of the boiler and possibly mechanical vibrations due to steam and water flow. A fatigue effect was certainly possible and the comparison of maximum stress levels alone was not the only factor affecting the significance of the defects in service.

Defect size. Micro examination of the tubing indicated that inclusions and micro cracks with a height of .5mm were common, but the length of the defects could not be accurately assessed and one would have to assume that, in the worst cases, these defects could be present throughout the full length of certain tubes.

Possible application of fracture mechanics. In attempting to decide on the suitability of these defects for future service, two types of failure must be considered:

1. During hydraulic test. With this type of failure the stress level can be estimated with some degree of certainty and estimates can be made of the worst defect sizes. The major difficulty would appear to be the estimation of relatively accurate fracture toughness values for the heat affected zone of the welded seam, particularly for a material exhibiting the non-metallic stringers previously discussed.

2. During normal operation. One could question the creditability of a fast unstable fracture at service temperature since the material would normally be expected to behave in a relatively ductile manner. Relevant work in this area by Oates, Price (1970) attempts to assess the acceptability for service of a water-tube boiler with known cracks in nozzle attachments. Reference is made to work at the Welding Institute (U.K.) which indicates that fast unstable fracture has been simulated in the laboratory at elevated temperatures. Crack growth due to fatigue and creep in service would, of course, increase the risk of sudden unstable failure.

Meanwhile the sugar was now ready for processing and a decision regarding the suitability for service had to be made. It was finally decided that the remaining tubes should be normalised at the ends and carefully expanded into the boiler. The boiler was subsequently hydraulically tested and operated successfully during the first season. At the next outage, however, the boiler was completely re-tubed at the user's insistence with alternative material which did not exhibit the defects referred to above. Clearly the user believed that the defects could affect the long-term life of the tubes in the boiler and was not prepared to accept at this stage any theories on the acceptance of the defects based on Fracture Mechanics theory.

Defects in Welded Vertical Storage Tanks.

Lest I give the impression that I am not impressed with the benefits to be derived from a proper application of Fracture Mechanics, I would like to give an indication of an area where Fracture Mechanics has been used successfully to accept defects which would otherwise be unacceptable to the relevant design and construction codes. The example concerns large vertical storage tanks of the type shown in Fig. 14.

Fig. 14. Large Vertical Storage Tank.

Naturally, these tanks are site welded and frequently the control of site welding is poor, resulting in the type of defect shown in Fig. 15.

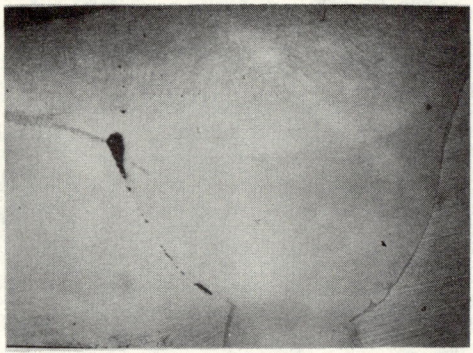

Fig. 15. Lack of Fusion Defect.

The porosity shown in the figure is probably insignificant, but the lack of fusion defects are crack-like in nature and in some of the worst examples extend virtually continuously in both the vertical and horizontal seams.

The material of construction is normally a semi-skilled carbon steel (say B.S.3604-43A) and there is usually no stress relief after welding. The welds are subject to the primary membrane stresses due to the internal pressure of the contents, to secondary stresses due to misalignment at joints and superimposed welded stresses.

In one specific example N.D.T. revealed that approximately 2mm lack of penetration occurred extensively in weld seams, particularly where the material was thicker in the lower courses of the tank. An estimate of the actual fracture toughness of the weld metal was obtained by performing Charpy Tests on the weld metal of actual samples removed from the tank and relating the absorbed energy values to a linear elastic toughness by an empirical relationship (suggested by U.S. Steel, Jackson (1977)). Using both a Linear Elastic and General Yielding fracture mechanics approach the calculated critical defect sizes were of the order of 5-6mm, i.e. an isolated through thickness crack of 10-12mm in length, or an imbedded weld metal crack (planar defect) of greater length. The effect of smaller imbedded planar defects is complex to assess as they are a function of defect height, plate thickness, stress residuals etc. and may also result in local bulging. Nonetheless, an attempt can be made, on a conservative basis, to quantify a height/length relationship, i.e. an equivalent stress concentration approach (considering both bending and membrane stresses) which indicated that an equivalent continuous 2mm defect on the neutral axis would be approximately 3.3 metres in length.

The tanks were also subject to internal corrosion in service and possible fatigue loading due to the filling/emptying cycles and both analyses had to allow for the effects of corrosion and defect growth due to fatigue during the life of the vessels. There were in fact 20 similar tanks involved in this particular assessment and thus, although the assessment was complex and costly, the tanks were eventually accepted for service subject to specific recommendations on in-service inspection and test, it being concluded that there was a greater likelihood of a through-thickness leak rather than an unstable fracture failure.

BRIEF REVIEW OF CODE APPROACH TO FRACTURE CONTROL

Design

To date the majority of design codes, including B.S.5500 and ASME VIII, have concentrated almost entirely on the avoidance of failure due to plastic bursting. The material yield or tensile strength being divided by a suitable safety factor to compensate for stress concentrations at nozzles, welding residual stresses, misalignment etc. Compliance with code design requirements ensures that at no point in the plain vessel shell or end will the yield stress be exceeded and, in the regions of stress concentrations, local yielding will cause "shake-down" to elastic behaviour. The design fabrication and testing sections of most codes concentrate extensively on avoiding defects in main welded seams although the failure of any vessel due to defective main seams is indeed rare, the majority of failures arising from stress concentration at nozzles, support brackets, arc marks etc. One cannot stress too highly that designers must concentrate more effort on good detail design to avoid stress concentrations at all vessel attachments and avoid welding wherever possible after stress relief. Good fatigue resistant designs will also greatly reduce the dangers of brittle fracture, since stress concentrations will be avoided.

Materials

Materials are at present selected basically on their ultimate tensile or yield strength and it is only at sub-zero temperatures that any attention is given to the toughness as demonstrated in the Charpy Impact Test. As I have described previously, it is possible to have semi-skilled steel which meets both the chemical and mechanical properties as required by the material specification and yet to exhibit alarming amounts of non-metallic inclusions and laminations with consequently poor toughness. Developments in furnace technology are reducing the extra costs required to produce fully-killed fine grain materials and these innovations must be welcomed as a major contributor to fracture control. In South Africa there is both an economic and strategic benefit in rationalising the materials used for Pressure Vessel and Boiler construction and this will undoubtedly involve an application of a "fitness-for-purpose" criteria and a more rational application of the mandatory requirements of some of our imported Pressure Vessel Codes.

Fabrication

Having heard much about the acceptable level of defects, one would perhaps expect a manufacturer to ask how quickly the design codes will increase the present level of defects permitted during fabrication. So far, however, code drafting committees have avoided increasing allowable defects sizes, presumably believing that the present level of acceptable defects represents a fair standard of workmanship and that it would be foolhardy to lower these levels unnecessarily and thereby encourage poor workmanship. It would appear likely therefore that there will be no significant increase in the present level of acceptable manufacturing defects, but rather that guidelines will be given on a basis of appraisal for actual manufacturing defects which exceed the present limits. Any appraisal of this kind will require an individual and detailed assessment of each defect and should be considered as an emergency situation, rather than an easy method of accepting defects which normal quality control could avoid.

Testing

Until recently non-destructive testing has concentrated on the somewhat negative activity of searching for every defect and, at first sight, it might appear that to fully utilise the fracture mechanics approach it is necessary to inspect the complete vessel with N.D.T. to locate all possible defects in the plate, welds and heat affected zones. The development of acoustic emission monitoring associated with fracture mechanics appears to have extremely exciting possibilities for the future and should increase significantly our understanding of a successful hydraulic test. The extension of the approach to in-service inspections and regular monitoring has also considerable potential.

CONCLUSIONS

The development of Fracture Mechanics Technology has certainly resulted in a more honest approach to the existence of defects in welded structures. It provides a much needed basis for the assessment of the significance of real defects in Engineering Structures. Even at the present stage of development it is clear that the application to Pressure Vessels will not be a simple process since successful application requires a knowledge of design stresses in different regions of the vessel, knowledge of the limitations of our present N.D.T. Techniques for sizing defects and a knowledge of the material properties including the effects of forming and welding. I hope that the illustrations given in the Paper have not deterred anyone from exploring the available literature on Fracture Mechanics, but I trust that I have also illustrated that it is not a simple matter of feeding variables into a formula and one must tread warily in applying it as a basis of acceptance.

If Engineers can be criticized as naive for assuming that the tensile and yield strengths of a material indicated its total resistance to failure, the current work in Fracture Mechanics would tend to suggest that the assessment of the true toughness of engineering materials may be more complex than was at first thought. The advocates of the Fracture Mechanics approach would suggest that one could select materials in which defects would be allowed to grow to through-thickness cracks and thus create a "leak before break" situation. If Harrisburg or the recent D.C.10 incidents have taught us anything, it is that such a philosophy must be approached with extreme caution since the accurate computation of the true safety margins for such a situation appears unlikely.

The considerable volume of research work into Fracture Mechanics is assisting the practicing Engineer, not only in providing a basis for defect assessment but in focusing our attentions on the critical criteria causing fractures and thereby indicating more clearly the direction in which one should proceed to produce safer and more reliable equipment.

REFERENCES

Griffiths, A.A. (1921). The phenomena of rupture and flow in solids. Phil. Trans. R. Soc., 163-221.
Irwin, G.R. (1960). Fracture Mechanics, Structural Mechanics. Naval Symposium (Pergamon Press).
Harrison, J.D., Burdekin, F.M. and Young, J.G. (1968). A Proposed Acceptance Standard for Weld Defects based upon Suitability for Service. 2nd Conf. on Significance of Defects in Welded Structures. Welding Institute, London.
Institute of Mechanical Engineers (1971). The Practical Application of Fracture Mechanics to Pressure Vessel Technology. Library of Congress Catalogue Card No. 71-184575.
Nicols, R.W. (1977). International Conference on Pressure Vessels, Tokyo. April 1977. CME September 1977.
Harrison, J.D. The Analysis of Fatigue Test Results for Butt Welds with lack of Penetration Defects using a Fracture Mechanics Approach. B.W.R.A. Report E/13/67
Jackson, W.J. (1977). Fracture Mechanics - Friend or foe? T.I.S. Conference Paper Ref. 771181/804.
Denham, J.B., Russel, J. and Wills, C.M.R. (1968). A comparison of Predicted and Measured Stresses in a Large Storage Tank. American Petroleum Institute.
Oates, G. Price, A.T. (1970). Application of Fracture Mechanics to Boiler Drum/ Nozzle Repairs. Institute of Mechanical Engineers, London.

FRACTURE TOUGHNESS CONSIDERATIONS IN THE DESIGN, MANUFACTURE AND USE OF HIGH STRENGTH COMPONENTS

G. T. van Rooyen

*Department of Material Science and Metallurgical Engineering,
University of Pretoria, R.S.A.*

ABSTRACT

Failures of drill string components used in raise boring were investigated. It is shown that there is a close interrelation in the satisfactory service performance of the final components and the design, the manufacturing of the components as well as the service conditions to which the components are subjected. The influence of chemical analysis and heat treatment necessary to obtain satisfactory fracture toughness coupled with optimum fatigue properties in the presence of a corrosive medium is considered at some length. As far as service behaviour is concerned under condition where service failure is due mainly to fatigue, adequate fracture toughness is important to ensure slow fatigue crack propagation to enable periodic non destructive testing of components to be carried out effectively before a catastrophic failure occurs.

INTRODUCTION

Raise boring since its introduction in South Africa has in many way revolutionised gold mining methods in South Africa. Instead of the tedius shaft sinking process, raise boring is accomplished by first drilling a pilot hole of about 300 mm diameter downwards. This hole in turn is reamed to a hole of 2,4 m in diameter. Reaming from the bottom upwards is accomplished by simultaneously pulling and rotating the reamer by a drill rod string which passes down the pilot hole. This is schematically shown in Fig.1A and Fig.1B. The drill pipes are internally threaded with a special tapered thread. The drill pipes in turn are screwed together with so called double pin subs shown in Fig. 2. One of these joints are usually made in a semi-permanent fashion by use of an expoxy resin adhesive. Failure of either a drill rod, double pin sub or stabilizer has very serious consequences in terms of damage to equipment, danger to operating personnel and to extended delays in production. It is therefore clear that the highest degree of integrity is required to eliminate any failure as far as humanly possible. In this regard not only the quality of the material used is important but factors such as details of the mechanical design, methods of construction and the operating conditions can also be of decisive importance.

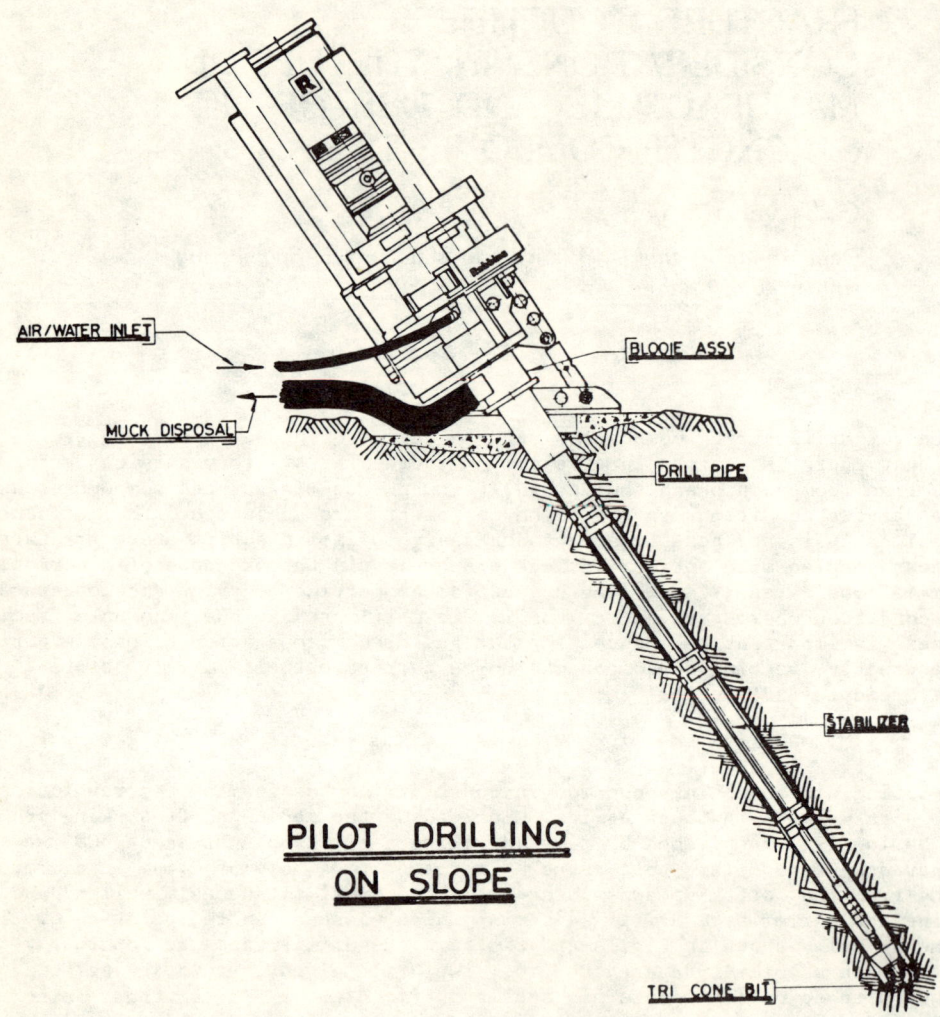

Fig 1a

Fracture Toughness Considerations

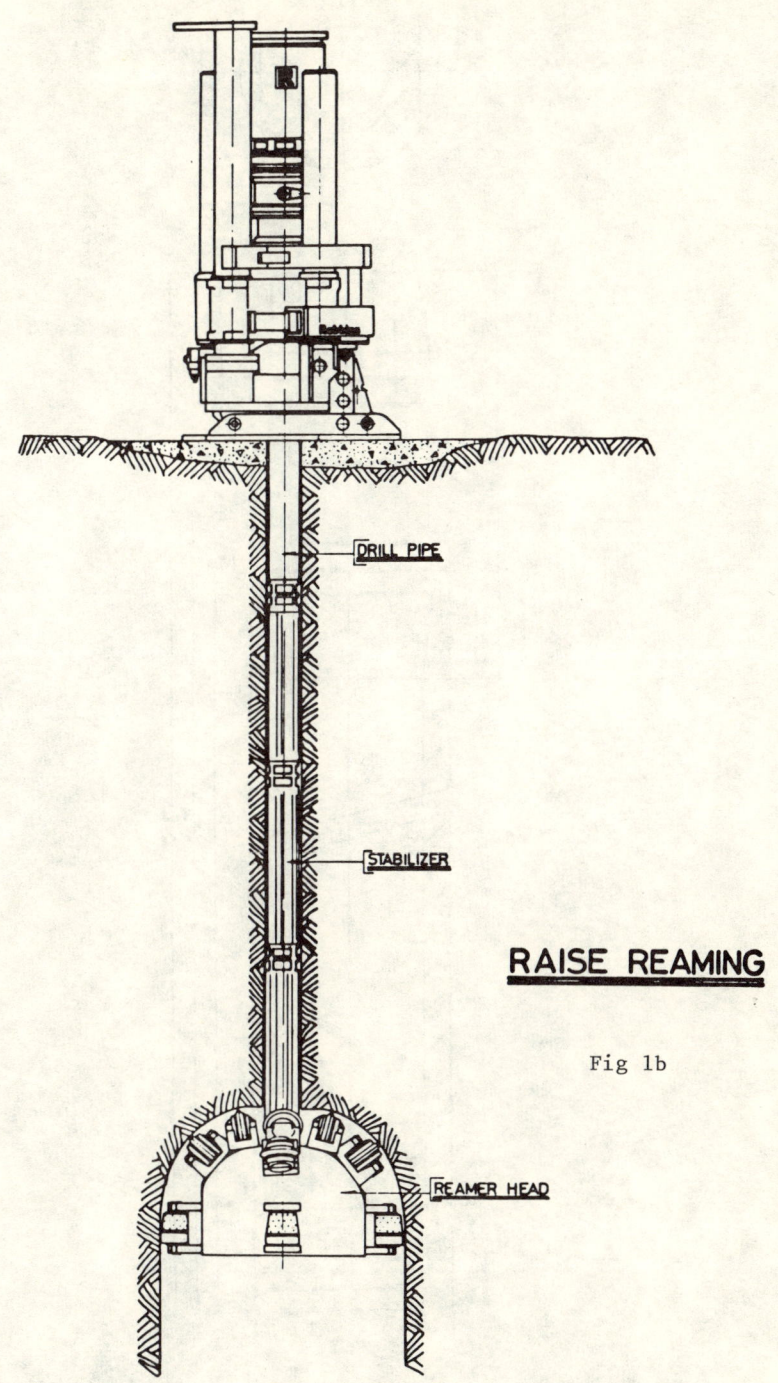

RAISE REAMING

Fig 1b

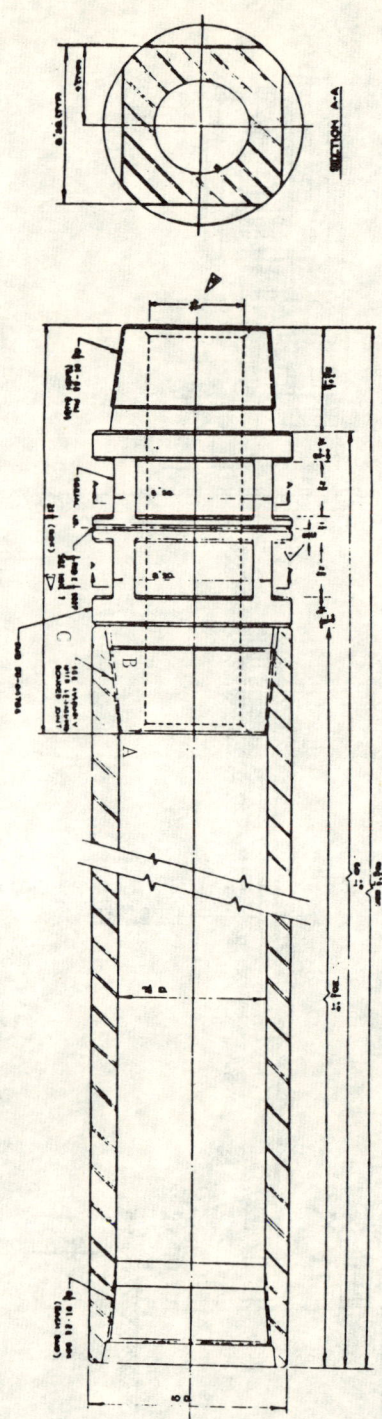

Figure 2. Drill pipe-double pin sub assembly

MECHANICAL DESIGN

Fig. 3. shows a photograph of a fractured double pin sub. Fig. 4. shows a section through the tapered thread which clearly reveals a number of fatigue cracks at the root of the threads.

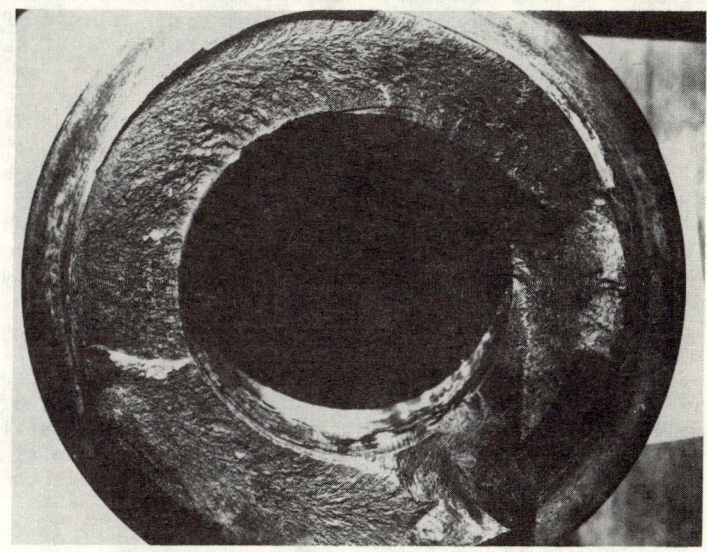

Fig. 3.   Brittle fracture of a double pin sub

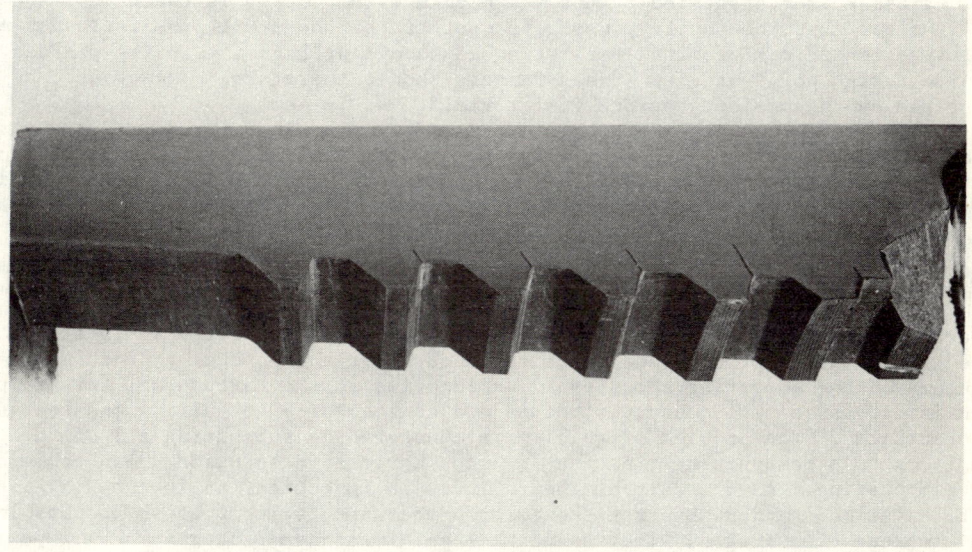

Fig. 4.   Tapered thread with fatigue cracks
outlined by magnetic particle testing

According to the instructions of the manufacturer the drill rods are to be
assembled by applying a torque of 244 kN.m. During operation the tensile loading
on the drill string is limited to 2900 kN. At the maximum specified loading this
results in a tensile stress of about 130 MPa superimposed on a shear stress of
35 MPa on the inside of the drill rod. (location A in Fig. 2) Using the Tresca
yield criterion this is equivalent to an effective tensile stress of 148 MPa. To
calculate the local stress at the root of the internal thread of the drill pipe
the influence of the stress raiser presented by the change of section at the
thread should also be taken into account.

At the location of the failure of the double pin sub shown in Fig. 3. (location B
in Fig. 2.) a stress calculation is complicated by the presence of the threaded
joint. The tensile stress at this location will be influenced by the stress
caused by the tightening of the joint as well as that caused by the axial load on
the drill string. The tightening stress is directly proportional to the tightening torque and inversely proportional to the tangent of the sum of the lead angle
of the thread and the friction angle. Depending on the actual value of the
friction coefficient this could vary considerably. Using a friction coefficient
of $\mu = 0,1$ and a tightening torque of 244 kN.m this results in a tightening stress
of approximately 750 MPa. The stress due to the axial loading (130 MPa) should
not be added directly to the tightening stress since the application of this
loading will invariably result in some reduction of the tightening stress. The
reduction will apart from the geometry of the joint also be affected by the
accuracy of the machining. For the purpose of a calculation of the combined stress
half of the stress due to axial loading was added to the tightening stress. The
shear stress exerted during drilling at location B will inter alia also be
dependant on the friction coefficient between the surfaces in contact.

Using the previous assumption calculation gives the effective tensile stress at
the point of fracture of the double pin sub equal to 820 MPa. If the presence of
the stress raiser presented by the sharp change in section is taken into account,
it is clear that the double pin sub in contrast to the drill pipe is indeed a
highly stressed member with local stresses in all probability in the plastic range.
From a design point of view stress raisers should therefore be alleviated as far
possible by the use of generous fillet radii.

From a design point of view it is advantageous to have double pin subs subjected
to higher stresses than the drill pipe. In this way the critical component is
much smaller and consequently can receive special attention in optimising its
mechanical properties.

MANUFACTURING PROCESS

Apart from the design considerations which have been enumerated it is clear that
the manufacturing process and machining accuracy can also have a decisive influence on the operating stresses. Inaccuracies in the thread form for example,
can lead to excessive contact stressed and consequently to galling and tearing of
the surfaces. The problem of galling is countered by subjecting all of the thread
surfaces to a phosphating treatment in phosphoric acid solution known commercially
as Kem-plating. As a result of the failures an investigation in the U.S.A. by
R.B. McLellan and F.R. Brotzen showed that the Kem-plating process results in the
development of hydrogen. They found that specimens Kem-plated absorbed hydrogen
which in turn could result in hydrogen embrittlement. To prevent thread galling
manufacturers also recommend the application of a heavy zinc base grease during
assembling of the drill string.

Inaccuracies in the axiality of the threads would result in a crooked drill string
which in a straight pilot hole would result in a bending stress added to any other

stresses present. Another machining inaccuracy which would affect the magnitude of any induced bending stresses in the screwed joint would be the extent to which the surfaces of the mating shoulders (position C in Fig. 2.) fit together. Calculations show that in the extreme the bending stresses in a close fitting joint, tightened so that mating surfaces always maintain contact, would be $2\frac{1}{2}$ times less than the bending stresses in a poorly fitted joint.

Fig. 5. Hard spot on surface due to the formation of martensite during machining

Metallographic examination of one of failures also revealed the presence of hard spots on the machined surfaces which consisted of shallow patches of white untempered martensite (Fig. 5 and 6.) in a matrix of tempered martensite. Some of these hardened spots with hardness values up to Vickers hardness HV 600 also contained small cracks. Enquiry revealed that the drill string components are normally first hardened and tempered to a Vickers hardness HV 400 before any machining is done. The hardened spots on the surface could therefore only have arisen during the machining operation. Machining of components with such a high hardness on a production basis is technically difficult and a blunt tool with an unstable built up cutting edge could presumably result in local temperatures which were occasionally high enough to result in austenitisation and consequently in hardening. The presence of such hardened spots of fresh martensite especially in the highly stressed fillet area of the double pin sub is an obvious source of cracks.

OPERATING CONDITIONS

The operating conditions in typical South African hard rock practice underground is particularly demanding with regard to load requirements. During use the drill string is also subjected to vibrations and fluctuations in loading due to variations in the geological formation. The resultant variation in stress could in time also result in metal fatigue. This would particularly be the case close to the reamer head. Additional cyclic stressing could also occur as a result of a crooked pilot hole. This would result in rotational bending fatigue stressing which would be superimposed on the relatively static stress due to tightening of the joint and the stressing due to reaming. Cyclic stressing of the double

Fig. 6. Cracking of untempered martensite formed during machining

pin sub at the threaded section in particular is to be avoided, since this position is also subjected to a high "static" tensile stress which would reduce the fatigue limit considerably.

During drilling water is passed down the drill string to aid in the reaming process. Typical mine water tends to be acidic and is quite aggressive as far as corrosion is concerned. Dissolved chlorides can be as high as 0,3%. Water with such a high chloride content in contact with highly stressed high strength steels can not only result in stress corrosion but also in corrosion fatigue where cyclic stresses are present.

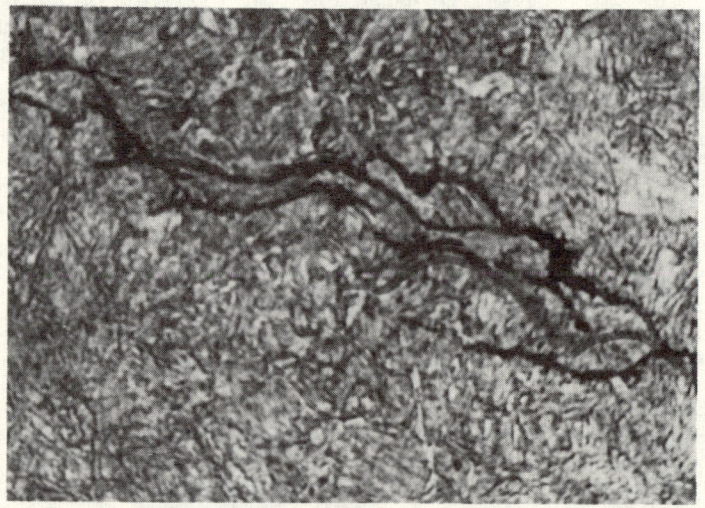

Fig. 7. Branching tip of a corrosion fatigue crack

Fig. 7 shows the crack tip of a fatigue cracks also shown in Fig. 4. at lower magnification. The branching nature of these cracks are typical of that usually present in cases of corrosion fatigue. Fig. 8. shows features of the resultant fatigue failure.

Fig. 8. Surface features of a drill pipe fatigue failure

MATERIAL REQUIREMENTS

From the above enumeration it is clear that material requirements for the double pin subs are very severe involving a combination of cyclic fatigue stress superimposed on very high tensile stresses which could be adversely affected by the manufacturing tolerances allowed. In addition added complications existed in the form of a corrosive medium, possible hydrogen embrittlement as a result of the Kem-plating process as well as the presence of hard spots of untempered martensite. The problem of finding a suitable material with optimum properties poses a problem. In general it can be said that the highest tensile strength would be required for resisting the high tightening stresses. As far as fatigue resistance is concerned optimum fatigue strength in the presence of stress raisers is obtained at a tensile strength of about 1200 MPa (see Fig. 9.) In the case of corrosion fatigue the situation is much more complicated. Depending on the aggressiveness of the corrosive medium and the exposure time, the fatigue strength can be seriously reduced. In general the percentage reduction in fatigue strength is more serious in the case of high strength steels and the tensile strength where optimum fatigue properties are realised are reduced to values much lower than the value of 1200 MPa quoted above. Qualitatively much the same can be said for the suscepibility of steel to hydrogen embrittlement as well as stress corrosion. From a fracture toughness point of view the inverse relationship of the plane strain fracture toughness K_{1C} and yield strength is well known (see Fig. 10.) Maximum resistance to catasthrophic crack propagation would apart from other factors, require a low strength highly ductile material. Alternately acceptable values of fracture toughness at high strength values can be obtained by optimising chemical composition and heat treatment.

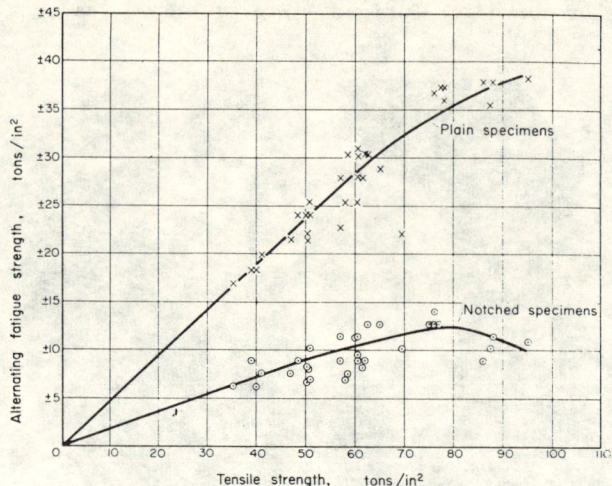

Fig. 9. The relationship between plain and notched fatigue strengths and the tensile strength of wrought steels (60° V groove, K_t = 4,4)
[From Pomp and Hempel[1], reprinted with permission from Addison Wesley]

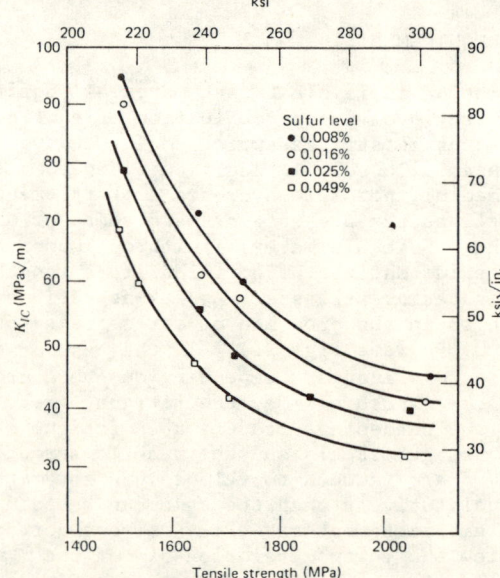

Fig. 10. Fracture toughness of 0,45 C-Ni-Cr-Mo Steel heat treated to different strength levels
[From Birkle et al[2], reprinted with permission of American Society of Metals]

Chemical Analysis

From an informative point of view Table 1 gives the chemical analysis of a number of different drill string components which failed.

TABLE 1   Chemical Analysis

	Sample A	Sample B	Sample C	Sample D	Sample E	SAE 4333
%C	0,46	0,21	0,36	0,29	0,30	0,30 - 0,38
%Mn	0,83	1,36	0,90	0,61	0,58	0,70 - 1,0
%P	0,015	0,027	0,017	0,011	0,013	0,035 max
%S	0,012	0,018	0,03	0,011	0,012	0,040 max
%Si	0,43	0,26	0,31	0,22	0,28	0,20 - 0,35
%Ni	2,22	1,28	2,11	2,30	2,23	2,0 - 2,5
%Cu	0,15	0,21	-	-	-	-
%Cr	0,97	0,53	1,02	0,96	0,88	0,8 - 1,2
%Mo	0,51	0,30	0,52	0,51	0,47	0,5 - 0,65
%V	0,02	0,06	0,005	0,11	0,11	-

From Table 1 it is clear that except for a few instances most of the element analyses conform to the SAE 4333 specification. SAE 4333 steel is an acceptable and widely used high tensile strength material with good fracture toughness properties. Although it is dangerous to generalise, the carbon content of low alloy steels seems to dominate the fracture toughness properties. By adjusting the tempering temperature in order to obtain the same strength level, steels with lower carbon contents generally have the highest fracture toughness values. On the other hand, the low hardenabilities associated with such "low carbon" steels usually necessitates the use of fairly high additions of other alloying elements (Mn, Ni, Cr & Mo). Experience has shown that in most instances where lack of fracture toughness was encountered in the low alloy steels with carbon contents in the range 0,3% to 0,4% it could be traced to either a lack of sufficient hardenability or otherwise in slack quenching during hardening. The resultant small amounts of either primary ferrite and or upper bainite together with martensite in the structure increases the brittle transition temperature of the tempered steel to such an extent that brittle failures can occur even after such steels have been tempered at very high temperatures. Calculation shows that the ideal critical diameter of 4333 steel could vary from 210 mm to 650 mm. In most cases this would be sufficient to allow full hardening of the solid stock (± 300 mm diameter) from which the double pin sub is machined, if water quenching is used. Oil quenching with an unfortunate combination of alloying elements all on the lower limit of 4333 specification will not result in full hardening. This would most probably result in a poor fracture toughness although the other tensile properties may be fully acceptable. From a practical point of view it may be advantageous to rough turn the double pin sub blanks in the softened condition before hardening and tempering. This will not only reduce the machining time but also improve the quenching due to the hole machined in the middle.

Tramp elements such as phosphorous, tin and antimony are also diletrious in that they may give rise to temper embrittlement and consequently also lead to low fracture toughness. In this respect the very low phosphorous contents with the exception of sample B is noteworthy.

Impurities in steel, especially inclusions such as sulphides (Mns) and oxides (SiO_2, Al_2O_3), can also influence the fracture toughness. During crack propagation decohesion along the particle-matrix interface can result in void formation ahead of the crack resulting in a reduction in fracture toughness. This is shown in Fig. 10. which shows the influence of MnS in high strength steels. For the drill string components vacuum degassed steel was specified which should result in a low inclusion content. Metallographic examination in some instances however showed a fair amount of inclusions. In general the low sulphur content of the components shown in Table 1 with the exception of sample C is highly desirable.

Heat Treatment

Metallographic structure in combination with strength has in important influence on the fracture properties of steels. The important parameters in this regard is the grain size as well as the size and distribution of second phase particles. In hardened and tempered steels the presence of even small percentages of primary ferrite or upper bainite can influence the fracture toughness beyond recognition. Table 2 for example shows heat treating data and the resultant mechanical properties of two 4m long 450 mm diameter axles. These axles were manufactured from 0,4%C 2,5%Ni 0,7%Cr 0,5%Mo (B26 M40) steel.

TABLE 2 Heat treatment and mechanical properties of B26 M40 steel axles 450mm diameter

	Fractured axle	Replacement axle
Austenitising Temperature (°C)	± 900	850
Quenching practice	time quench	direct quench
Tempering Temperature (°C)	620 – 630	645 – 650
Yield Strength (Re) MPa	735	755
Tensile strength (R_m) MPa	1000	910
Elongation (A) %	15	21
Red in area (Z) %	35	59
Charpy Impact Energy at room temperature (J)	8	104
25J Brittle Transition temperature	+110°C	-86°C
Brinell Hardness (HB)	282	278

The time quenching operation involved the interruption of the water quench for a certain period before the shaft was finally quenched to room temperature. Interruption of the quench was considered essential to prevent quench cracking. During the direct quench water quenching was maintained until the average temperature was about 200°C whereafter the axle was directly tempered. The incidence of cracking during the direct quenching in this instance was prevented by the fact that a small percentage austenite was retained before tempering commenced. The retained austenite was sufficiently ductile to alleviate the resultant transformation stresses.

The comparison shown in Table 2 indicates the equivalence of most of the mechanical properties with the exception of the large difference in the Charpy impact values. The very large difference in brittle transition temperature -82°C in comparison with +110°C is significant, Metallographic examination showed that the time quenched axle contained about 15 to 20% of upper bainite together with tempered martensite whereas the directly quenched axle had a fully hardened and tempered structure. From experience it appears as if it is an accepted custom amongst heat treatment shops to use a slack quenching procedure followed by an adjustment of the tempering temperature to obtain the specified hardness. This represents an unacceptable practice which could result in a mixed structure coupled with a low fracture toughness.

MECHANICAL PROPERTIES

Table 3 gives the mechanical properties of the samples of which the chemical analyses were given in Table 1.

TABLE 3 Mechanical properties of drill string components which failed

	Sample A	Sample B	Sample C	Sample D	Sample E
Rockwell Rc Hardness	45	39	40	39	38
Yield strength Re (MPa)	1330	-	1120	-	-
Tensile strength Rm (MPa)	1420	-	1240	1100	1150
Elongation A(%)	8,5%	-	14	16%	15%
Red in Area Z(%)	30%	-	48%	-	-
K_{IC} (MPa -$\sqrt{m}$)	-	-	126	-	-
Charpy Impact (J)	22,28,32	-	-	90	80
Falure mode	Brittle	Fatigue	Fatigue	Fatigue	Fatigue

It is significant that the only component which failed in a brittle fashion (sample A) was significantly harder that the other drill string components investigated. The Charpe impact energy value for this sample was similarly only a third of that of the other samples.

FRACTOGRAPHY

Microscopic examination of the specimens shown in Table 3 in all instances revealed a correctly hardened and tempered structure. The drill string components B C D and E which included both double pin subs as well as drill pipes(stabilizers) all clearly failed by a corrosion fatigue (see Fig. 11.) mechanism. Fatigue crack propagation was usually extensive. From a fracture point of view this represents a satisfactory situation. Such an extensive propagation of a fatigue crack is expected to take a fairly long time and presumably allow detection by non-destructive testing before fracturing in a unstable manner. In contrast the

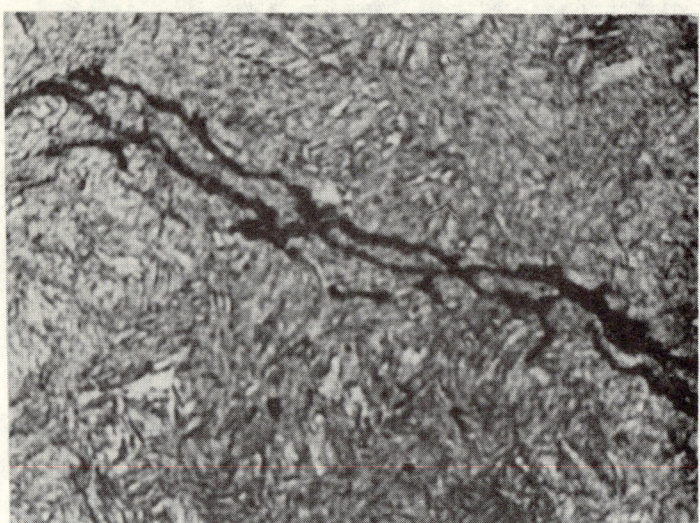

Fig. 11. The crack tip as a result of a corrosion fatigue crack showing the branching nature of the crack propagation

component A (double pin sub) failed in a brittle manner with no signs of metal fatigue. The fracture started at a small defect and propagated in two directions around the circumference following a hellical path along the two star thread (see Fig. 12.)

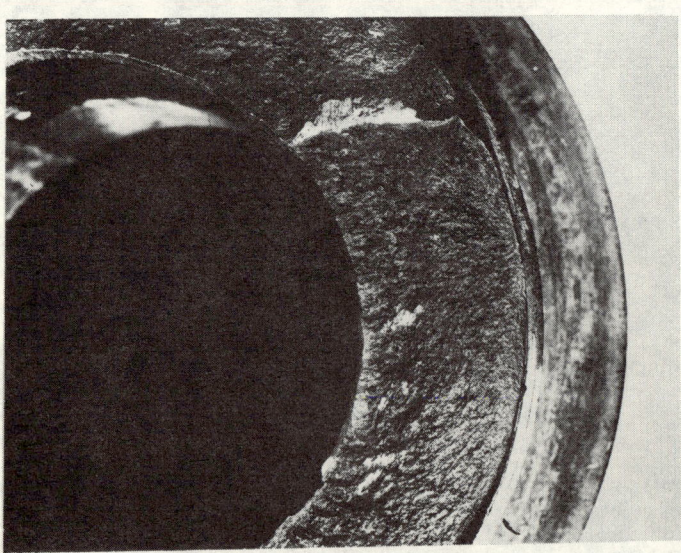

Fig. 12. The origin of the brittle fracture.

This portion of the fracture propagated as a result of a tensile or bending loading and was accompanied by the characteristic chevron marks which indicate the direction of crack propagation. The fracture surface of the final rupture was more complex and shear like in nature, and connected the two initial fracture surfaces at an angle across the two threads in between. The virtual absence of shear lips on the inside and outside diameters of the fracture surface is further evidence of a relatively brittle failure. Macroscopic examination at the fracture origin showed a number of subsidiary dracks at the origin. Metallographic examination Fig. 13. showed the presence of a branching netword of fine cracks characteristic of stress corrosion cracking. Scanning electron microscopy examination of the fracture surface was inconclusive as a result of general corrosion of the fracture surface.

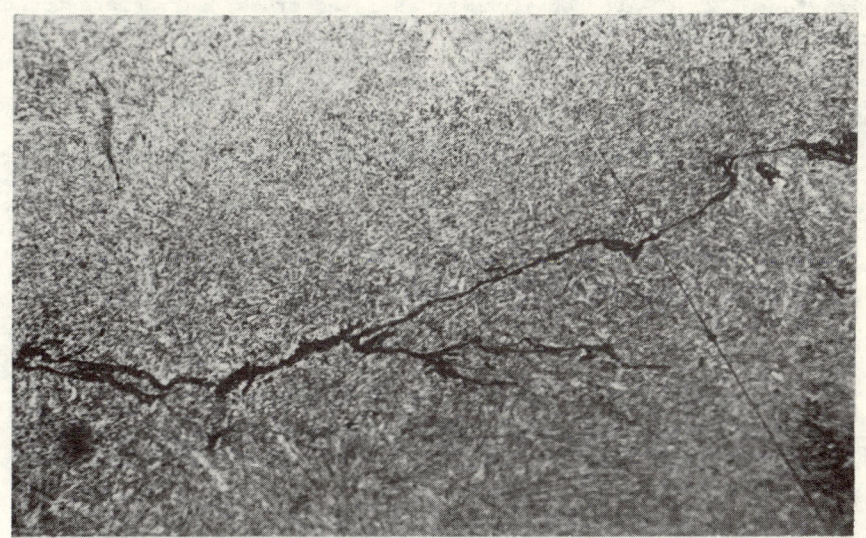

Fig. 13. Branching crack pattern as a result of stress corrosion at the origin of the brittle fracture

FRACTURE ANALYSIS

From the investigation it is clear that a combination of adverse conditions such as cyclic stresses superimposed on high static stresses in the presence of stress raisers could arise. In addition embrittlement due to hydrogen, surface defects due to untempered martensite spots and the presence of a corrosive environment complicates the issue. It would be unrealistic to expect to find a material which would operate satisfactory under all of these adverse conditions. The mechanical properties such as represented by the components B C D and E (Table 3) are considered very close to the optimum attainable for this type of application. The value of K_{1C} = 126 MPa $\sqrt{m}$ determined according to B S D D 3 : 1971 for component C is quite representative of published values for similar steels heat treated to the same hardness. Indirectly this also indicates the absence of temper brittleness. The higher hardness of component A (Rc = 45) coupled with the resultant lower Charpy values are considered less desirable. In this regard the fact that this component is the only one which failed in a brittle manner is probably significant.

Using the Rolfe and Barsom relationship $\frac{(K_{1C})^2}{E} = 2(CVN)^{1,5}$ to calculate the equivalent K_{1C} from the average Charpy energy a value of 80 MPa $\sqrt{m}$ is obtained for component A. Due to the difficulty in determining the true stress intensity factor, an upper bound value of the probable defect size for the brittle fracture of this component can be calculated from the quation $K_{1C} = \sigma\sqrt{\pi a}$ (σ = applied stress a = half crack length). By assuming stresses equal to the yield stress (1330 MPa) and K_{1C} = 80 MPa $\sqrt{m}$ a crack with a length a = 13 mm is obtained From the fractographic analysis however the defect size responsible for this fracture appeared smaller. From this it can be concluded that the fracture toughness was probably adversly affected by the presence of the corrosive mine water. No estimate could however be made of K_{1EAC} the environment assisted cracking fracture toughness. In general it can be said that the fracture toughness of high strength steels are affected much more by a corrosive environment (see Fig. 14.)

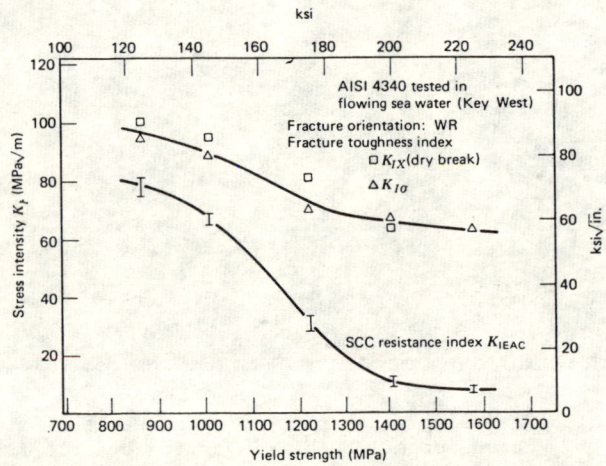

Fig. 14. Effect of yield strength on K_{1C} and K_{1EAC} (in water) for 4340 steel |From M.H. Peterson[3], reprinted with permission of National Association of Corrosion Engineers|

As far as the fatigue failures are concerned it is of interest to note that most of these occurred in the threads of the joints which were screwed together with an epoxy resin. The formation of extensive fatigue cracks around the root of so many threads as shown in Fig. 4. is quite uncommon in pure fatigue, unless the cyclic stresses are far in excess of the fatigue limit. The extensive propagation of the fatigue cracks in this instance is further proof of the action and presence of corrosion fatigue which was substantiated by the metallographic examination. The heavy zinc base grease which is used on the one set of joints most probably prevented the ingress of water affording protection. Why the same protection was not afforded by the epoxied joint is not clear. Positive protection from stress corrosion as well as corrosion fatigue should be possible by plasma spraying the thread surfaces with zinc. In this way galvanic protection could be attained.

The elimination of hardspots due to the formation of martensite during machining could be eliminated by controlling the hardness in the vicinity at a Rockwell hardness HRc = 40 and by the use of numerical controlled machining of the thread surfaces. Whether hydrogen embrittlement was important in the instances of failures which occurred was never positively established. At present the Kem-plating process is still being used apparently without any detrimental effects.

In essence it can be concluded that adequate fracture toughness in the case of drill string components which mainly fail as a result of corrosion fatigue is necessary to allow sufficient fatigue crack propagation without the danger of catastrophic failure. Extensive fatigue crack propagation associated with a high fracture toughness will then enable the replacement of such critical components before complete failure occurrs.

From a practical point of view the investigation highlights the complexity of problems which confronts the engineer at the design stage. In spite of the advances made in Material Science and Fracture Mechanics, good judgement and practical experience is still of paramount importance. Otherwise large safety factors (ignorance factors) will have to be used.

REFERENCES

Pomp, A., and M Hempel (1950). Arch. Eisenhuttenw 21, 53.
Birkle, A.J., R.P. Wei and G.E. Pellisier, (1966). Trans ASM 59, 981
Peterson, M.H., B.F. Brown, R.L. Newbegin and R.E. Groover,(1967).Corrosion,23 142.

CRACKING IN WELDMENTS OF STRUCTURAL STEELS

R. E. Dolby

The Welding Institute, Abington, Cambridge, England

ABSTRACT

Three types of cracking that can occur during the welding of structural steels are reviewed; hydrogen induced cracking in the HAZ and weld metal, solidification cracking and lamellar tearing. Emphasis is placed on ways of avoiding cracking during fabrication and on the significance of current trends in steel compositions.

INTRODUCTION

Structural steels can exhibit several forms of cracking when welded. The cracks are examples of fracture occurring locally to the weldment and the purpose of this paper is to discuss three common cracking problems with emphasis on their avoidance rather than on mechanisms, details of which can be found in the literature. (Coe, 1973; Farrar, 1974; Bailey & Jones, 1978). The three types of cracking to be discussed are associated with the fabrication stage of an engineering structure rather than in-service performance. In principle, the risk of the cracks forming during fabrication could be assessed by fracture mechanics techniques and indeed some attempts have been made to apply this approach, (Cane, Dolby & Baker, 1973; Makhnenko & co-workers, 1979) but particular difficulties lie in the quantitative assessment of the acting stress or strain and the production of suitable specimens containing the weld regions of interest. For these reasons cracking risks are usually assessed by direct welding tests, or by correlation of small scale test results with practical experience.

HAZ AND WELD METAL HYDROGEN INDUCED CRACKING

A frequent question posed by fabricators is the following; "Is preheat necessary to avoid hydrogen induced cracking and if so what level of preheat should be used?" From economic considerations the first question is usually the most important one to be settled. A decision on the need for preheat must be based on :

   i) the anticipated check analysis range of the steel supplied

  ii) the plate/section thicknesses in each joint

 iii) possible restrictions on welding position due to access, etc

  iv) the expected weld hydrogen level for the consumables chosen

v) joint restraint

Preheating decisions for C, C-Mn and C-Mn microalloyed steels to avoid HAZ cracking of the type shown in Fig 1 have been greatly aided in recent years by the development of nomograms such as that in Appendix E of BS 5135 'Specification for metal arc welding of carbon and carbon-manganese steels'. However, there is increasing evidence that the nomogram is not accurate for C levels less than 0.12% and at the root of the problem is the IIW carbon equivalent formula.[1] There should be no surprise that this formula appears inadequate for low carbon steels in view of the higher carbon steel compositions used for its original derivation. In essence it is a formula describing the hardenability of the HAZ but works well to describe the risk of cracking within a limited range of composition. Outside this range, where hardenability and susceptibility to cracking are not necessarily related in the same way, we need new formulae to describe the risk of cracking.

The best approach for assessing low carbon steel behaviour at the present time appears to be that due to Ito & Bessyo (1968)

$$P_{cm} = C + \frac{Mn}{20} + \frac{Si}{30} + \frac{Ni}{60} + \frac{Cu}{20} + \frac{Cr}{20} + \frac{Mo}{15} + \frac{V}{10} + 5B$$

This was originally developed from examination of C-Mn and low alloy steels having carbon contents in the range 0.07-0.22% using the y-groove welding test and weights carbon more strongly than other elements compared to the IIW formula. The Welding Institute approach has been to use restrained fillet welds for assessing cracking risks and these tests in fact form the basis for the nomograms in BS 5135. Whilst the merits of particular welding tests can be debated at length, recent findings, confirming that plate composition can influence the residual stress in a weld (Hart, 1976), point strongly to the need to use some form of self-restrained welding test for ranking steels with regard to hydrogen cracking. Thus more confidence can be placed in data from y-groove and fillet weld tests than in data from tests where stress is an independent variable, as would be the case in a fracture mechanics approach or in the implant method. (Granjon, 1969)

In general, the improvements in steel processing and lower carbon contents in recent years have brought about an improved resistance to HAZ hydrogen induced cracking and have encouraged less rigorous, more economic, welding procedures. For example, improved controlled rolling techniques, incorporating reheating above Ac_3 to produce fine grained austenite at an intermediate stage in processing now enable plates in 12.5mm thickness to be produced with a yield of $380N/mm^2$ having a carbon level of 0.06%, and an IIW carbon equivalent of only 0.26%. (Sumitomo Metal Industries, 1976) Significant reductions in the carbon equivalent of rolled sections have also been achieved by improved rolling practices. (Lessells, Randerson & Smith, 1978) Many fabricators are agreed that a contribution of major significance would be the development of steels of $350N/mm^2$ yield strength up to 50mm thickness which could be welded without preheat. Such 'crack-free' steels have been reported in the literature, specifically from Japan. Suzuki (1976) details a steel of $380N/mm^2$ yield (HT60CF) which reportedly can be welded in thicknesses up to 50mm without preheat and has the following composition:

0.06C, 1.31Mn, 0.16Cr, 0.16Mo, 0.03V, P_{cm} = 0.16%, CE(IIW) = 0.36%

---

[1] $CE = C + \frac{Mn}{6} + \frac{Ni + Cu}{15} + \frac{Cr + Mo + V}{5}$

Fig. 1. Typical HAZ hydrogen induced cracks.

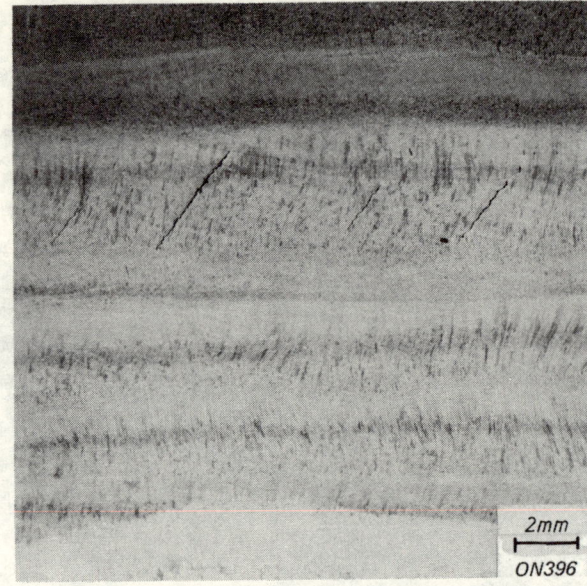

Fig. 2. Weld metal hydrogen induced cracks in a multipass submerged arc deposit.

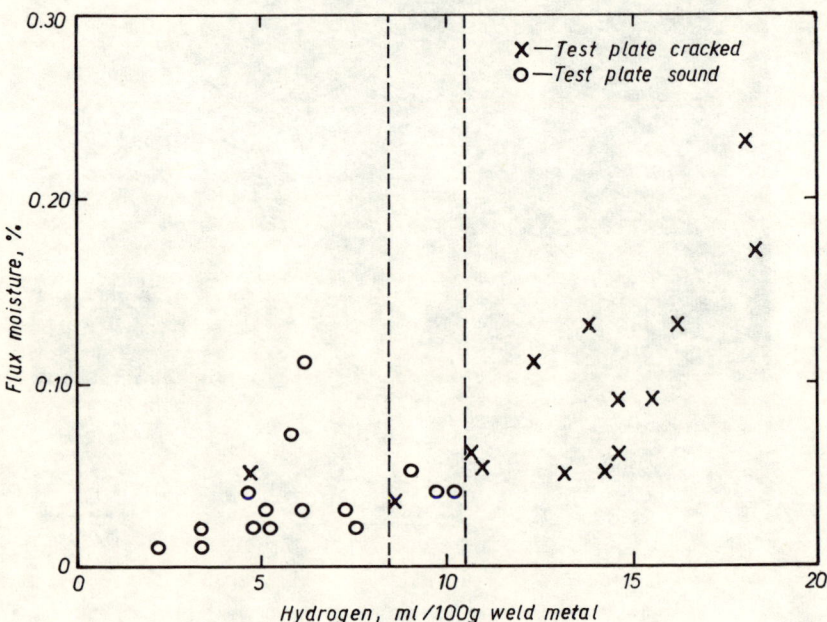

Fig. 3. Relationship between flux moisture, weld metal hydrogen content and the incidence of weld metal cracking in butt weld tests (after Wright and Davison).

Whilst the data presented show that plate up to 38mm can be satisfactorily welded without preheat using y-groove tests, some weld metal hydrogen cracking was found at 50mm thickness. This highlights a major problem area currently being encountered in structural and pressure vessel fabrication, namely an increased incidence of weld metal cracking.

Found with manual metal arc (MMA), submerged arc and electroslag processes, cracking in the weld deposit usually disappears on taking remedial measures normally employed for overcoming hydrogen cracking. Typically, in MMA and submerged arc welds, it is located as a series of buried small fissures angled at approximately 45° to the surface in the weld deposit (Fig 2) and has been termed chevron cracking. Small fissures can also be found in electroslag weld metal and are reported as being intergranular. (Konkol & Domis, 1979)

Because of the importance of the problem to many industries, several organisations have researched this topic and the results of some of this work were reported at recent conferences.(Mota, Apps & Jubb, 1978; Wright & Davison, 1978, Konkol & Domis, 1979) In one study into the submerged arc process, practical experience was confirmed in that agglomerated basic fluxes showed a greater tendency to fissuring than lower basicity and/or fused fluxes. (Wright & Davison, 1978). The effect of weld hydrogen and flux moisture on the incidence of cracking, taken from the same study, is reproduced in Fig 3. The weld hydrogen limit of 10.5ml/100g weld metal for avoiding cracking must be viewed with some caution since there is no agreed standardised method for measuring submerged arc weld metal hydrogen levels and the data were obtained for a specific welding condition. However, the work reinforces the results of a number of similar investigations and practical experience which have shown that hydrogen is a major controlling factor and that anti-hydrogen measures, such as increased preheat and interpass temperatures, higher flux and electrode baking temperatures, etc, are generally successful in eliminating fissuring. The use of clean filler wires can be just as important in avoiding fissuring as the choice of flux type, since weld metal hydrogen arises, in part, from grease, oil, etc, arising from wire drawing.

Eventually a nomogram procedure for predicting risks of weld metal cracking should emerge, and it is obviously desirable that any new predictive procedure should be combined in some way with nomograms in current use, e.g. Appendix E in BS 5135. The latter needs to be improved not only because of its inadequacy to cope with weld metal cracking, but also because of (i) the more widespread use of higher strength quenched and tempered C-Mn steels where reliable data are scarce, and (ii) the limited accuracy of existing methods for predicting safe procedures when using the MIG, flux cored wire and submerged arc process. Some account may have to be taken of the evidence that cleaner (low S) steels are less tolerant to normal welding procedures than normal S steels and may show a greater susceptibility to HAZ hydrogen cracking. This has been recently explained by the higher HAZ hardenability of low inclusion content steels. (Hart, 1978).

WELD METAL SOLIDIFICATION CRACKING

This form of cracking is encountered sporadically as a centreline defect in submerged arc welded root runs or other runs involving high dilution of parent steel into the weld. Typical cracks in a butt and fillet weld are shown in Fig 4. The cracking may be sub-surface and difficult to detect. It can be also found in vertical welding processes, such as electroslag, where dilution of the parent plate into the weld is normally high. Minimising the cracking usually involves a reduction of welding speed and adjustment of other welding parameters and consequently deposition and production rates have to be lowered.

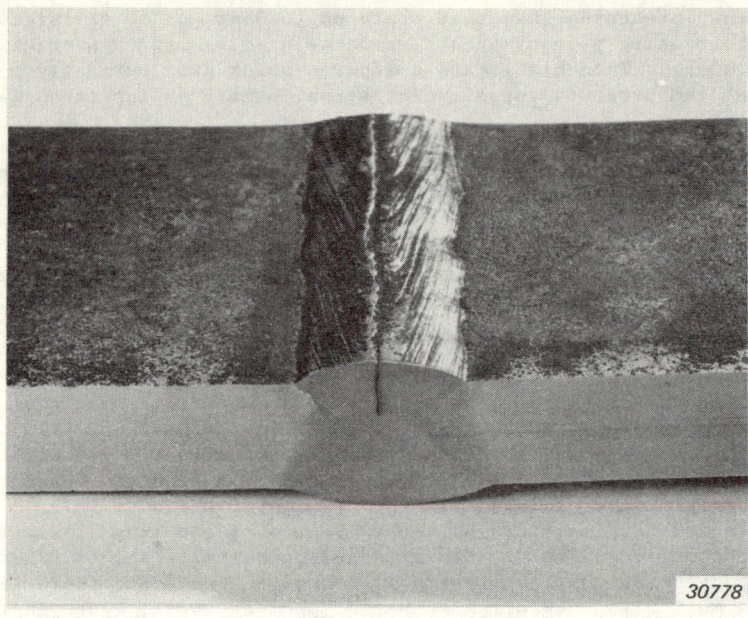

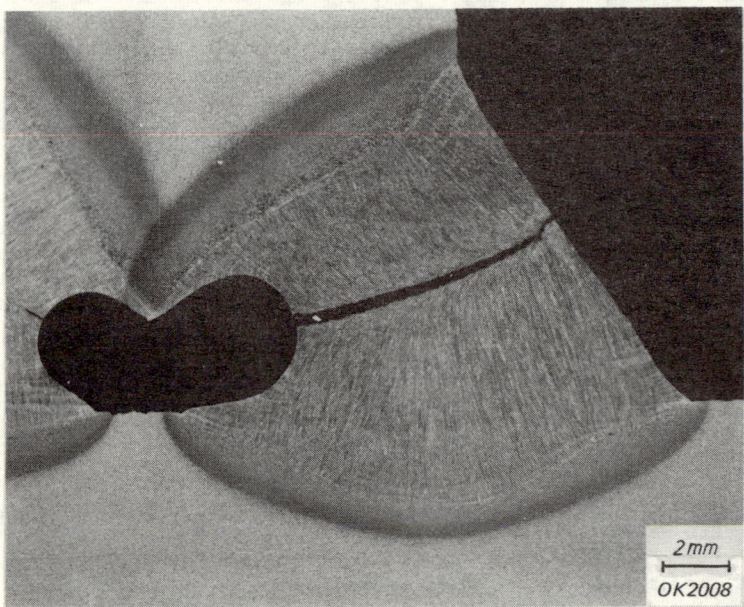

Fig. 4. Typical solidification cracks in butt and fillet welds of C-Mn steels made with the submerged arc process.

Recent work at The Welding Institute (Bailey & Jones, 1978) has allowed a closer definition of the important compositional factors in submerged arc welding and the following regression equation based on weld metal composition has been produced. Crack susceptibility is expressed by arbitrary units (UCS) established using the Transvarestraint test:

$$UCS = 230C^* + 190S + 75P + 45Nb - 12.3Si - 5.4Mn - 1$$

For C <0.08%, C* is taken as equal to 0.08%C

This gives guidance as to the key elements to control and monitor in plate, wires and fluxes, e.g. carbon, but the relative importance of composition, weld pool shape and strain acting across the joint still needs clearer definition. Decreasing the depth to width ratio of the weld bead can often solve a problem of severe cracking but in some situations, the strain factor can be dominant and changes in consumables and welding parameters can have little effect in mitigating cracking. By direct experiment and confirmed by experience, UCS values of greater than 19 and 25 for T fillet and butt welds respectively, normally indicate a crack sensitive composition in submerged arc weldments.

Undoubtedly, the development of controlled process HSLA steels having generally leaner chemistry and lower impurity elements than conventional steels has been helpful in reducing the incidence of solidification cracking, enabling high speed welding to be exploited widely in submerged arc panel lines, and in pipe and fabricated beam production. This is mainly a result of reductions in C and S and the fact that the levels of microalloying elements commonly employed, which may act to increase the risk of cracking, do not have a serious detrimental effect. Nevertheless, occasional problems still arise and will continue to occur in specific situations, despite the development of lower carbon steels, e.g. where excessive displacement of the weld root occurs caused by distortion, inadequate tacking or insufficient number or size of strongbacks.

LAMELLAR TEARING

Avoidance of lamellar tearing in welded construction (Fig 5) requires knowledge of material through thickness (short transverse) ductility and some indication of joint restraint. The usual measure of ductility is %RA in short transverse (ST) tensile tests using specimens of between 5 and 12mm diameter. The test is not aimed at reproducing restraint in HAZ regions but only at obtaining a measure of the non-metallic inclusion population in the steel and at reproducing the void coalescence fracture mechanism of lamellar tearing. Correlation with fabrication history has revealed limiting values of 10%, 15% and 20% STRA appropriate to avoiding tearing[2] in joints of low, medium and high restraint (Fig 6). German and French standards have since emerged enabling steel to be procured to %STRA minima of 15, 25 and 35% and an increasing number of steelmakers now offer grades of plate with guaranteed minimum ductility.

Welding Institute experience shows that mean STRA values of 20% are quite adequate to avoid tearing in joints even of high restraint, but in using the test for specification or plate release purposes some allowance must usually be made by user organisations for factors such as inclusion segregation. Thus minimum values often have been set at 25% and over. However, once specific steelmaking routes are confirmed as giving consistently high ST ductility, mechanical testing can usually be discarded.

---

[2] Stahl-Eisen Lieferbedingingen 096   1974

Norme Francaise NF A 36-202   1977

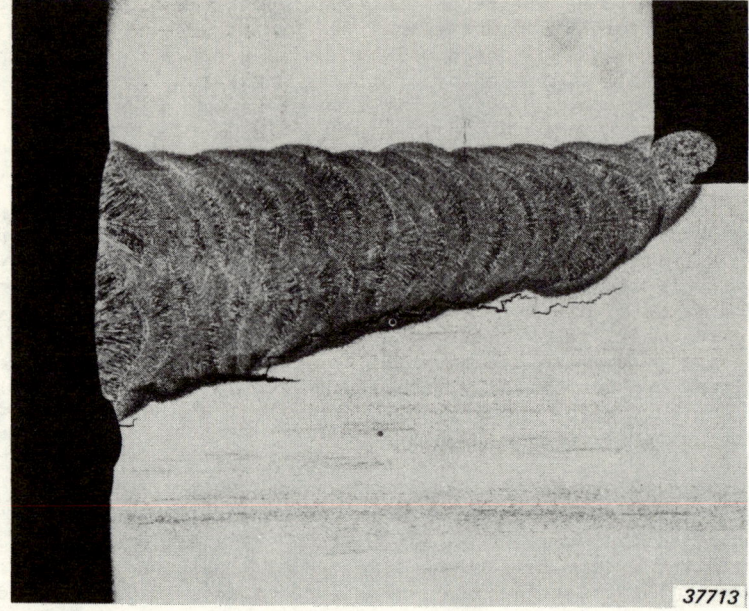

Fig. 5. Lamellar tear in flange plate of a box column.

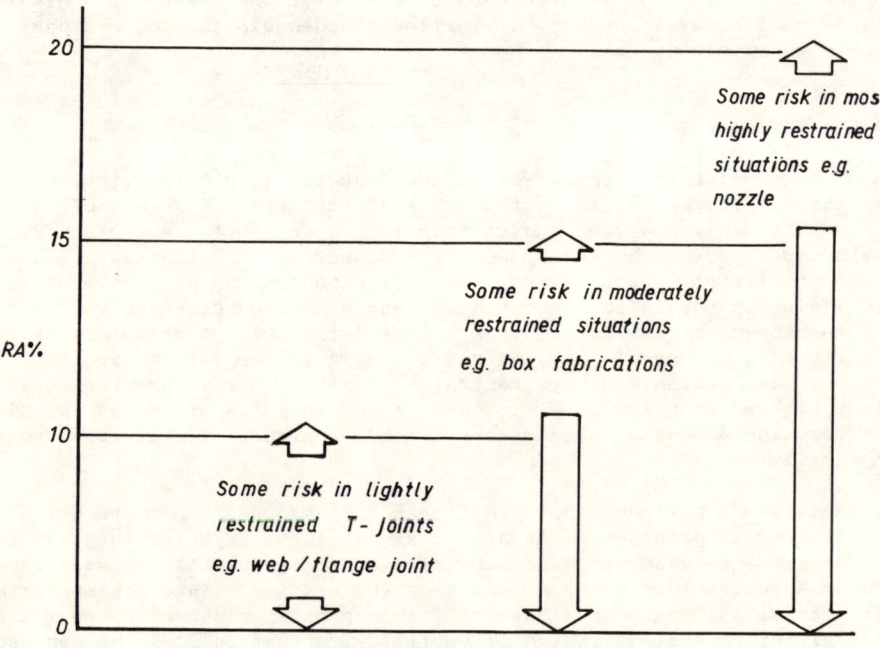

Fig. 6. Correlation of % STRA with practical experience of lamellar tearing.

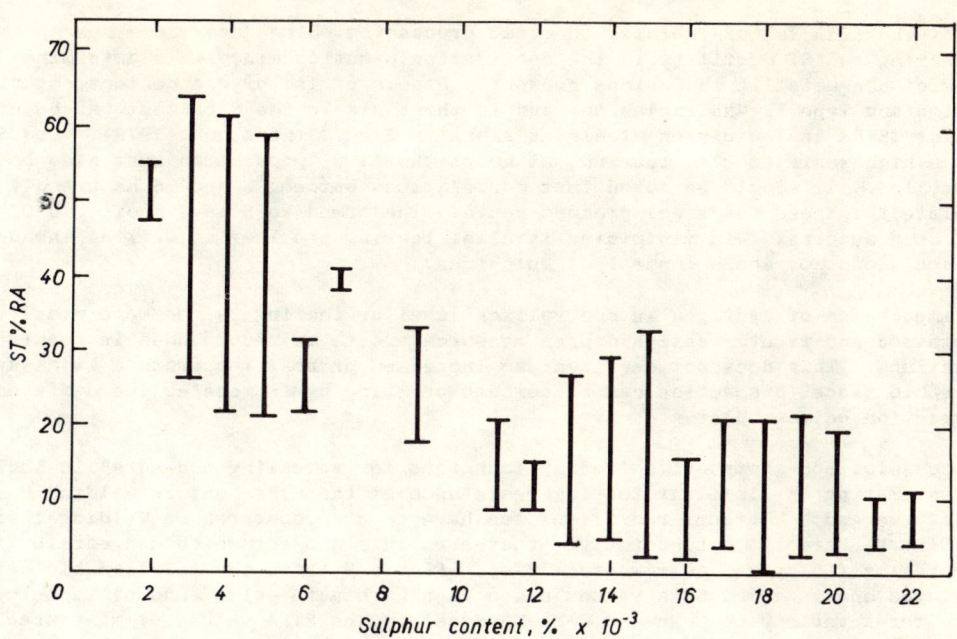

Fig. 7. Relationship between % STRA and sulphur content (after Kanazawa).

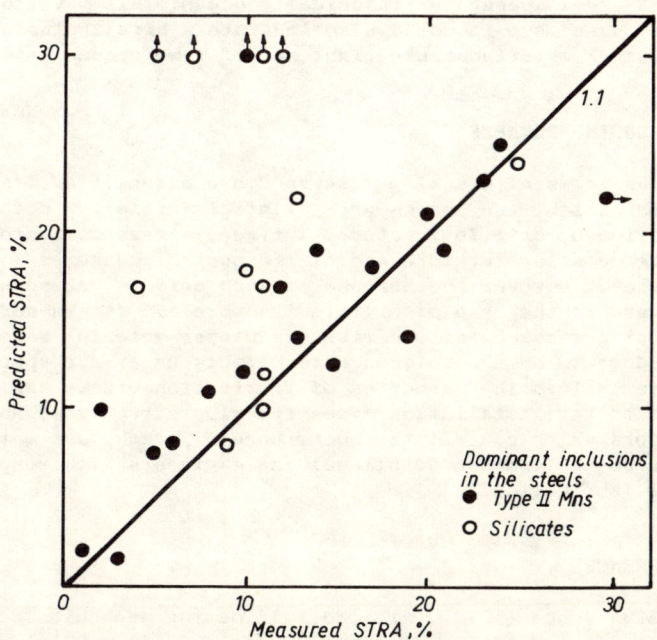

Fig. 8. Relationship between predicted and measured short transverse reduction in area values for various C-Mn and C-Mn-Nb steels. Those with Type II MnS inclusions dominant are Al-treated. Predicted values obtained using the ultrasonic techniques described by Dolby and co-workers.

One of the main factors related to steel processing which determines susceptibility to tearing and ST ductility is the deoxidation practice since this determines the types of non-metallic inclusions present. Steels of low oxygen content contain alumina and Type II MnS inclusions and in these steels the S content can be used to predict %STRA in low oxygen steels as shown in Fig 7. (Kanazawa, 1974). Low S steels have a high resistance to tearing, although the Al_2O_3 population must also be closely controlled. It should be noted that correlations between S and %STRA are a function of plate thickness and steel process route. The trend to S levels of <0.010% has thus been successful in minimising lamellar tearing problems as well as the use of REM additions for shape control of sulphides.

The importance of hydrogen in controlling lamellar tearing is now more widely recognised and tighter anti-hydrogen measures are to be recommended in repair situations. This does not mean that an increased preheat temperature is always desirable since this action can exacerbate cracking by increasing the differential contraction between plates.

A successful non-destructive testing technique for revealing non-metallic inclusions and estimating the lamellar tearing resistance of the plate before welding has not so far emerged. Particular difficulties have been encountered in Welding Institute/ NDT Centre, Harwell, collaborative studies on this subject which had earlier looked encouraging (Dolby and co-workers, 1974). Figure 8 shows the relationship between predicted and measured STRA values based on a high gain pulse echo ultrasonic technique for a variety of C-Mn and C-Mn-Nb steels. The STRA values of many steels are reasonably predicted but several appear clean ultrasonically, although showing low measured STRA values. Investigation showed that the inconsistency in STRA prediction appears to have its root in the variable impedance of inclusions to ultrasound, a factor which can be influenced by plate rolling history and heat treatment. In broad terms, if a steel appears ultrasonically clean, this may indicate a low inclusion content steel but it could also indicate a high inclusion content material where inclusion/steel interfaces are tight and of low impedance to ultrasound.

CONCLUDING REMARKS

Whilst the various forms of cracking discussed are examples of fractures that can occur during fabrication, the cracks are potential initiating defects for subsequent extension in service by brittle fracture, fatigue, stress corrosion,etc. Avoidance of fabrication defects is therefore one of the control measures for minimising failures in service. However the absence of such defects can never be totally assured and in the case of brittle fracture and stress corrosion, the major control for reducing service failure risks is proper material selection and choice of welding procedure to ensure tolerance to defects under the appropriate design conditions. Nevertheless the detection of fabrication cracks can cause serious and expensive delays to the installation of engineering structures, and a broad understanding of factors which control the occurrence of cracks and methods for their avoidance should be the concern of all welding engineers with responsibility for structural steel fabrication.

REFERENCES

Bailey, N. and S.B. Jones (1978). The solidification mechanics of ferritic steel during submerged arc welding. Welding Journal 57, 217s-231s

Cane, M.W.F., R.E.Dolby and R.G. Baker (1973). A slow bend test for HAZ hydrogen cracking susceptibility and its correlation with welding experience. Weld Research Int'l 3, (2)

Coe, F.R. (1973) Welding steels without hydrogen cracking. The Welding Institute.

Dolby, R.E., J.C.M.Farrar, M.G.Silk and B.H.Lidington (1974). An ultrasonic method for assessing susceptibility to lamellar tearing. Conf on Quality Control & Non-Destructive Testing in Welding. London, Nov 1974. The Welding Institute

Farrar, J.C.M., (1974). Inclusions and susceptibility to lamellar tearing of welded structural steels. Welding Journal, 53. 321s-331s

Granjon, H (1969). The implant method for studying the weldability of high strength steels. Met.Con 1, 509-515.

Hart, P.H.M. (1976). Some limitations of the implant cracking test for predicting hydrogen cracking behaviour in welds. Conf on Welding of HSLA (microalloyed) structural steels. Rome, Nov. 1976. American Society for Metals

Hart, P.H.M. (1978). Low sulphur levels in C-Mn steels and their effect on HAZ hardenability and hydrogen cracking. Procs of Conf on Trends in steels and consumables for welding. London, Nov. 1978. The Welding Institute

Ito, K and K. Bessyo (1968). Cracking parameter of high strength steels related to HAZ cracking. Jnl JWS 37. 983-991 and 38. 1134-1144.

Kanazawa, S., K.Kawamura, K. Yamato, and T. Haze (1974). Lamellar tear resisting steels and the direction for use of them. IIW Doc. IX-873-74

Konkol, P.J. and W.F. Domis, W.F (1979). Causes of grain boundary separations in electroslag weld metals. 60th AWS Annual Meeting, Detroit. April 1979. To be published in Welding Journal.

Lessells, J., K. Randerson and C.I.Smith (1978). The impact of welding on the development of plates and sections. Procs of Conf on Trends in steels and consumables for welding. London, Nov. 1978. The Welding Institute

Makhnenko, V.I., Y.A. Sterenhogen, N.I. Pivtarak and V.E.Pochinok (1979). The fracture mechanics application for evaluation of crystalline crack initiation in the weld metal. IIW Doc. IX-1141-79

Mota, J.M.F., R.L.Apps & J.E.M.Jubb (1978). Chevron cracking in manual metal arc welding. Procs of Conf on Trends in steels and consumables for welding. London Nov. 1978. The Welding Institute.

Sumitomo Metal Industries (1976). Bulletin 002K-76.

Suzuki, H. (1976). Recent Japanese high strength steels for large welded structures. Paper to Public Session. IIW Annual Assembly, Sydney, Australia.

Wright, V.S. and I.T. Davison (1978). Chevron cracking in submerged arc welds. Procs of Conf on Trends in steels and consumables for welding. London, Nov 1978. The Welding Institute.

SECTION 4

Advances in Fracture

FACTORS INFLUENCING THE IMPACT TOUGHNESS OF MULTIPASS SUBMERGED-ARC WELDS IN MICRO-ALLOYED STEEL

J. I. J. Fick

Vecor Heavy Engineering, Vanderbijlpark, R.S.A.

ABSTRACT

This paper describes some of the microstructural components of weld metals in low carbon structural steel. The influence of the various structures on cleavage crack propagation is illustrated with fractographs. A few interesting aspects of weld metal toughness testing are discussed.

KEYWORDS

Impact toughness, submerged-arc, weld metals micro-alloyed steel, microstructures, fractographs.

INTRODUCTION

This paper describes some results of a programme of research that was carried out by the author at the School of Welding Technology of the Cranfield Institute of Technology. The purpose of the work was to establish whether low root area toughness occurred in thick weldments made by the submerged-arc process using plate and consumable material similar to those used in the fabrication of off-shore oil and gas production platforms.

A series of welds were made in 70 mm thick structural steel plate conforming to BS 4360 Grade 50D employing both C-1,5% Mn and C-1,5% Mn - 0,5% Mo electrode wires with a fully basic flux. The experimental programme was designed to determine the effects of electrode type, dilution, the thermomechanical effect of subsequent runs, the grain refining effect of subsequent runs and stress relief heat treatment on the impact toughness and mechanical properties of the root area compared with those of the subsurface.

It will be beyond the scope of this paper to discuss all the results obtained. It is proposed, however, to concentrate on those aspects of the results which illustrate the very interesting relationships between microstructure and impact toughness of weld metals and the various factors which influence the microstructure. A full account of the results may be obtained in reference 1.

Microstructural Components of Weld Metals

The first step in the formation of the weld metal microstructure is the solidification process. A major feature is the growth of a columnar grain structure from the fusion boundry into the weld pool along the maximum thermal gradient.

At the moment that solidification is complete the structure consists of columnar grains of delta-ferrite stretching inward from the fusion boundry.

On further cooling the delta-ferrite transforms to austenite by epitaxial growth from austenite grains already present in the base material and from nuclei formed on the high angle delta-ferrite grain boundaries.

On cooling below the A_3 temperature pro-eutectoid ferrite is nucleated on the columnar austenite grain boundaries. These grow into relatively large equiaxed ferrite grains to form largely continuous veins marking the prior austenite grain boundaries. Depending on composition and cooling rate parallel plates of ferrite may grow from the pro-eutectoid ferrite on the columnar boundaries into the remaining austenite within the columnar grains. This structure should be refered to as ferrite side plates and not as upper-bainite. Fig. 1 shows a typical microstructure of weld metal in low carbon steels with pro-eutectoid ferrite and ferrite side plates. A microstructure displaying more typical features of upper-bainite was observed by the author in welds made on thinner plates (32 mm) and thus having a slower cooling rate. See Fig. 2.

At some temperature below the A_1 temperature the austenite remaining within the columnar grains transforms to a fine-grained non-polygonal ferrite the exact nature and name of which is still the subject of much debate and controversy. The accepted terminology for this constituent is acicular ferrite. Figure 3 shows at high magnification the typical tweed-like structure of acicular ferrite. The more random and interlocking the appearance of these grains the better the cleavage strengths.

A distinguishing feature of multi-pass welds is the bands of fine grained equiaxed structure formed in the areas where weld metal is heated into the austenite range by subsequent runs. Figure 4 shows the typical structure of relatively equiaxed ferrite grains with small areas of pearlite.

Microstructure versus Mechanical Properties

The influence of microstructure on fracture behaviour is illustrated very strikingly in Figs. 5 and 6. These micrographs were obtained by nickel plating the fracture face of a charpy specimen broken at low temperature and then preparing a metallographic section perpendicular to the fracture face. Figure 5 shows that fracture occurred preferrentially along the bands of pro-eutectoid ferrite marking the columnar grain boundaries. The stepped appearance of the fracture was due to the fact that the plane of fracture and the axis of the columnar structure was at an angle. Figure 6 shows in detail how the crack propagated by easy cleavage across an area of pro-eutectoid ferrite and ferrite side plate structure into an area of acicular ferrite. Arrest and blunting of the crack occurred in the acicular ferrite structure. The cracks then linked across the columnar grain back into the plane of fracture by a ductile mechanism which is evident as plastic shear of the acicular ferrites.

The observed easy cleavage through grains of pro-eutectoid ferrite helps to explain the reported observation that the resistance to cleavage crack initiation decreases with increasing fraction of pro-eutectoid ferrite in the microstructure. (2). This decrease in toughness occurs despite that an increase in pro-eutectoid ferrite almost always goes together with a decrease in yield stress of the weld metal. The major factor contributing to the embrittling effect of this constituent is its large grain size, typically 200-300 micron.

A fine acicular ferrite with interlocking grains and a minimum of carbide and pro-eutectoid precipitation has been found to promote optimum strength-toughness properties. (2,3,4,5). The high yield strength of this structure is as a result of the extremely fine grain size and a highly developed dislocation substructure. (6,7). High angle grain boundaries between the individual grains increase the resistance to cleavage crack propagation by necessitating reinitiation and a change of direction of the propagating crack front.

Figure 7 shows the propagation of a cleavage crack through refined weld metal structure. Refined weld metal almost always has a lower yield strength and a better impact toughness than the weld metal from which it formed. T.E.M. metallography reveals that the refined structure has very low dislocation densities (1) and that the grains are separated by high angle grain boundaries. Comparison of Figs. 3 and 4 shows that "refined" weld metal can actually have a coarser grain structure than that of the acicular ferrite in the as deposited weld metal. The great benefit of refinement lies in the complete destroying of the pro-eutectoid ferrite structure outlining the columnar structure.

Some Practical Results of the Study

Figure 8 shows schematically how the test welds were made and how charpy impact specimens were extracted to determine the root and subsurface toughness. When the root sample was extracted roughly equal proportions of refined and as-deposited structures were present in the plane of fracture. To afford an unbiased comparison the subsurface test-pieces were purposely sited to also sample the two structures in equal proportions.

A result of one of the comparisons between root and subsurface toughness is shown in Fig.9. Also included is the result of a test weld where the preparation was buttered with weld metal before the joint was filled, the purpose being to eliminate the effect of dilution of the weld metal by parent metal. The results are interpreted as follows:- If dilution had no effect the results of the full root and the buttered root would be identical. On the other hand, excluding dilution by buttering, if the straining and ageing effect of subsequent welding over the root had no effect on the toughness, the results for the buttered full root and the subsurface would be identical. It was concluded therefore that both dilution and the thermo-mechanical effect of subsequent welding had a detrimental effect on the impact toughness of the root.

However, in none of the test welds were the specified limits of 61J at $-10^{\circ}C$ not met.

The effect of the proportion of refined structure in the plane of fracture was determined by shifting the position for the extraction of the subsurface test pieces slightly to the right of that shown in Fig.8. So that the plane of fracture contained 0% refined structure. The results of the comparison are shown in Fig.10 and it may be observed that an appreciable difference in measured toughness results.

This principle is wittingly or unwittingly exploited in fabrication practice by refining the weld metal just below the level of the plate surface (from which most fabrication codes specify the impact specimens should be taken) by means of so called capping passes or temper beads. These final weld runs stand proud of the plate surface and comprise large areas of unrefined structure which are then not included in the qualifying test pieces but which are subject to loads during service and could initiate brittle failure.

With the materials and procedures used in this investigation it was found that if the impact specimens were sited to sample the least tough zone in the sub-surface i.e. a complete unrefined weld bead, the root would almost certainly have matching or higher impact toughness than the sub-surface.

Looking at weld metal impact properties only, post weld stress relieving had very little advantage. In most cases a reduction in impact toughness resulted and where improvement occurred this was accompanied by an unfavourable loss of tensile and yield strength. However, this does not imply that post weld stress relieving should forthwith be dispensed with. Fracture mechanics theory predicts that residual tensile stresses would lower the applied load at which a crack of certain length would propagate catastrophically in a stressed member.

One way to prove this would be by using a full scale wide plate test. This is very expensive and has to the author's knowledge not been extensively applied to ascertain the beneficial effects of stress relieving. This is an aspect that could be persued because industry could save large amounts of money if it were proved that in some cases stress relieving after welding caused more metallurgical damage than it brought benefits from a lowering of residual stresses.

REFERENCES

(1) Fick, J. I. J., (1978). Impact Toughness of Submerged-arc Weld Metal in Microalloyed Structural Steels. Ph. D. Thesis. Cranfield Inst., of Technology.
(2) Widgery, D. J. (1969). The Influence of Microstructure on Fracture Initiation in Mild Steel Weld Metals. Weld Inst. Members Report. M/46/69.
(3) Farrar, R. A., S. S. Tuliani and S. R. Norman (1974). Relationship between Fracture Toughness and Microstructure of Mild Steel Submerged-arc Weld Metal. Welding and Metal Fab., Vol.42. (2) P68.
(4) Almqvist, G., S. Polgary, C. H. Rosendahl and G. Valland (1972). Some basic factors controlling the properties of weld metal. Proc. Conf. Welding Research related to Power Plant. Marchwood, England. Paper 16, P.204.
(5) Ito, Y. and M. Nakanishi (1976). Study on Charpy Properties of Weld Metal with Submerged-arc Welding. The Sumitomo Search. No. 15. IIW Doc: XII A-113-75.
(6) Wheatley, J. M. and R. G. Baker (1963). Microstructural Factors Governing the Yield Strength of a Mild Steel Weld Metal deposited by the Metal Arc Process. Vol. 10. (1), pp 23-28.
(7) Wheatley, J. M., R. G. Baker (1962). Mechanical Properties of a Mild Steel Weld Metal deposited by the Metal Arc Process. Vol.9. Brit. Weld J. (6) p.378.

Fig. 1. Typical microstructure of weld metal in low carbon steel showing equiaxed pro-eutectoid ferrite, ferrite side plates and intracolumnar acicular ferrite. X 250

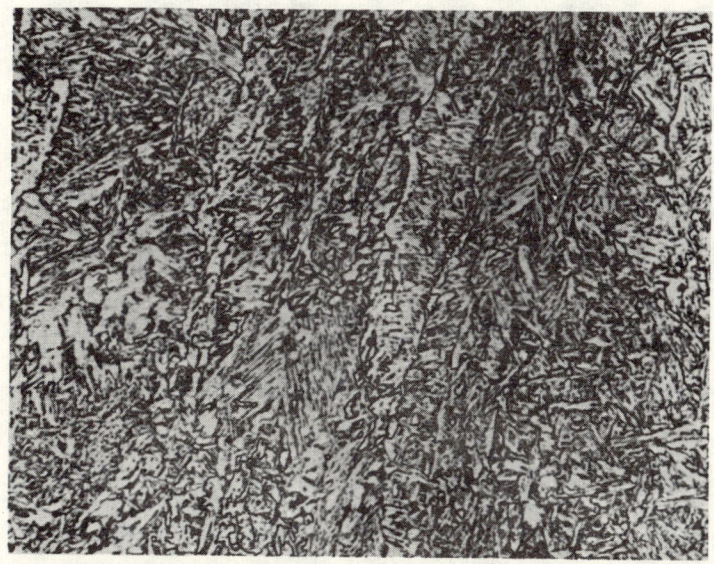

Fig. 2. Bainitic weld metal microstructure observed in welds on 32 mm thick plates. X 250

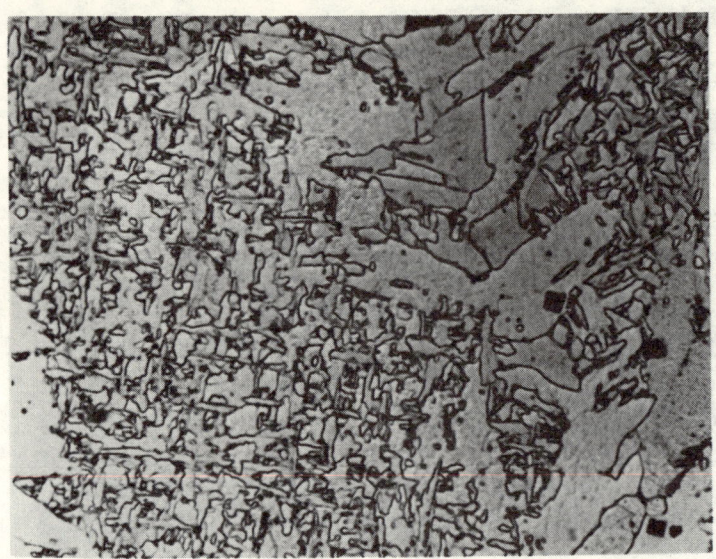

Fig. 3. Detail micrograph of non-polygonal or acicular ferrite. X 1000

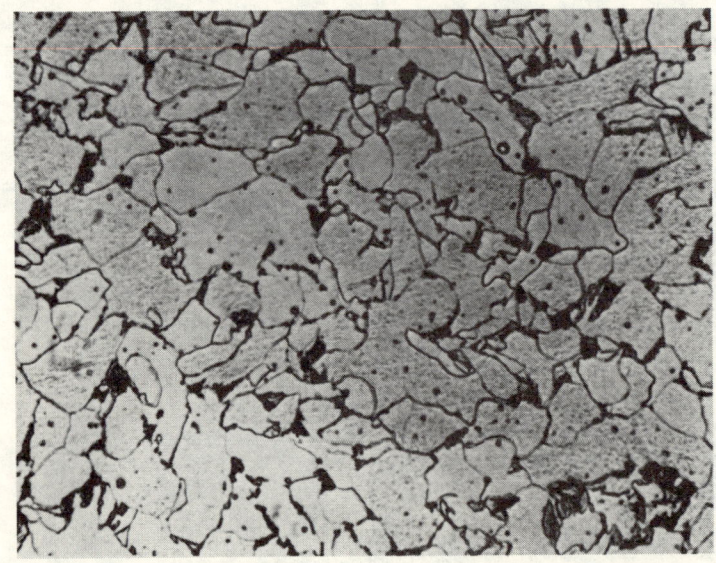

Fig. 4. Typical structure of refined weld metal. X 1000

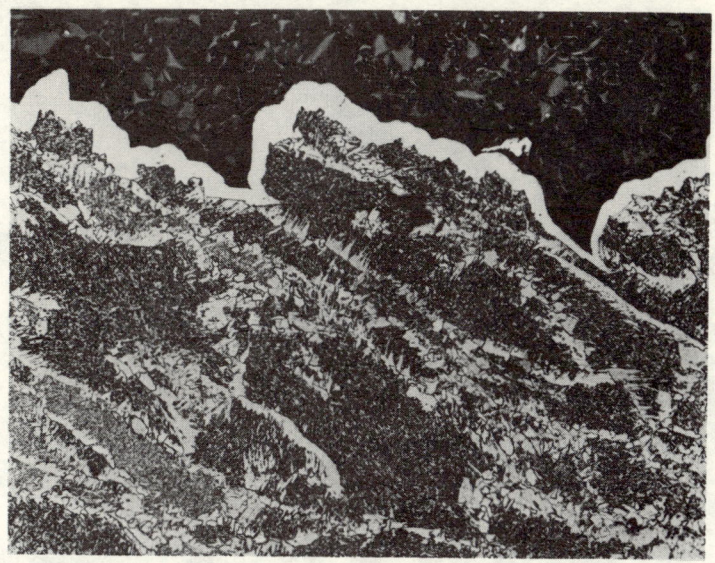

Fig. 5. Showing crack propagation through as deposited weld metal. X 80

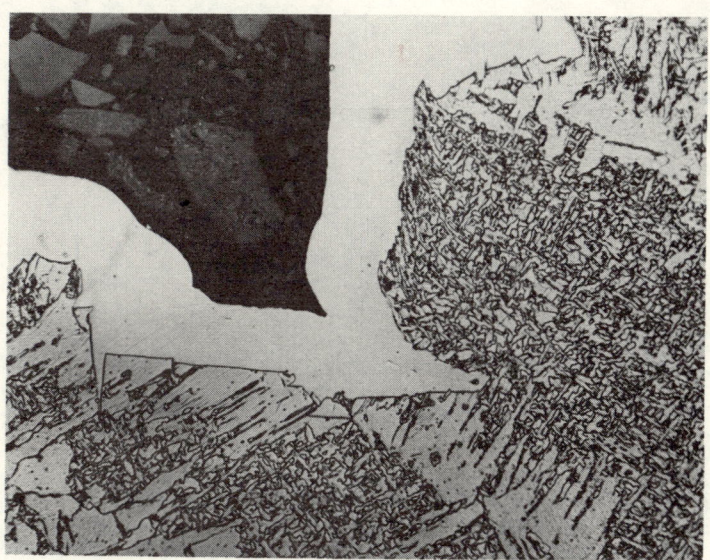

Fig. 6. Detail of cleavage through ferrite side plates and arrest in acicular ferrite. X 400

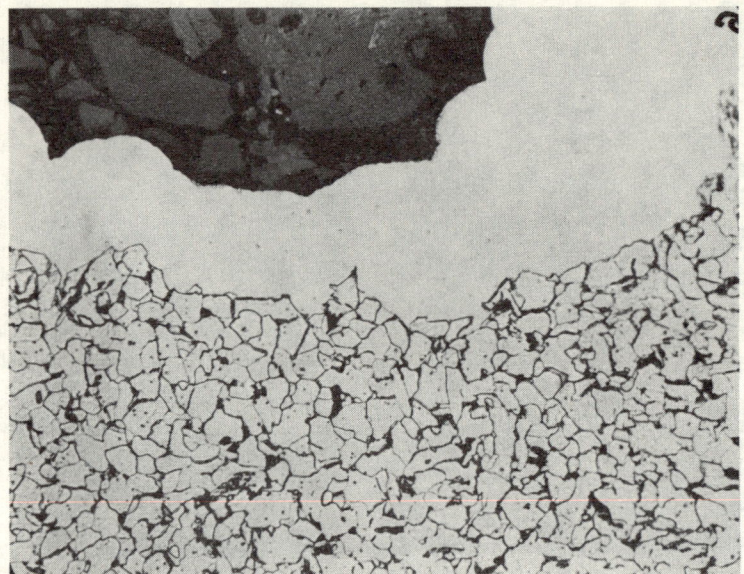

Fig. 7. Showing cleavage crack propagation through refined weld metal. X 500

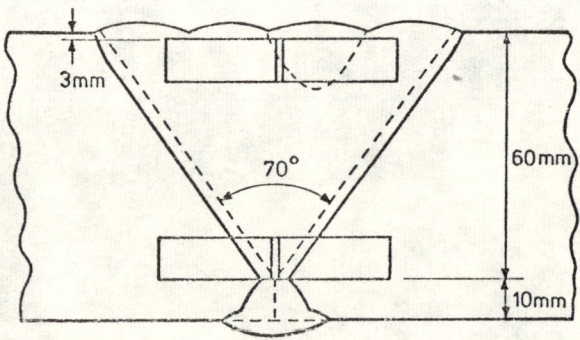

Fig. 8. Details of weld preparation and charpy extraction in test welds to compare root and subsurface toughness.

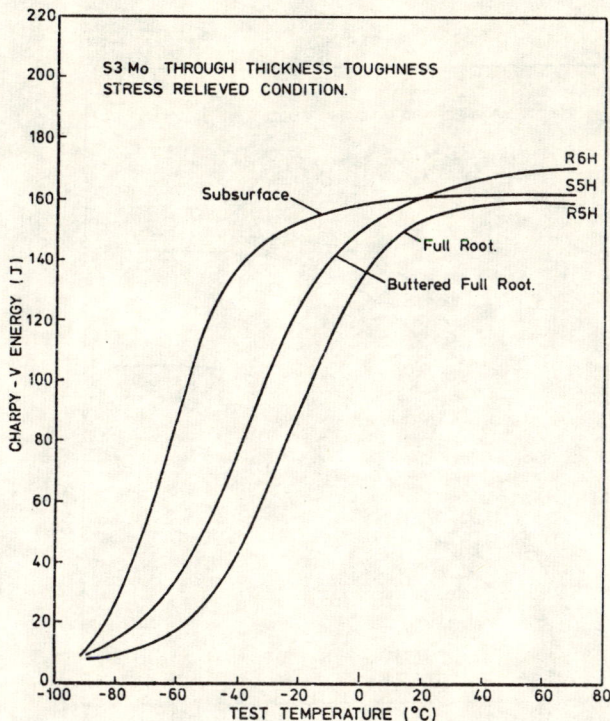

Fig. 9. Charpy transition curves showing the variation in toughness between root and subsurface.

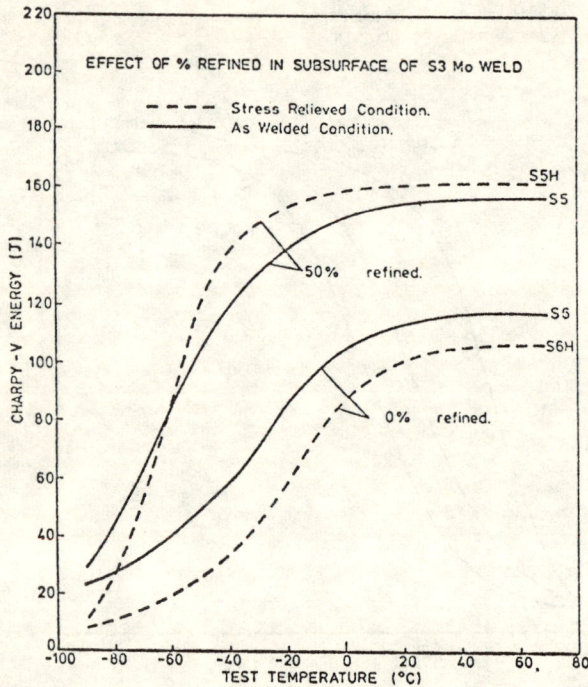

Fig. 10. Charpy transition curves showing the marked effect that % refined structure can have on impact results.

A STUDY OF CRACK ARREST RELATED TO NUCLEAR PLANT INTEGRITY

D. L. Marriott*, R. P. G. Anderson** and G. G. Garrett**

*Licensing Branch, Atomic Energy Board, Pretoria, R.S.A.
**Department of Metallurgy, University of the Witwatersrand, Johannesburg, R.S.A.

ABSTRACT

This paper describes a recent program, carried out jointly by the Atomic Energy Board, Licensing Branch and the Department of Material Science and Metallurgy of Cape Town University, to investigate the phenomenon of Crack Arrest. The program was part of an international co-operative effort organised by the ASTM to standardise a test for measurement of crack arrest material properties. Results of the study are given. Two methods proposed for a standard crack arrest test are discussed in relation to approximate design calculations.

KEYWORDS

Crack arrest, dynamic fracture toughness, thermal shock, fracture testing.

THE CRACK ARREST PROBLEM

There are circumstances, such as thermal shock, in which it is difficult to prevent the initiation of fast crack growth. Component integrity would be difficult to demonstrate if it were not for the fact that, given certain conditions, moving cracks will halt or "arrest". The problem is that both the stress intensity at the crack tip, and the material fracture toughness, are rate sensitive and strictly require a dynamic analysis. Recent work at Battelle (Hoagland et al, 1977) and Materials Research Laboratories (Crosley and Ripling, 1977) has been aimed at simplified approaches to crack arrest leading to similar test methods, but with different objectives.

The Battelle approach is based on the principle that the dynamic fracture toughness, K_{ID}, has a minimum value, K_{IM}, at a certain critical crack velocity, V. Fig. 1. This quantity, K_{IM}, combined with a dynamic stress analysis, gives a safe bound on crack arrest, i.e. overestimates cracks jump, and is simpler to perform than the full rigorous analysis.

** Formerly of University of Cape Town.

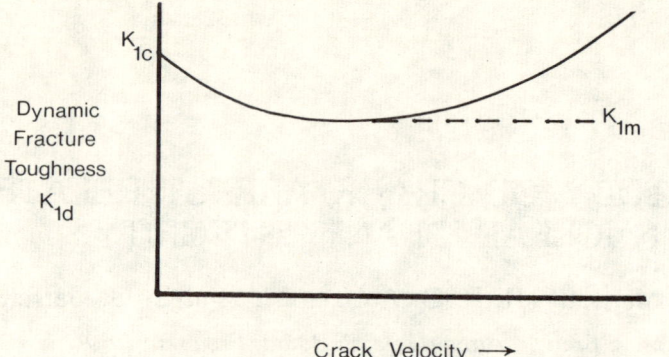

Fig. 1. Schematic variation of K_{Ia} with crack velocity

The MRL approach postulates a "material property" K_{Ia}, the crack arrest toughness, which is the stress intensity at an arrested crack tip after the dynamic effects have died away. K_{Ia} is not a real material property, since it depends on the component and crack shapes. However, use of K_{Ia} in a static analysis is relatively simple, and may be justified on the grounds of ease of computation.

THE CO-OPERATIVE TEST PROGRAM

The ASTM International Co-operative Program on Crack Arrest (Hahn et al, 1977) is aimed at testing the relative practicality of the Battelle and MRL alternatives. To achieve this 30 laboratories world-wide were supplied with four each of the Battelle and MRL test specimens machined from a single controlled plate of A533B steel (a nuclear pressure vessel material), and detailed instructions on the two test procedures. The results have since been collated to determine the variability expected from one laboratory to another, and a preliminary analysis has been published recently (Hahn et al, 1978a).

Both Battelle and MRL tests use a transverse wedge loaded compact tension fracture specimen (Fig. 2). A controlled deflection of the wedge leads to a stable crack in this configuration because the stress intensity at the crack tip decreases with crack growth. To produce a dynamic crack jump a crack starter is placed at the initial crack tip which artificially elevates the initiation toughness above the dynamic toughness to a value K_Q. Battelle use a blunt notch in a section of high strength low toughness steel, electron beam welded to the test metal. MRL use a notch in a weld bead. Measurements of the crack opening at the wedge are made just before and after the crack jump. After the test the crack surface is heat tinted and the specimen broken to determine the length of crack jump.

From this point on the Battelle and MRL techniques diverge.

(a) Battelle Test. This uses measures of K_Q, the initiation stress intensity and the crack jump distance, together with calibration curves, to determine K_{ID} and V indirectly.

(b) MRL Test. This calculates K_{Ia} directly from the crack jump and the measurement of wedge opening at the instant of the jump.

Fig. 2. General View of Wedge loaded test specimen

THE TEST PROGRAM

Description of Test Specimens

Ten specimens were supplied to each laboratory for testing, four each of the Battelle and MRL designs in A533B material, and two trial specimens in AISI 4340 steel. Both designs conform approximately with those of a standard 50 mm ASTM E 399 fracture test specimen, as shown in Fig. 3. The specimens are side-notched

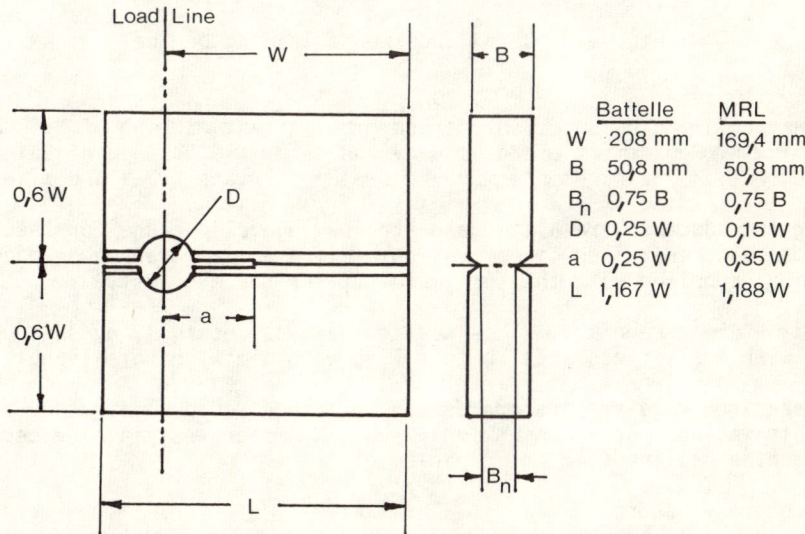

	Battelle	MRL
W	208 mm	169,4 mm
B	50,8 mm	50,8 mm
B_n	0,75 B	0,75 B
D	0,25 W	0,15 W
a	0,25 W	0,35 W
L	1,167 W	1,188 W

Fig. 3. Dimensions of crack arrest test specimens

30% to control cracking direction. In both cases wedge loading is used to achieve sufficient load train stiffness for the load condition to be effectively fixed displacement. The load train is shown in Fig. 2 assembled, and separated in Fig. 4. The MRL wedge (shown in the Figure) has a diameter of 25 mm, and the Battelle model is 50 mm.

Fig. 4. Details of wedge loading device

Testing Equipment

Tests were performed on an E.S.H. servohydraulic test machine of 250 KN capacity (Fig. 5) in the Department of Metallurgy and Materials Science of University of Cape Town (UCT). Measurements were recorded on Bryans X-Y-Y and X-X-t recorders.

Measuring transducers were a standard strain gauge clip gauge for the Battelle test and an a.c. energised linear variable differential transducer made by R.D.P. Electronics, supplied with the test machine, for the MRL test.

Testing at $0^{o}C$ was required. This was achieved successfully by surrounding the specimen with a stainless steel bath filled with a mixture of alcohol and dry ice.

The MRL specimen required precompression to induce residual stresses at the notch tip. This was done on a conventional 3000 KN Amsler test machine because the E.S.H. machine had insufficient capacity.

Test Procedure - General

In the following respects both tests are similar. The test specimen is placed flat in the test machine, transverse to the load line, as shown in Fig. 6.

Fig. 5.   General view of test rig

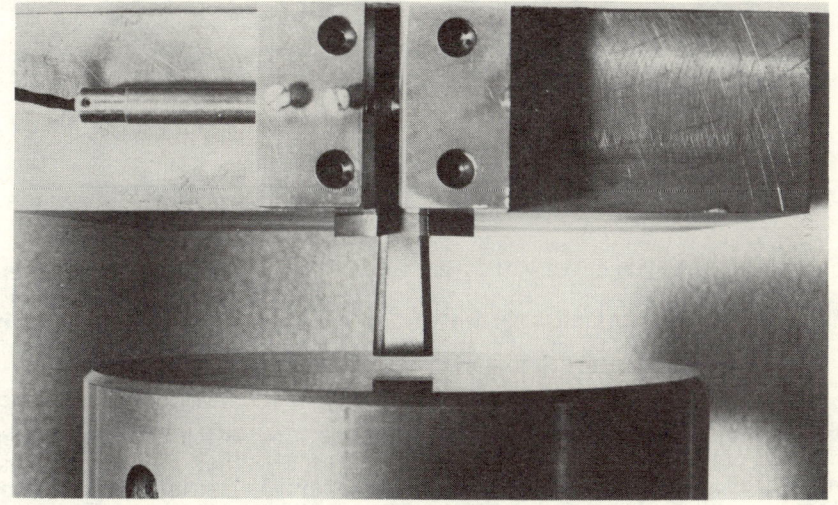

Fig. 6.   Wedge loading arrangement in test rig

Load is applied through a controlled displacement of the wedge while the opening displacement across the wedge is measured. This displacement can be later related to the stress intensity through calibration curves produced by the test originators (Crosley and Ripling, 1977; Hoagland et al, 1977). The equation for stress intensity is

$$K = f_\eta \frac{E}{\sqrt{W}} \sqrt{\frac{B}{B_n}} \delta$$

where K = stress intensity
       E = Young's Modulus
       W = Specimen length, measured from load line to end of specimen (See Fig. 2).
       B = Specimen width
       B_n = Nett specimen width (See Fig. 2)
       δ = Crack opening displacement at load line (Battelle) or outside face (MRL)
       f_n = Coefficient depending on crack depth shown in Fig. 7.

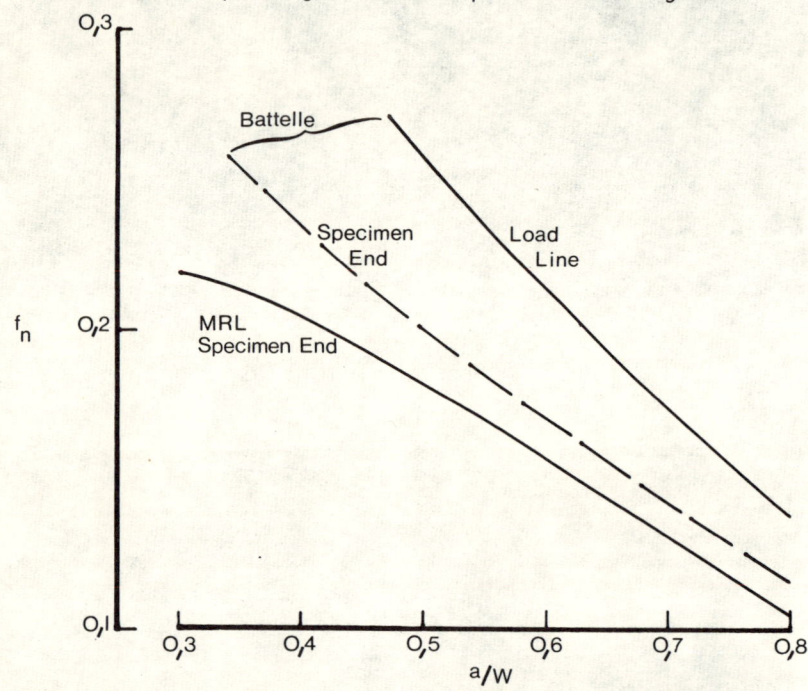

Fig. 7. Compliance calibration curves for crack arrest specimens

Since the conditions described will result in slow stable crack growth in a uniform material, it is necessary to create an artificial crack starter with an initial critical stress intensity, K_Q, significantly greater than the static facture toughness K_{IC}, in order to achieve a dynamic crack jump. This is achieved firstly by putting a machined notch at the crack tip, which is significantly blunter than a fatigue type crack, and hence requires a higher critical stress intensity for initiation. Unfortunately, in tough materials, this is not sufficient because the resulting elevation of stress intensity is accompanied by an increase in the plastic zone size, given by

$$r_y = \frac{1}{6\pi}\left(\frac{K_Q}{\sigma_y}\right)^2$$

to an extent that widespread yielding of the specimen could precede crack initiation.

Example. In A533B K_{IC} is about 200 MPa√m and K_{IM} is about 150 MPa√m at ambient temperature. A suitable K_Q value is therefore about 250 MPa√m. The yield stress σ_y is about 500 MPa. The plastic zone size r_y is therefore

$r_{y\ A533B}$ = 13 mm

Compared with a through thickness dimension of 50 mm this means that deviation from plane strain conditions is likely, leading to a much larger <u>plane stress</u> plastic zone of about 40 mm.

To avoid problems with large plastic zones the crack starter therefore incorporates a section of high strength material, with a yield of about 1000 MPa, in which the notch is machined. This gives a crack tip plastic-zone size of 3 mm, which is small enough.

When the stress intensity of the crack tip reaches K_Q the crack propagates.

Since the wedge displacement is held sensibly constant, the advancing crack tip stress intensity decreases until it falls below some critical material value, at which point the crack will arrest. The specimen is subsequently heat tinted and broken apart at liquid nitrogen temperature to reveal the crack jump (Figs. 8 and 9).

Fig. 8. Fracture surface (Battelle)   Fig. 9. Fracture surface (MRL)

The details of the Battelle and MRL tests differ. These are described in the following two sections.

Test Procedure - Battelle Test

The Battelle test specimen has a test section of AISI 4340 steel electron beam welded to the test section. Crack initiation is achieved at a predetermined value of K_Q by cycling the wedge load from zero up to the load corresponding K_Q; in effect initiating fracture by low cycle fatigue.

The object of the Battelle test is to obtain K_{IM}, the minimum value of K_{ID}, the dynamic fracture toughness (Fig. 1). This cannot be done directly. The originators of the test (Kanninen, 1974) have made a dynamic analysis of the test specimen and have devised calibration curves which enables K_{ID} to be obtained from measured values of K_a and the crack jump length a. Such a curve is shown in Fig. 10 which also illustrates the procedure for calculating K_{ID}. Strictly

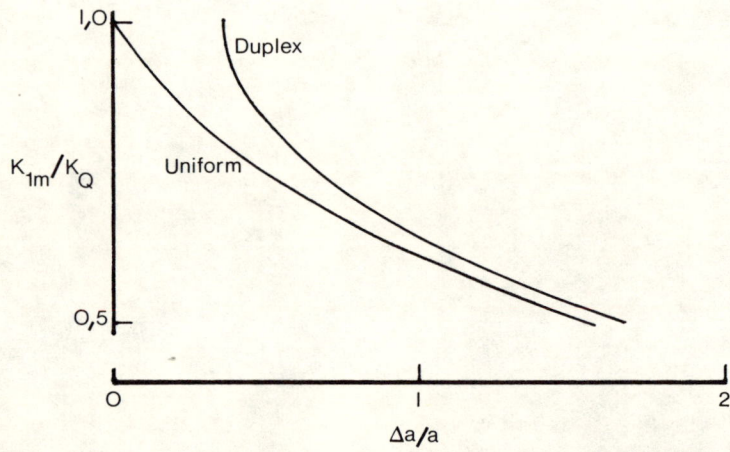

Fig. 10. Calibration curves for K_{IM} - Battelle Test

speaking K_{IM} can only be obtained by plotting a full curve of K_{ID} vs V as shown in Fig. 1. In practice this has not been necessary because, for the geometry chosen, the minimum value, K_{IM}, is obtained if the crack jump falls between limits which are defined in the detailed procedure (Hoagland et al, 1977). All that is described here is the basic procedure. The full Battelle test includes a large number of validity tests which can be referred to in the original proposal (Hoagland et al, 1977). Typical output plots are given in Figs. 11 and 12.

Test Procedure - MRL Test

The MRL test specimen is relatively simple having a weld bead deposited at the base of a "U" notch for the crack starter section. A simple machined notch in the weld bead was found by Crosley and Ripling (1977). to have too high a value of K_Q. The test procedure therefore involves precompressing along the load line, causing plastic deformation at the crack starter, and leading to a residual tensile stress field on unloading which reduces the initiating K_Q.

Loading and measurements were as for the Battelle test. The crack jump is measured by the normal method of heat tinting and fracture at low temperature.

The MRL crack arrest philosophy is based on static analysis using K_{Ia}, the value of arrest stress intensity after static equilibrium has been achieved. Therefore the only measurement required is the value if K_{Ia}, which can be obtained from the crack length at arrest and the displacement of the wedge at the same instant. Since the inertia of the wedge causes it to move beyond the initiation position it is necessary to obtain a continuous recording of displacement with time, from which the displacement at initiation can be identified by the displacement jump of the run-arrest event as shown in Fig. 13. A load-deflection curve is given in Fig. 14. Although not required for the MRL test the initial displacement just prior to initiation was used to calculate K_Q.

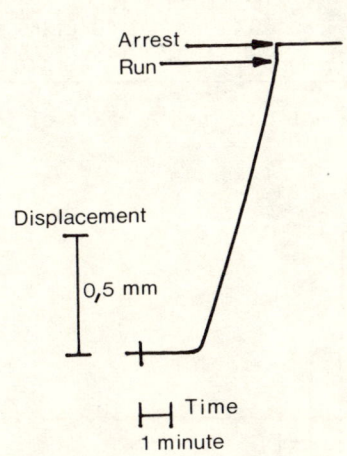

Fig. 11. Battelle δ-t plot

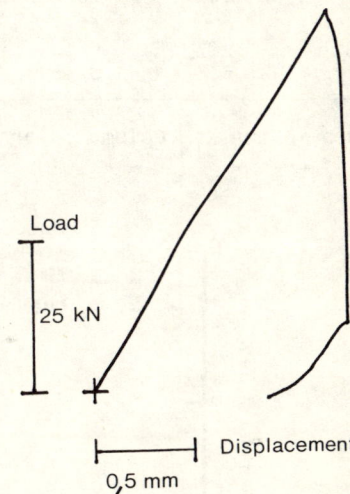

Fig. 12. Battelle Load -δ plot

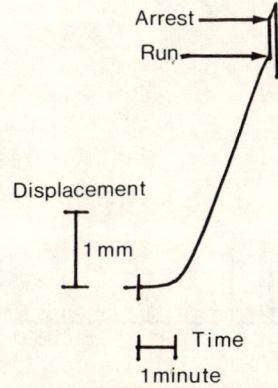

Fig. 13. MRL δ-t plot

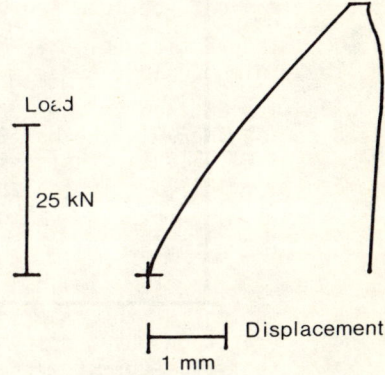

Fig. 14. MRL Load -δ plot

Results of Test Program

The results of the tests performed at UCT are given in Table 1.

Table 1 Results of Crack Arrest Tests Performed at UCT

Test	Type	Temp °C	K_Q MPa√m	K_{Ia} MPa√m	K_{IM} MPa√m	Material
6341/44	MRL	21	150	81	-	AISI 1018
6341/64	"	22	128	71	-	"
105C	"	21,5	179	111	(141)	A533B
501C	"	21,5	149	123	(135)	"
405A	"	0/5	142	93	(115)	"
410A	"	-4/-1	134	69	(36)	"
RG144	Battelle	25	247	108	166 (163)	"
RD31	"	25	224	117	166 (162)	"
RH84	"	0/+4	222	83	140 (136)	"
RC103	"	0/+5	216	98	145 (145)	"

The values in the K_{IM} column in brackets were not obtained as part of the test

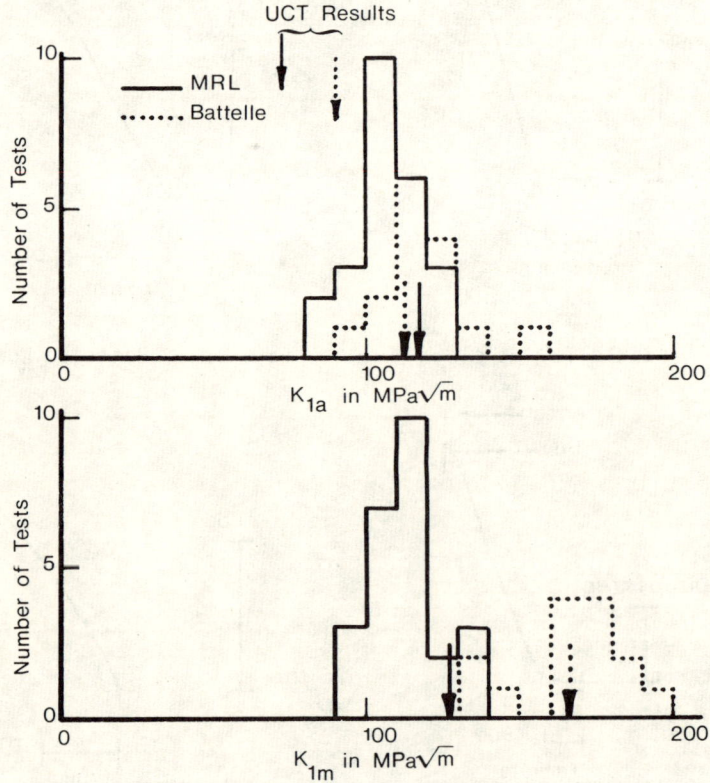

Fig. 15. Comparison of UCT results with results of entire program

program, but were calculated using an alternative approximate method mentioned later in this paper. Fig. 15 shows results of the UCT tests superimposed on the combined results of the co-operative test program. Compared with the variability of the test program results as a whole the UCT values fall very close to the mean values, indicating a satisfactory grasp of the proposed methods. A full analysis of the Co-operative Test Program has been published recently (Hahn et al 1978a) and can be referred to for further details. The essentials are summarised in the following section.

Discussion of Test Program Procedures and Results

The experience from participating in the test program can be considered under two headings

(i) Comments on Test Procedures.

(ii) Comments on Results.

Comments on Test Procedures. Both the Battelle and MRL test procedures were relatively simple to perform given the appropriate equipment. Concerns expressed before the tests about dynamic response of machines and instruments were unfounded. The crack-arrest event was easily discerned and the simple clip gauge was found to be satisfactory for displacement measurement. It is possible to make the following observations on comparison of the two methods.

(i) As a standard test specimen the MRL design is to be preferred since it can be manufactured by standard fabrication techniques.

(ii) The Battelle specimen has the advantage of providing a residual stress-free starter for which K_Q can be calculated accurately. Both welding residual stresses and precompression in the MRL specimen lead to uncertainty over the value of K_Q, which is not independent of residual stress.

(iii) The Battelle method of crack initiation by cyclic loading is more controlled than the MRL precompression method.

(iv) In summary, it may be concluded that the MRL specimen may be more practical, but the test procedure should be the Battelle cyclic loading technique. There remains the problem of weld residual stresses in the MRL crack starter. These will have some effect on the initiation K_Q. However, weld stresses are relatively randomised compared with the precompression residual stresses, and should have a minor effect on K_Q. Improvements on the MRL test can be achieved by using a low heat input multirun TIG weld which will randomise residual stresses still further and enable a greater choice of starter materials to be used.

Comments on test results. It can be seen from Table 1 and Fig. 15 that there is a significant difference between the values of K_{IM} obtained from the two techniques, which cannot be explained by normal experimental scatter. The reason has not been clearly identified but it is highly probable that the discrepancy derives from erroneous calculation of K_Q in the MRL test because the effects of residual stresses have been ignored. The value of K_{Ia} obtained from the two tests do not show any significant difference. The uncertainty over K_Q in the

MRL test does not affect that test as such because its philosophy is based on the use of K_{Ia} alone. However, the discrepancy is important if it is decided by ASTM that K_{IM} should be adopted as the standard crack arrest parameter, because an accurate measure of K_Q is essential to the calculation of K_{IM}. Furthermore it is likely that K_{IM} will become the preferred measure. K_{Ia} is not a real material parameter. Analysis of the co-operative test program results (Hahn et al, 1978a) shows that K_{Ia} is dependent on crack jump, decreasing with increasing crack jump. This is a predictable result, as will be shown in the next section of this paper.

THE ANALYSIS OF CRACK ARREST

Under dynamic conditions both the stress intensity at a crack tip, and the dynamic fracture toughness, depend on the crack tip velocity. This makes for an extremely difficult problem if a rigorous analysis is called for. Fortunately some approximations can be made which considerably simplify the problem.

The problem of varying stress intensity will be considered first. For similar geometries the stress intensity at a crack tip decreases with crack tip velocity as shown in Fig. 16. The reason is that the energy released by a slowly advancing crack is all available to form new crack surface. Some of the energy released by a fast moving crack must go to provide kinetic energy. Therefore, if the stress intensity is calculated under static conditions, this should overestimate the actual dynamic value, providing a safe estimate for design purposes. For short crack jumps, typical of thermal cracking, the error is small anyway.

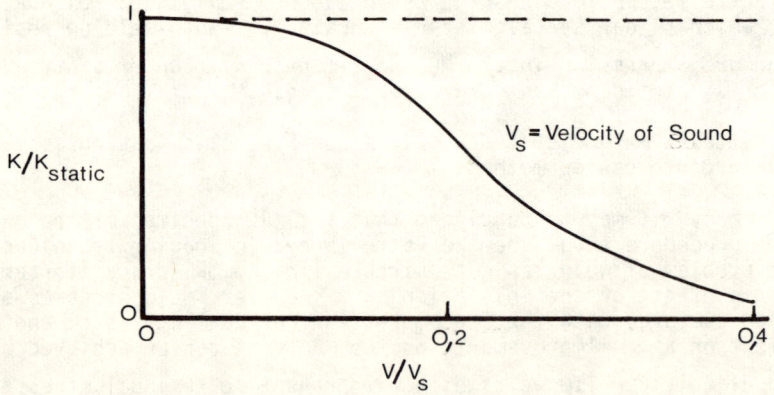

Fig. 16. Schematic variation of stress intensity with crack velocity

The question of dynamic fracture toughness is more difficult to deal with. Fracture toughness varies with strain rate in accordance with the strain rate sensitivity of the material. In most practical cases, e.g. steel at temperatures near its brittle-ductile transition, the effect is to reduce the fracture toughness with increasing velocity. Very high crack velocities, or temperatures well onto the upper shelf of toughness, increase fracture toughness above the static values. Nevertheless it is obvious that it is not satisfactory to use the static value K_{IC}. At present there are two alternative philosophies for dealing with this problem, corresponding to the two alternative test proposals evaluated by the ASTM Co-operative Test Program on Crack Arrest.

The K_{Ia} Method based on Static Analysis

This approach assumes that the quantity K_{Ia}, the static stress intensity after arrest, is approximately a material property which can be used in a static analysis in exactly the same manner as static fracture toughness K_{IC}. Figure 17 illus-

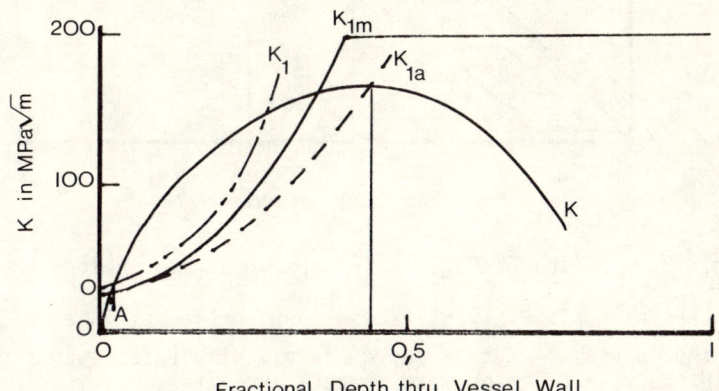

Fig. 17. Variation of stress intensity and fracture toughness through a pressure vessel wall during a thermal transient

trates the stress intensity variation through a nuclear pressure vessel wall following an injection of cold water. Methods for calculating the variation of stress intensity with crack depth in a problem such as this have been comprehensively reviewed in a paper presented at this conference (Van der Walt, 1979). Superimposed on the stress intensity variation in Fig. 17 are the temperature varying values of K_{IC} and K_{Ia}. It is assumed that any crack initially greater than OA will propagate until the stress intensity drops below the K_{Ia} line, at which point it will arrest. This method has been used extensively for analysis of thermal shock (Buchalet and Bamford, 1975). The disadvantage of the method is the uncertainty of whether the results are conservative or not, since K_{Ia} is not in fact a constant, and may be smaller under conditions in the structure than in the original test specimen.

The K_{IM} Method of Crack Arrest Analysis

The quantity K_{IM} is the minimum value of the dynamic fracture toughness K_{ID}. It may be adopted as a material property with the same degree of confidence as the static toughness measure K_{IC}. Use of K_{IM}, together with a static analysis for stress intensity K, provides a consistently conservative estimate of crack arrest. However, for an analysis of crack arrest further information is required which must be obtained from energy balance considerations. Referring to Fig. 17, it can be seen that, following crack initiation, there is an excess of stress intensity over dynamic toughness, so that static equilibrium is not satisfied. The difference represents an excess of energy released, which is more easily appreciated by replotting Fig. 17 as Fig. 18 in terms of energy release rate $G = K^2/E$.

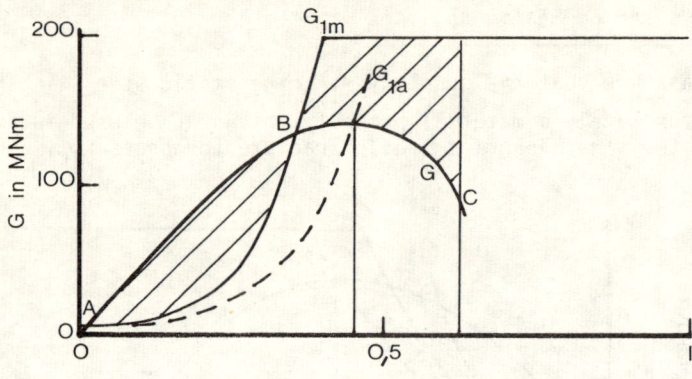

Fig. 18. Replot of Fig. 17 in terms of Energy Release Rate G

At point B, where K equals K_{IM}, the excess energy released is equal to the shaded area between the curves. If this energy is all dissipated away from the crack, for example in the form of an outward propagating stress wave, then arrest will occur at point B. If, on the other hand, all the energy is reflected back to the crack tip, as may occur if the crack jump is large and the component dimensions are small, arrest will only occur after an area has been accumulated between the K and K_{IM} curves, representing energy absorbed, equal to the energy released.

Arrest under 100% reflection is given by point C. The crack arrest stress intensity K_{Ia} can be seen to vary with circumstance. There are two extreme cases.

(a) 100% energy dissipation away from crack. This is equivalent to a small crack jump in a large component

$$K_{Ia} = K_{IM} \tag{1}$$

(b) 100% energy absorbtion at the crack tip. This is equivalent to a long crack jump and is given by the equal area construction shown in Fig. 18. For the particular case of constant K_{IM} it has been shown (Hahn et al, 1978b) that the arrest stress intensity is given by

$$K_{Ia} = \frac{K_{IM}^2}{K_Q} \tag{2}$$

This condition is satisfied by either the Battelle or MRL test specimens. It is an alternative method of calculating K_{IM} which is simpler than the calibration curve proposed by the Battelle Institute. In Table 1 estimates of K_{IM} calculated from eqn (2) are shown in brackets. The correspondence with K_{IM} calculated from the calibration curve is very good for the Battelle test. Estimates of K_{IM} can also be obtained for the MRL test by this method. Unfortunately the residual stress error in calculating K_Q for the MRL test is reflected in low values of K_{IM} compared with the Battelle test.

Comparison of K_{Ia} and K_{IM} Methods of Crack Arrest Analysis

Both methods use a static analysis for stress intensity. Thereafter the K_{IM} method is slightly more complicated, but the extra computational effort is not significant. Furthermore it is possible to obtain a consistently conservative estimate with the K_{IM} approach by assuming 100% energy absorbtion at the crack tip. The only advantage of the K_{Ia} approach is the greater ease of determining this value from tests compared with K_{IM}. However, the use of eqn (2) to calculate K_{IM} from K_a and K_{Ia} direct virtually cancels this advantage. It would appear therefore that the method based on K_{IM} is in all respects superior to the K_{Ia} approach.

CONCLUSIONS

1. The test methods for determining crack arrest properties have been evaluated. It is found that for both design and material test purposes the quantity K_{IM}, representing the minimum value of dynamic fracture toughness, is a preferred measure.

2. From experience of the test program it is proposed that a preferred crack arrest test should incorporate the following features.

 (i) The MRL weld embrittled specimen modified to take the large Battelle loading wedge, and using multipass TIG welding to produce a weld bead with minimal residual stress effect on K_Q.

 (ii) The Battelle crack initiation technique of load cycling to produce fracture at a controlled value of K.

(iii) Use of the simpler Battelle type clip gauge for displacement measurement.

 (iv) Dispensing with calibration curves and using eqn (2) to calculate K_{IM} directly.

3. Approximate analysis of crack arrest using K_{IM} and 100% energy absorbtion at the crack tip is no more difficult than a static analysis and gives a conservative estimate of crack propagation before arrest, i.e. the crack jump is overestimated. For very short jumps, e.g. residual stress cracking in welds, the 100% energy dissipation model applies which requires only a static analysis with arrest defined by $K_{Ia} = K_{IM}$.

REFERENCES

Buchalet, C.B. and W.H. Bamford (1975). Method for fracture mechanics assessment of nuclear reactor vessels under severe thermal transients. Paper 75WA/PVP 3, ASME PVP Annual Winter Meeting.

Crosley, P.B. and E.J. Ripling (1977). Guidelines for measuring K_{Ia} with a Compact Specimen. Report to ASTM. Crack Arrest Task Group, Prospectus for a Co-operative Test Program on Crack Arrest Toughness Measurement conducted under auspices of ASTM E24.03.04, December 1977

Hahn, G.T. (program co-ordinator) (1977). Prospectus for a co-operative test program on crack arrest toughness measurement, ASTM E24.03.04 Sub-committee on Dynamic Testing, Dynamic Initiation - Crack Arrest Task Group.

Hahn, G.T., R.G. Hoagland, A.R. Rosenfield, C.R. Barnes, (1978a). A co-operative program for evaluating crack-arrest testing methods, ASTM Symp. on Crack Arrest Methodology and Applications, Philadelphia.

Hahn, G.T., A.R. Rosenfield, C.W. Marshall, R.G. Hoagland, P.C. Gehlen, and M.F. Kanninen (1978b). Crack arrest concepts and applications, Fracture Mechanics, Proc. 10th Symp. Naval Struct. Mech. Washington D.C., University Press of Virginia, Charlotteville, 205 - 227

Hoagland, R.G., P.C. Gehlen, A.R. Rosenfield, M.F. Kanninen and G.T. Hahn (1977). Proposed tentative method of test for fast fracture toughness and crack arrest toughness. Report to ASTM Crack Arrest Task Group, Prospectus for a Co-operative Test Program on Crack Arrest Toughness Measurement conducted under auspices of ASTM E24.03.04, December 1977.

Kanninen, M.F. (1974). An analysis of dynamic crack propagation and arrest in the DCB test specimen. Int. J. Fract., 10, 3, 415 - 430

Van der Walt, W. (1979). Simplified stress intensity evaluation of a nuclear reactor vessel under a given accident condition. Fracture 79, Proc. Joint SAIW/SAIP Conf. on Fracture, Johannesburg, 6 - 9 Nov.

STRESS CORROSION AND CORROSION FATIGUE IN LIGHT WATER REACTOR ENVIRONMENTS

D. de G. Jones

Licensing Branch, Atomic Energy Board, Pelindaba, Private Bag X256, Pretoria 0001, R.S.A.

ABSTRACT

The main factors in stress corrosion are briefly reviewed with regard to:

(i) generation of chemical environment, where it is shown that consideration must be given to boiling and crevice situations as well as bulk solutions;

(ii) interaction of (i) with metals under tensile stresses where the main factors are:

(a) tensile stresses, which can be applied or residual, steady or varying but must be above yield;

(b) corrosion effects with different types of chemical solutions with either corrosion or non-corrosion resistant metals.

The stress corrosion process consists of initiation and propagation stages and the various factors which affect these, including electrochemical polarisation and fracture mechanics, are described. The effect of microstructural effects on crack morphology is briefly discussed.

The above arguments are extended to corrosion fatigue where it is shown that there are two basic forms of corrosion fatigue:

Type I, where cyclic stresses and corrosive environment result from the same se= quence of events;

Type II, where cyclic stress and corrosive environment arise independently.

Stress corrosion incidents in the pressure vessel and piping of boiling water reactors and steam generators of pressurised water reactors are described to= gether with related laboratory studies. It is concluded that these are con= sistent with general stress corrosion theories and are caused predominantly by chemical factors.

The main factors which have been found in laboratory studies to affect the corrosion fatigue of pressure vessel steels in LWR environments are described and related to the possible risk of corrosion fatigue in service.

INTRODUCTION

The aim of this paper is to

1. briefly review some of the main aspects of stress corrosion and corrosion fatigue theories; and

2. discuss their application in light water reactors.

Stress corrosion may be defined as the interaction of essentially static stress and corrosion together where each separately would not cause any significant effect. It is localised and often fast and catastrophic, with cases of multiple failures occurring in austenitic superheaters, 5 mm thick, of coal fired power plant half an hour after a malfunction of the associated water chemistry plant.

Corrosion fatigue may be defined as the interaction of alternating stress and corrosion to produce an additive effect rather than the synergistic effect observed in stress corrosion.

STRESS CORROSION THEORIES

The two main factors which have to be considered in stress corrosion are:

1. generation of chemical environment;

2. interaction of 1 with metal under tensile stress.

Generation of Chemical Environment

Chemical environments which cause stress corrosion are normally ionic solutions in which electrochemical corrosion processes can take place.

However, when stress corrosion in plant is considered, not only the bulk stable environments should be considered but also how they can be altered by boiling or crevices.

Boiling. Concentration processes which can occur under boiling conditions have been reviewed (Jones and others, 1971). Figure 1 shows how concentrated solutions can be formed by boiling under a deposit, such as scaling or bulk corrosion, by "wick boiling" in which the boiling is stabilised by the deposit. Concentration factors of up to 10^6 have been demonstrated.

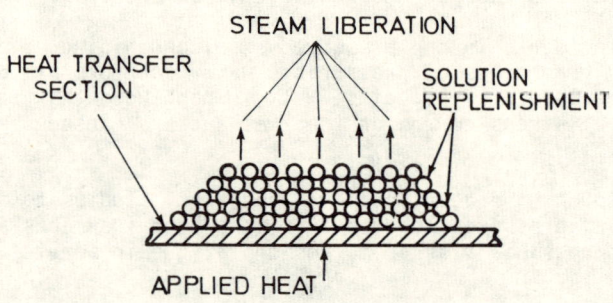

Fig. 1. Solution concentration in wick boiling.

Crevices. Brown (1971) has postulated and demonstrated experimentally how both concentrations and pH alter in crevices and cracks. Figure 2 shows a simple crevice system involving the corrosion of iron.

In the initial stage of active corrosion at the crack tip, iron passes into solution as ferrous ions

$$Fe \longrightarrow Fe^{2+} + 2e$$

The ferrous ions subsequently hydrolise

$$3Fe^{2+} + 4H_2O \longrightarrow Fe_3O_4 + 8H^+ + 2e$$

to give a solution whose pH is determined by the equilibrium pH at the solubility limit of ferrous ions.

Values calculated from the above have been confirmed by pH measurements on precracked compact tension samples. (Brown, 1971). This effect does not occur in buffered solutions, since these can compensate and overcome the hydrolytic effect.

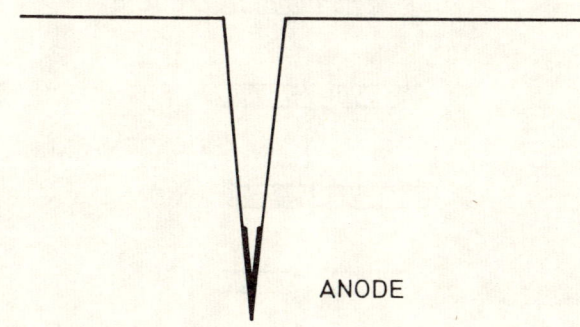

STAGE 1 ACTIVE CORROSION AT ANODE
$Fe \longrightarrow Fe^{++} + 2e$
STAGE 2 HYDROLYSIS
$3Fe^{2+} + 4H_2O \longrightarrow Fe_3O_4 + 8H^+ + 2e$

Fig. 2. Crevice chemistry in the corrosion of Iron

Interaction of Chemical Environment with Metal under Tensile Stress

A number of comprehensive reviews of stress corrosion has been published. (Logan, 1966; Parkins, 1972; Scully, 1971; Staehle and others 1969, 1977). From these, the main factors in stress corrosion are

(a) tensile stresses which may be

(i) applied or

(ii) steady or varying

but must be locally or generally above yield;

(b)   corrosion under two main conditions

(i)   aggressive solutions with corrosion-resistant metals such as stainless steels, etc.; or

(ii)   inhibiting solutions with less corrosion-resistant metals such as mild steel, etc.

The range of stress corrosion behaviour can vary from corrosion dominant systems (such as carbon steels in nitrate solutions) to stress dominant systems (such as high strength steels in water).

<u>Stress</u>. There are two stages in the stress corrosion process:

(i)   initiation which basically consists either of pitting due to chemical/electrochemical breakdown of the oxide film on the metal or film rupture under stress (shown in Fig. 3);

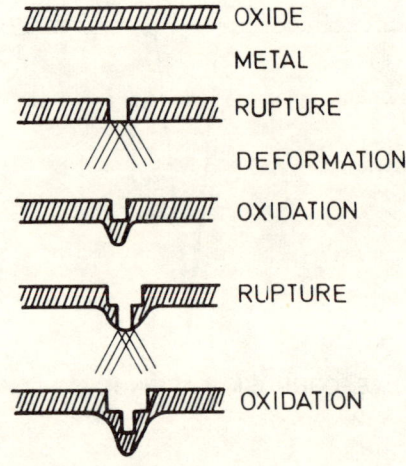

Fig. 3. Oxide film rupture mechanism.

(ii)   propagation which is an extension of the initial crack by competition at the crack tip between corrosion and stress. If the stress is too high, ductile failure from general yielding will occur. If the corrosion effect is too high, crack blunting and ductile failure will occur. In the balance condition between stress and corrosion, localised failure will occur (Scully, 1975). In some cases this balance is only achieved under controlled strain-rate conditions and Fig. 4 shows the effect initially demonstrated by Parkins (1972). This effect has some implications with regard to corrosion fatigue.

Crack-propagation rates in stress corrosion are measured using pre-cracked samples (Spiedel, 1971) and data shown in Fig. 5 is typical of the data produced relating crack growth rate and calculated stress intensity at the crack tip. In Zone 1 the crack growth is stress dependent, with an activation energy of $\sim$20k cal/mole while in Zone 2 the crack growth is independent of stress intensity. Calculations and measurements by Beck (1971) have shown that Zone 2 is diffusion controlled. From the data in Fig. 5, the stress intensity at which no measurable crack growth occurs is defined as K_{ISCC} but the result will obviously depend on the sensitivity of the crack growth technique and the time allowed for measurements. Values as low as 5 - 10 MN m$^{-3/2}$ have been measured for stress corrosion on stainless and low alloy steels.

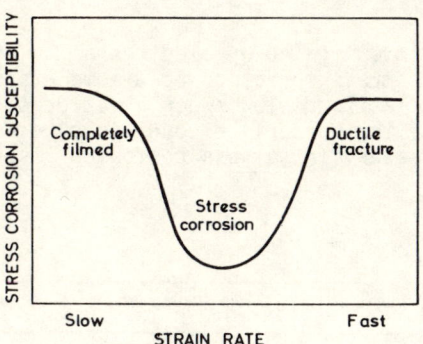

Fig. 4. Effect of strain rate on stress corrosion crack propagation (after Parkins).

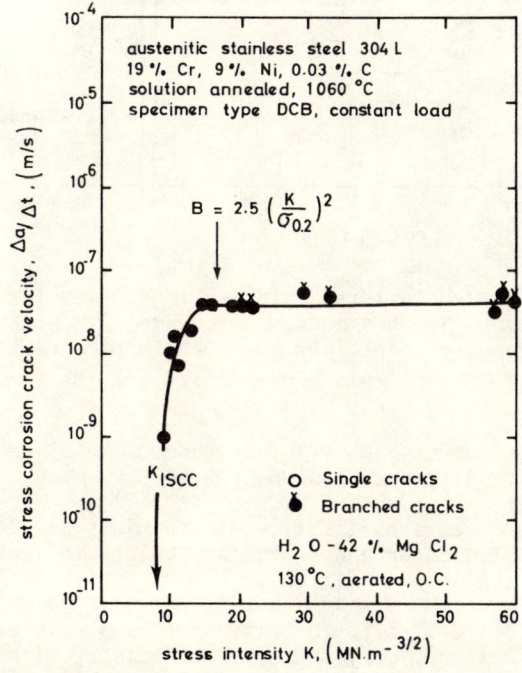

Fig. 5. Crack propagation as a function of stress intensity in pre-cracked samples (after Speidel).

Corrosion. The conditions necessary for stress corrosion are the borderline between general corrosion and immunity or passivity. The two main cases are

(i) aggressive solutions with corrosion resistant materials, i.e. stainless steel in oxygenated chloride solutions;

(ii) inhibiting solutions with non-corrosion-resistant metals, i.e. mild steel in bicarbonate solutions.

Both have been extensively investigated by electrochemical techniques (Parkins, 1972) which have shown that stress corrosion occurs in areas where there is the greatest difference between fast and slow sweep electrochemical experiments; this is a good indication of the borderline conditions between active corrosion and passivity which are necessary for stress corrosion (see Fig. 6).

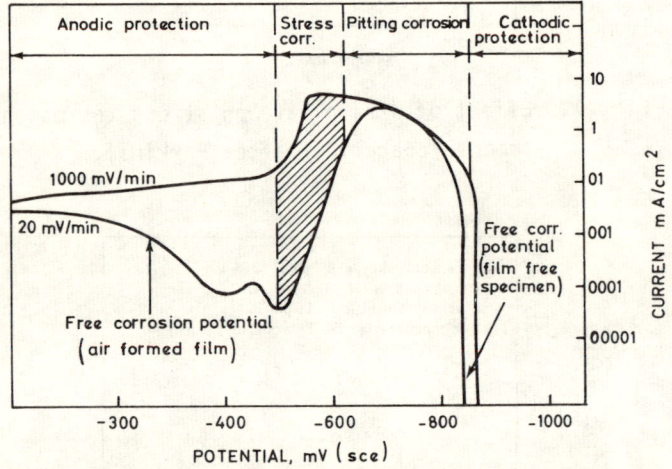

Fig. 6. Electrochemistry polarisation curves for mild steel in 2N ammonium carbonate, showing the types of behaviour observed in different potential ranges (after Parkins).

Morphology. The previous comments apply to homogeneous metal behaviour which tends to give transgranular stress corrosion.

Microstructural effects such as sensitisation of stainless steels and temper em= brittlement in low-alloy steels can change the morphology to intergranular stress corrosion under the same chemical conditions.

Summarising therefore the stress corrosion situation is a complex but understand= able interaction of physical chemistry and electrochemistry with physical metallur= gy and fracture mechanics which is illustrated in Fig. 7 (Staehle, 1969).

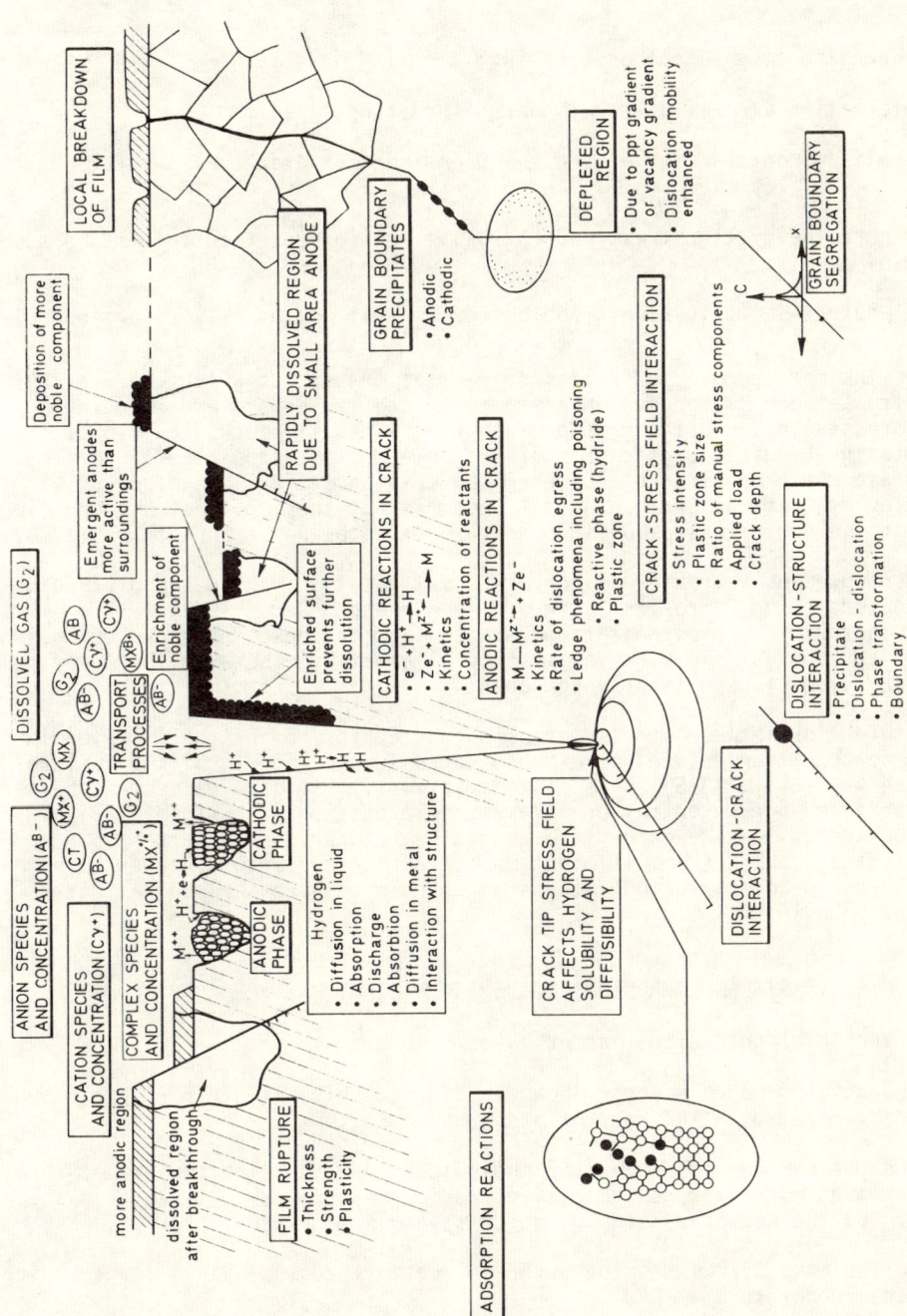

Fig. 7. Montage showing important processes operating which affect stress corrosion (after Staehle).

CORROSION FATIGUE THEORIES

The two main factors which have to be taken into account in corrosion fatigue are

1. generation of chemical environment;

2. interaction of 1 with metal under alternating stress.

Two generalised sources of the stress and environment are necessary for corrosion fatigue.

<u>Type I</u> where the cyclic stress and corrosive environment result from the same sequence of events.

<u>Type II</u> where the cyclic stress and corrosive environment arise independently.

Type I is the more common and relies on thermal transients which produce thermal cyclic stresses and corrodent concentration, the latter as described in "Boiling". Cyclic stresses in Type II corrosion fatigue are usually mechanical in origin, i.e. rotating bending, cyclic pressurisation and start-stop centrifugal loading. In this case the generation of the chemical environment will be as described in "Generation of Chemical Environment". Examples of the two types of corrosion fatigue in the power generation industry are given by Jones and others (1971).

The most comprehensive review of corrosion fatigue is given in the proceedings of a NACE conference on Corrosion Fatigue (Staehle and others, 1972). From this the main factors in corrosion fatigue are similar to those in stress corrosion except of course that the stresses must be alternating; however the damaging part of any cycle is the tensile component.

The possible link between stress corrosion and corrosion fatigue is given in the balance condition between stress and corrosion to give localised failure (described by Scully, 1975). As stated previously, in some cases this balance is only achieved under controlled strain rate conditions (Parkins, 1972) which may be considered to be a borderline condition between stress corrosion and corrosion fatigue. Perhaps it might be better therefore to consider both stress corrosion and corrosion fatigue as aspects of "Environment affected mechanical behaviour".

STRESS CORROSION INCIDENTS IN LIGHT WATER REACTORS

The two types of light water reactors are

(i) the boiling water reactor shown in Fig. 8 which is a direct cycle system with steam generated in the reactor pressure vessel;

(ii) the pressurised water reactor shown in Fig. 9 in which the primary coolant, overpressurised with hydrogen to prevent radiolysis, is pumped through a steam generator, on the secondary side of which steam is generated.

Stress corrosion problems have occurred in reactor pressure vessels and piping and in steam generators.

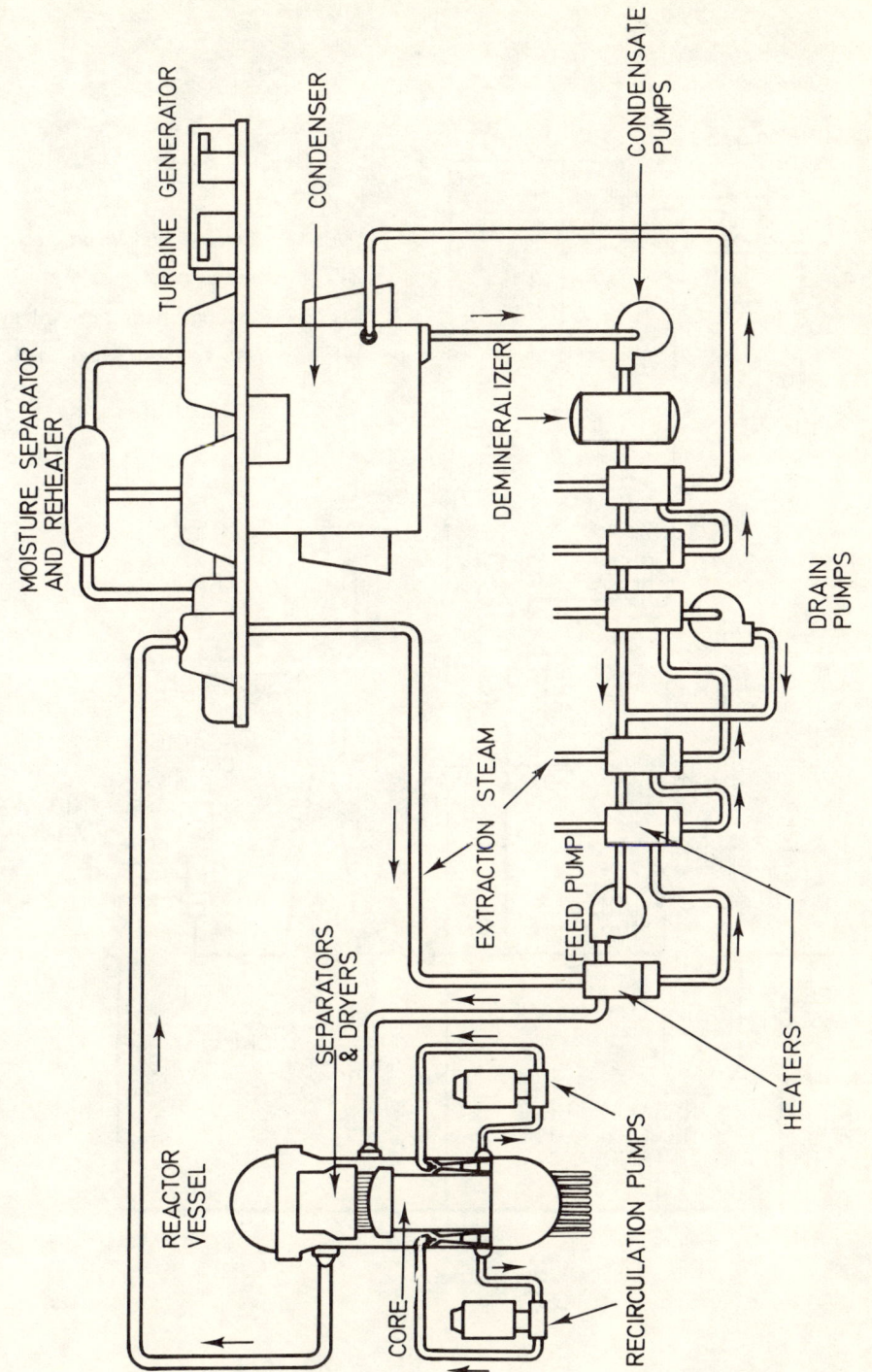

Fig. 8. Boiling Water Reactor system.

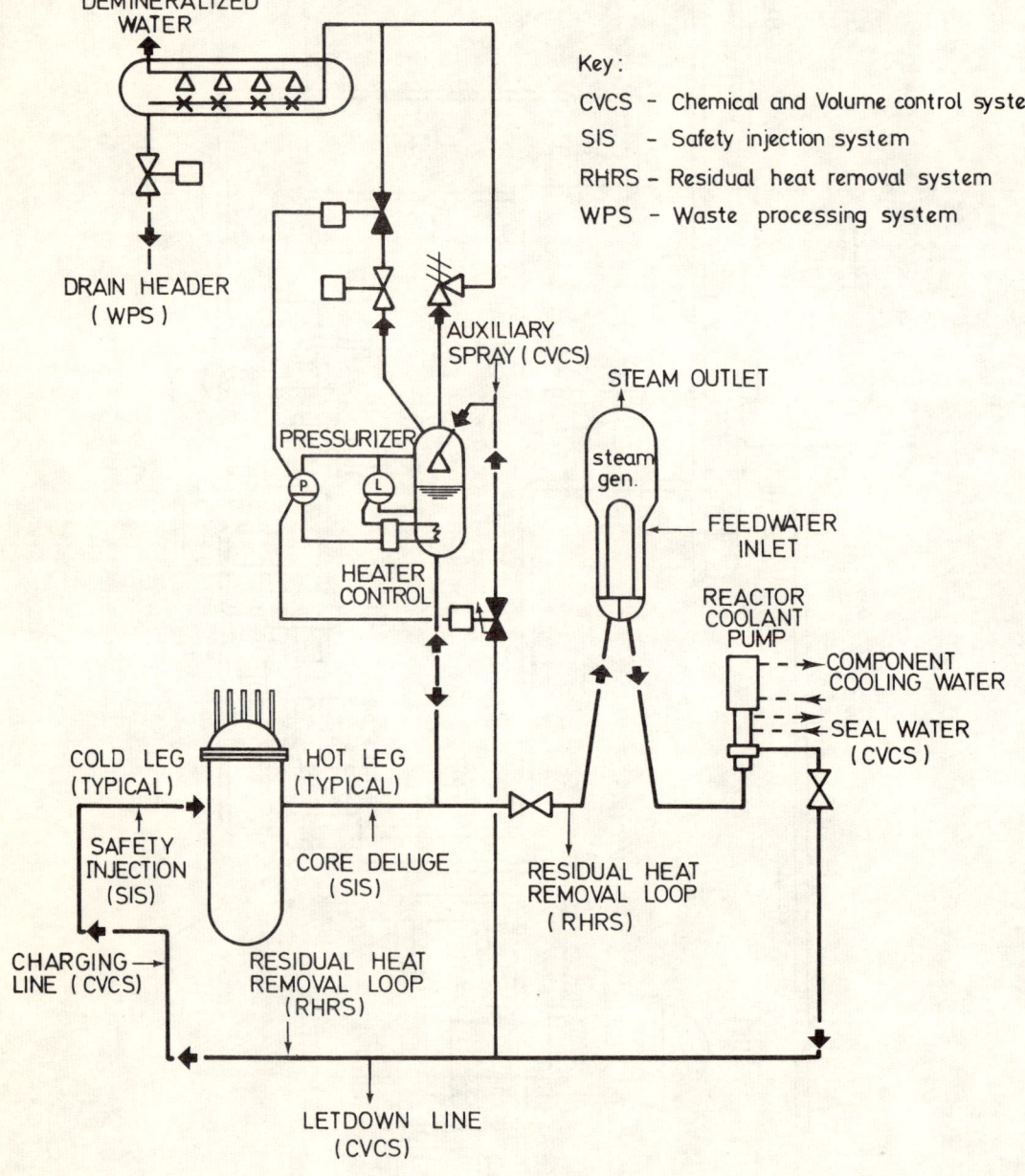

Fig. 9. Pressurized Water Reactor system.

Reactor Pressure Vessels and Piping

The incidents in BW reactors have been summarised by Berge (1971), Bush and Dillon (1977), and the Nuclear Regulatory Commission of the USA (1975) and are listed in Table 1. The main recommendation is that in future reactors only stabilised or low carbon austenitic steels should be used so as to minimise or eliminate sensitisation of the steels during welding and stress relief heat treatments.

TABLE 1. BWR Pressure Vessel and Piping Stress Corrosion.

POSITION	REACTOR	DIAGNOSIS
1. SS stub tube welds	Preservice in Oyster Creek and Tarapur	Intergranular SCC Chem - F from flux Stress - residual in welds Structure - sensitised by RPV stress relief and welding
2. SS cladding nozzle safe ends, bypass lines, core sprays, piping	Post service - Dresden, etc.	Intergranular SCC Chem - O_2 - HT, HP water Stress - residual weld Structures - sensitised by RPV stress relief and welding
3. Core structure	Post service - Millstone Point	Transgranular SCC Chem - condenser leakage chloride Stress - residual

No problems have occurred so far in the pressure vessels of PWRs. Since the design and construction of the two vessels is similar (except for position of control rods and size of the RPV) then other factors must be involved in the very different stress corrosion behaviour.

The most likely difference is the water chemistry of the two reactor systems, where

(i) in the boiling water reactor, radiolysis of the pure water, used as feed water, gives rise to both oxygen and hydrogen;

(ii) in the pressurised water reactor primary circuit, boric acid ($\sim$5 000 ppm) is used as a soluble poison to help control reactor operation and an over= pressure of hydrogen ($\sim$100 cm^3/kg water) is used to prevent radiolysis.

High temperature electrochemical studies have shown that oxygen stimulates the corrosion of stainless steels in high temperature water (Jones and others, 1976) whilst borates act both as inhibitors for mild and low alloy steels and are buffer solutions.

A possible danger area is the reduction of boric acid levels during the coast down conditions during the end of core life, before refuelling.

Steam Generators

Figure 10 shows the design of the U tube type of steam generator used in some PWR's including Koeberg.

Stress corrosion problems in these have been described in a number of papers (Hare, 1976, 1977; Stevens-Guille and Hare, 1975; Stevens-Guille, 1975) and appears to be due to the use of phosphate dosing for the secondary circuit of the PWR.

The complex chemistry of breakdown of the phosphate under deposits near the tube plate to give rise to free hydroxide (which causes stress corrosion of the high-nickel Inconnel 600 tubing) has been described (Fletcher and Malinowscki 1976; Weeks, 1976). The use of phosphate dosing has also caused both general and "denting" corrosion in PWR's generally near tube supports. In all cases the use of "all volatile treatment" (AVT) from the start of operation appears to have solved the corrosion problems, although if phosphate treatment is used initially, subsequent use of AVT does not solve the problem (Fletcher and Malinowscki, 1976; Hare, 1976, 1977; Stevens-Guille and Hare, 1975; Stevens-Guille 1975; Weeks, 1976) since the reaction products are already established within the crevices and deposits.

CORROSION FATIGUE IN LWR ENVIRONMENTS

Although there has not yet been any direct experience of corrosion fatigue in PWR's or BWR's, a number of authors have observed in laboratory studies that low cyclic strains can cause corrosion fatigue of low alloy steels in both PWR and BWR environments at high temperatures ($\sim 300°C$). (Kondo and others, 1976; Slama and others, 1977).

This is consistent with observations from "Stress Corrosion Theories" that

(i) borate and oxygenated water act as inhibitors for low alloy steels; and

(ii) that low strain rates cause stress corrosion crack propagation in mild and low alloy steels.

The main parameters which have been found to affect the rate of corrosion fatigue are

(i) frequencies below 10 cpm;

(ii) R ratio$(\frac{K\ min}{K\ max}) > 0.6$; and

(iii) low Δ Ks.

Typical results are shown in Fig. 11 where the growth rate seems to display two effects; at low values of ΔK, the growth rate increases very rapidly with ΔK values, whereas at medium and high ΔK values, a lower growth rate is observed. At the low ΔK end it can be seen that under certain test conditions the results are above the curve recommended by ASME for water reactor environments (ASME 1974).

To integrate research work in this area an International Co-operative Group on Cyclic Crack Growth Rate has been set up comprising workers from the USA, Europe

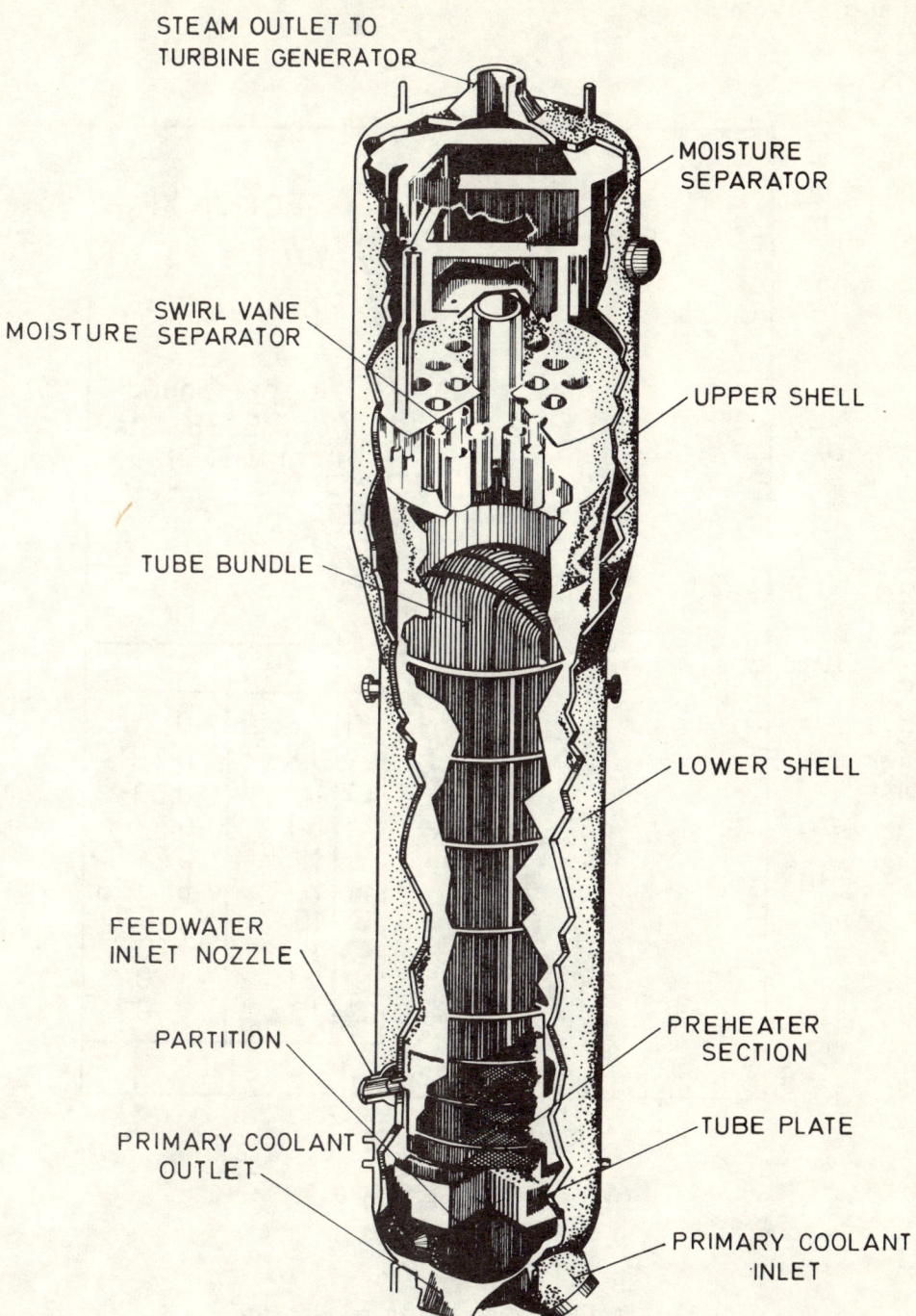

Fig. 10. PWR U tube steam generator.

and Japan. Their planned program includes examination of the effect on corrosion fatigue rates of factors such as

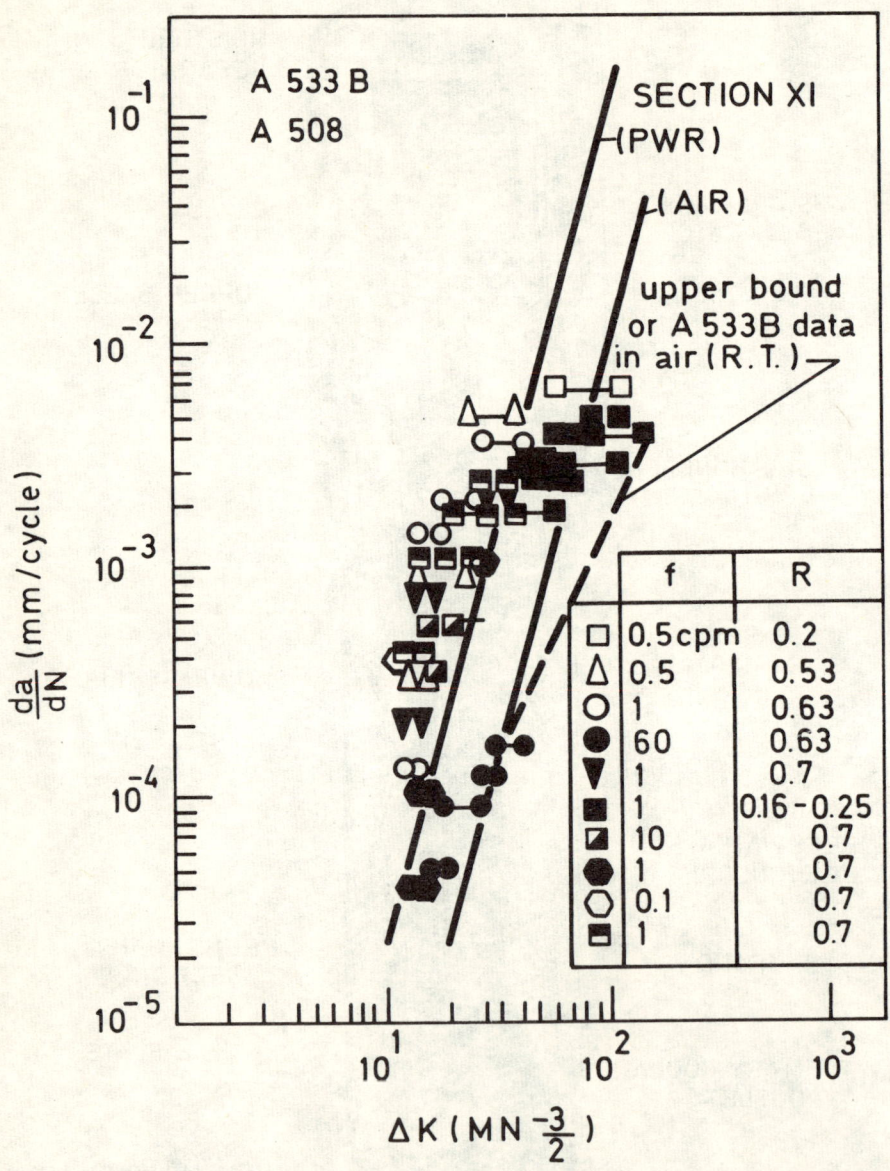

Fig. 11. Fatigue crack growth results for A533B and A508 RPV steels in PWR environment at 290° C (after Bamford).

R ratio
Frequency and wave form
Starting effects
Threshhold Δ K effect
High Δ K
Pressure vessel materials
Effect of cladding
Other material variables
Water chemistry
Temperature
Crack branching
Irradiation
Water chemistry at the crack tip and electrochemical effects
Size effect

In addition derivation of mechanisms of corrosion fatigue and application of data to practical cases will form part of the program.

Until the results of the above program are available, it would be conservative to use the ASME "wet" and "dry" lines rates (which differ by 15 times) for risk assessments. These assessments involve

1. measurement of defect sizes by UT or estimation of largest defect size from hydrotest;

2. estimation of extension of cracks in 1 taking into account the various tran= sients likely in water reactor operation and the ASME XI "wet" line data;

3. comparison of those with critical defect sizes calculated for normal transient and accident conditions for the RPV;

4. if it appears that the corrosion fatigue extended cracks are likely to exceed the critical defect sizes, further in service UT and hydro tests will be neces= sary to clarify the situation.

Even though at present the corrosion fatigue of low alloy steels is similar in BWR and PWR environments, the PWR vessel may be less likely to undergo corrosion fatigue since its stainless steel cladding is less likely to be penetrated by stress corrosion than the BWR's due to the difference in chemical activity between the PWR borate and BWR's oxygenated water environments (see "Reactor Pressure Vessel and Piping").

CONCLUSIONS

1. The stress corrosion incidents in LWR reactors are consistent with recent theories of stress corrosion.

2. The predominating factor in the stress corrosion of BWR pressure vessel and piping and PWR steam generators is the chemical environment.

3. Observations of corrosion fatigue in LWR environments are consistent with recent theories of corrosion fatigue.

4. Although the corrosion fatigue propensity of low alloy steels is similar in BWR and PWR environments, in practice PWR should be less susceptible to stress corrosion than that in BWR's.

REFERENCES

American Society of Mechanical Engineers. (1974). Rules for in-service inspection of nuclear power plant components. Boiler and pressure vessel code, Section XI, ASME, New York. Section A4300 and Fig. A 4300-1.
Beck, F. (1971). In J.C. Scully (Ed.), The Theory of Stress Corrosion in Alloys, North Atlantic Treaty Organisation, Brussels.
Berge, Ph. (1971). A review of stress corrosion cracking in water cooled reactors. In Effects of Environment on Materials Properties in Nuclear Systems, Institute of Civil Engineers, London pp 73 - 78.
Brown, B.F. (1971). In J.C. Scully (Ed.), The Theory of Stress Corrosion in Alloys, North Atlantic Treaty Organisation, Brussels.
Bush, S, and P. Dillon. (1977). In R.W. Staehle, J. Hochmann and J. Slater (Ed.), Stress Corrosion Cracking and Hydrogen Embrittlement of Iron Base Alloys, National Association of Corrosion Engineers, Houston.
Fletcher, W.D. and D.D. Malinowscki. (1976). Operating experience with Westinghouse steam generators. Nuclear Technology, 28, 356.
Hare, M.G. (1976). Steam-generator tube failures : World experience in water-cooled nuclear power reactors in 1974. Nuclear Safety, 17, 231 - 242.
Hare, M.G. (1977). Steam-generator tube failures: World experience in water-cooled nuclear reactors in 1975. Nuclear Safety, 18, 355 - 364.
Jones, D. de G., G.M.W. Mann, H.G. Masterson, J.F. Newman, and B. Hearn (1971). Stress corrosion and control of the working environment. In Effects of Environment on Materials Properties in Nuclear Systems, Institute of Civil Engineers, London, pp 79 - 89.
Jones, D. de G. and I.L. Mogford (1972). Corrosion fatigue studies in the power generation industry. In R.W. Staehle, A.J. McEvily and O. Devereux (Ed.), Corrosion Fatigue : Chemistry Mechanics and Microstructure, National Association of Corrosion Engineers, Houston, pp 30 - 39.
Jones, D. de G., R.W. Staehle and J. Slater. (1976). (Ed.), High Temperature, High Pressure Electrochemistry in Aqueous Solutions, National Association of Corrosion Engineers, Houston.
Kondo, T., H.E. Watson, and W.H. Bamford. (1976). Papers presented at US Nuclear Regulatory Commission Conference on Water Reactor Safety Research.
Logan, H.L. (1966). The Stress Corrosion of Metals. John Wiley, New York.
Nuclear Regulatory Commission, (1975). Investigation and Evaluation of Cracking in Stainless Steel Piping of BWR Plants, NUREG 75/067. US Nuclear Regulatory Commission, Washington.
Parkins, R.N. (1972). Stress corrosion spectrum, British Corrosion Journal, 7, 15 - 28.
Scully, J.C. (1971). (Ed.), The Theory of Stress Corrosion in Alloys, North Atlantic Treaty Organisation, Brussels.
Scully, J.C. (1975). Stress corrosion crack propagation : A constant charge criterion, Corrosion Science, 15, 207 - 224.
Slama, G., B. Haussin, and J.L. Bernard. (1977). Effect of nuclear environment (PWR) on the deterioration of materials in service, paper presented at symposium on "Deterioration of Materials in Service" organised by S.A. Institute of Physics at Cape Town.
Spiedel, M. (1971). In J.C. Scully (Ed.), The Theory of Stress Corrosion in Alloys, North Atlantic Treaty Organisation, Brussels.
Staehle, R.W., A.J. Forty, and D. van Rooyen. (1969). (Ed.), Fundamental Aspects of Stress Corrosion Cracking, National Association of Corrosion Engineers, Houston.
Staehle, R.W., A.J.McEvily, and O. Devereux. (1971). (Ed.), Corrosion Fatigue : Chemistry, Mechanics and Microstructure. National Association of Corrosion Engineers, Houston.

Staehle, R.W., J. Hochmann, and J. Slater (1977). (Ed.), <u>Stress Corrosion Cracking and Hydrogen Embrittlement of Iron Base Alloys</u>, National Association of Corrosion Engineers, Houston.
Stevens-Guille, P.D. (1975). Steam-generator tube failures : World experience in water-cooled nuclear power reactors during 1972. <u>Nuclear Safety</u>, <u>16</u>, 354 - 364.
Stevens-Guille, P.D., and M.G. Hare. (1975). Steam-generator tube failures : World experience in water cooled nuclear power reactors in 1973, <u>Nuclear Safety</u> <u>16</u>, 603 - 614.
Weeks, J.R. (1976). Materials performance in operating P.W.R. steam generators. <u>Nuclear Technology</u>, <u>28</u>, 356.

A REVIEW OF THE USE OF ISOPARAMETRIC FINITE ELEMENTS FOR FRACTURE MECHANICS

F. J. Heymann

The Research and Process Development Department, The Iron and Steel Industrial Corporation of South Africa, R.S.A.

ABSTRACT

Analytical solutions are only available for relatively few idealized geometries. It is therefore required to resort to approximate methods such as the finite-ele= ment method, to solve problems of a more general nature. Many standard elements are required to model the singularity of the crack tip accurately and resort has often been taken to special crack-tip elements. However, Henshell and Shaw (1975) and Barsoum (1976a) have shown that the parabolic isoparametric element produces the correct singularity at the crack tip when the mid-side nodes are shifted to the quarter points. This type of element is discussed here as well as the method proposed by Lynn and Ingraffea (1978) to improve the strength of the finite-ele= ment singularity.

Some isoparametric special elements are mentioned and finally the determination of solution parameters is discussed.

KEYWORDS

Finite element; fracture mechanics; quarter point; crack; isoparametric.

INTRODUCTION

In order to solve general fracture-mechanics problems, resort has often to be taken to numerical methods. Currently the most popular and most powerful method, with general application, is the finite-element method (see Zienkiewicz (1977)). Here a continuum is divided into discrete elements where the contributions of each element is obtained separately, see Fig. 1.

Boundary conditions are imposed and finally element contributions are added to form a set of linear equations which are then solved in an appropriate manner.

The displacement approach to finite-element analysis, can be seen as the minimi= zation of the total potential energy of the system.

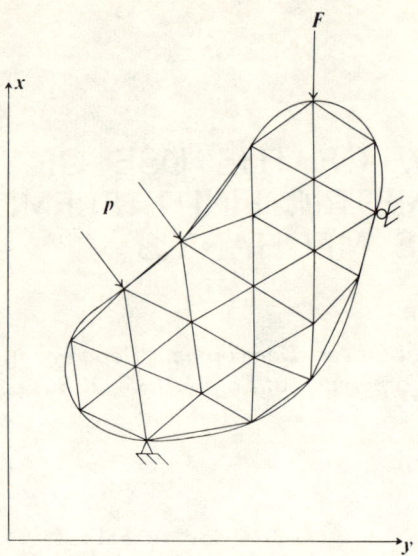

Fig. 1

As $\Pi = U + W$, it follows that a good representation of the displacements in areas of high strain gradient is required. One way in which this can be achieved is by refinement of the finite-element mesh, however, Chan, Tuba and Wilson (1970) have shown that a tremendous mesh refinement is required to describe the conditions at a crack tip adequately. Many special elements have thus been developed to model crack-tip behaviour in an acceptable manner, e.g. Byskov (1971), Wilson and Thompson (1971), Blackburn (1972), Tracey (1971), Tong, Pian and Lasry (1973), Stern and Becker (1978), Akin (1976), Fawkes (1978) and Swedlow (1979).

Since their introduction by Taig (1961), isoparametric elements have become popular and have proven to be well suited to fracture-mechanics applications - Henshell and Shaw (1975) and Barsoum (1976a), (1976b).

THE ISOPARAMETRIC FORMULATION

The term 'isoparametric' is used for elements for which the shape (interpolation) functions have been formulated in a curvilinear co-ordinate system, termed the 'natural' co-ordinate system, and are subsequently mapped onto a cartesian co-ordinate system via a suitable transformation, see Fig. 2.

$$\begin{bmatrix} x \\ y \\ z \end{bmatrix} = f \begin{bmatrix} \xi \\ \eta \\ \zeta \end{bmatrix}$$

Furthermore, the same shape functions that describe the geometry also describe the displacements:

$$x = \sum_i N_i x_i$$

and $\quad u = \sum_i N_i u_i$

where $\quad \sum_i N_i = 1$

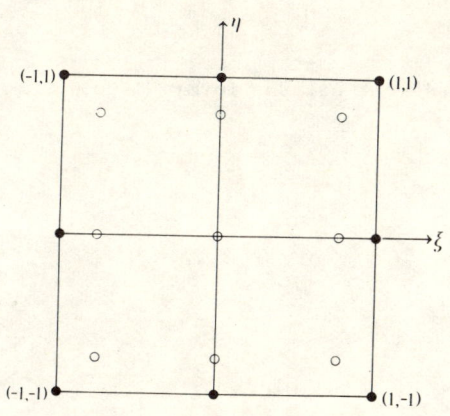
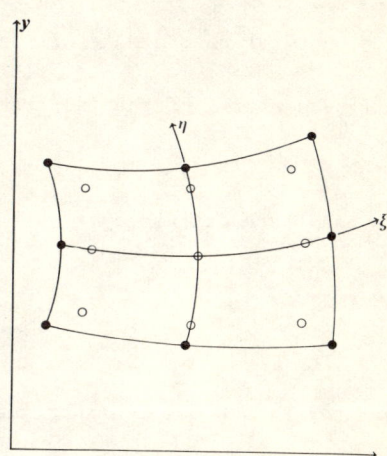

Fig. 2

The shape functions of the 8-node isoparametric plane quadrilateral are:

$$N_i = \tfrac{1}{4} (1 + \xi_0) (1 + \eta_0) (\xi_0 + \eta_0 - 1) \eta_i^2 \xi_i^2$$

$$+ \tfrac{1}{2} (1 - \xi_i^2) (1 - \xi^2) (1 + \eta_0) \eta_i^2$$

$$+ \tfrac{1}{2} (1 - \eta_i^2) (1 - \eta^2) (1 + \xi_0) \xi_i^2 \tag{1}$$

and for the 12-node cubic plane element:

$$N_i = \tfrac{1}{256} (1 + \xi_0) (1 + \eta_0) [-10 + 9 (\xi^2 + \eta^2)][-10 + 9 (\xi_i^2 + \eta_i^2)]$$

$$+ \tfrac{81}{256} (1 + \xi_0) (1 + 9 \eta_0) (1 - \eta^2) (1 - \eta^2)$$

$$+ \tfrac{81}{256} (1 + \eta_0) (1 + 9 \xi_0) (1 - \xi^2) (1 - \xi_i^2) \tag{2}$$

where

$N_i = i^{th}$ shape function

$\xi_0 = \xi\xi_i$

$\eta_0 = \eta\eta_i$

and

ξ_i = natural co-ordinate of the ith node

η_i = natural co-ordinate of the ith node

The stiffness matrix of the isoparametric element has the form:

$$K = \int_V B^T D B \, dV$$

Where B and D are defined by:

$\varepsilon = Bu$

and

$\sigma = D\varepsilon$

The operator B is a matrix (in plane analysis)

$$\begin{bmatrix} \frac{\partial N_i}{\partial x} & 0 \\ 0 & \frac{\partial N_i}{\partial y} \\ \frac{\partial N_i}{\partial y} & \frac{\partial N_i}{\partial x} \end{bmatrix}$$

but as the N_i are defined in the natural co-ordinate system:

$$\frac{\partial N_i}{\partial x} = \frac{\partial N_i}{\partial \xi}\frac{\partial \xi}{\partial x} + \frac{\partial N_i}{\partial \eta}\frac{\partial \eta}{\partial x}$$

This can be written as:

$$\begin{Bmatrix} \frac{\partial N_i}{\partial x} \\ \frac{\partial N_i}{\partial y} \end{Bmatrix} = J^{-1} \begin{Bmatrix} \frac{\partial N_i}{\partial \xi} \\ \frac{\partial N_i}{\partial \eta} \end{Bmatrix}$$

where

$$J = \begin{bmatrix} \frac{\partial x}{\partial \xi} & \frac{\partial y}{\partial \xi} \\ \frac{\partial x}{\partial \eta} & \frac{\partial y}{\partial \eta} \end{bmatrix}$$

is the Jacobian operator.

The stiffness matrix becomes

$$K = \int_{-1}^{1} \int_{-1}^{1} B^T DB \det J \, d\xi \, d\eta$$

and

$$\varepsilon = J^{-1} B(\xi, \eta) u$$

Now consider Fig. 3

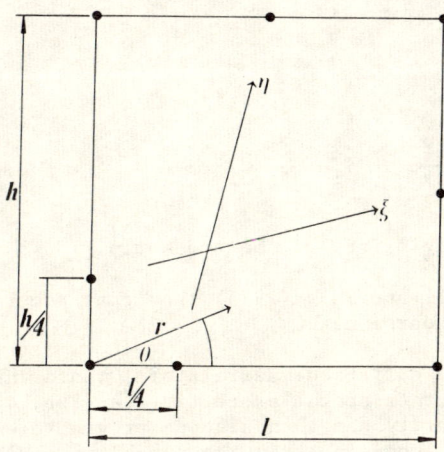

Fig. 3

for the edge where $\eta = -1$; $x = 0, p, 1$ and $y = 0$ we determine the value for p which makes J singular at $x = 0$.

$$x = \sum_i N_i x_i$$

$$= -\tfrac{1}{2}(1-\xi)\xi x_1 + (1-\xi^2) x_2 + \tfrac{1}{2}(1+\xi)\xi x_3$$

$$= (1-\xi^2) p + \tfrac{1}{2}(1+\xi)\xi$$

and

$$\xi = \frac{-\tfrac{1}{2} + \tfrac{1}{2}\sqrt{1 - 8(1-2p)(p-x)}}{1 - 2p} \tag{3}$$

Considering only the positive root:

$$\frac{\partial \xi}{\partial x} = \frac{2}{\sqrt{1 - 8(1-2p)(p-x)}} \tag{4}$$

and is singular when

$$1 - 8(1-2p)(p-x) = 0$$

to be singular at $x = 0$ $\quad p = \tfrac{1}{4}$

and

$$\frac{\partial \xi}{\partial x} = \frac{1}{\sqrt{x}} \tag{5}$$

$$u = \sum_i N_i u_i$$

and

$$\varepsilon_{xx} = \frac{du}{dx} = \frac{\partial u}{\partial \xi}\frac{\partial \xi}{\partial x} = \left(\sum_i \frac{\partial N_i}{\partial \xi} u_i\right)\frac{1}{\sqrt{x}} \tag{6}$$

therefore the $r^{-\tfrac{1}{2}}$ singularity exists at the crack tip.

Numerical results by Henshell and Shaw (1975) show that accurate results can be obtained with a relatively coarse mesh.

Barsoum (1976a) proposed the collapsed quadrilateral where three nodes of one side are constrained to move together at the crack tip (see Fig. 4). In a similar manner as above, the singularity can then also be proven to exist along the ξ axis and consequently along rays emanating from the crack tip - Hibbit (1977).

An analysis made by Hibbit (1977), he shows that for a one-dimensional polynomial interpolation such that:

$$u(\xi) = a_0 + a_1 \xi + a_2 \xi^2 + \ldots + a_u \xi^n$$

with

$$x = A_0 + A_1 \xi + A_2 \xi^2 + \ldots A_n \xi^n$$

a particular choice of nodal positioning delivers for $n \geq 2$

$$A_1 = A_2 \ldots A_{n-1} = 0; \quad A_n > 0$$

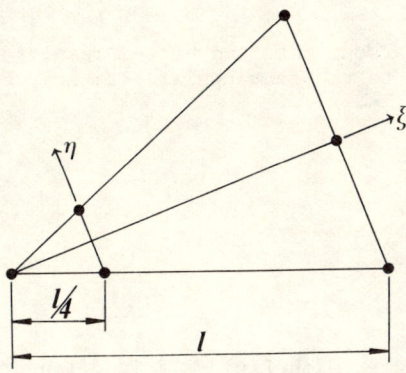

Fig. 4

so that $x = A_0 + A_n \xi^n$

With $A_0 = 0$ we have

$$\frac{du}{dx} = \frac{1}{nA_n} \xi^{1-n} (a_1 + 2a_2 \xi + \ldots + na_n \xi^{n-1}) \quad (7)$$

$$= \frac{1}{nA_n} \{a_1 (\frac{x}{A_n})^{(1-n)/n} + 2a_2 (\frac{x}{A_n})^{(2-n)/n} + \ldots + na_n\} \quad (8)$$

The strain singularity as $x \to 0$ is thus determined by the interpolation order thus from Hibbit (1977).

Interpolation order n	Leading strain term as $x \to 0$
2	$O(x^{-1/2})$
3	$O(x^{-2/3})$
4	$O(x^{-3/4})$

etc.

Hibbit (1977) then suggested that this phenomenon can be used to model the changing singularity order as per Rice (1968) where the singularity is $x^{-1/(1+N)}$ with N a hardening exponent varying between 0 and 1.

A simple method thus accrues from the above to vary the order of the singularity by using a high-order element and by shifting the nodes to positions to produce the required singularity order. This can, of course, be done during an elasto-plastic computation.

Hibbit also explained why Barsoum obtained better results with the collapsed qua= drilateral element compared to the rectangular element. For the triangular ele= ment:

$$x = \bar{x} \, r^n$$
$$y = \bar{y} \, r^n \, \eta$$

where
$$0 \leq r \leq 1$$
$$-1 \leq h \leq 1$$

The singular term is associated with $R(r)$ - (R function of r)

$$\frac{\partial u}{\partial R} = 0 \, (R^{(1-n/n)})$$

The area mapping

$$dA = \bar{x} \, \bar{y} \, n \, r^{2n-1} \, dr \, d\eta$$

The strain energy associated with the singular strain is:

$$E \propto \int_{-1}^{1} \int_{0}^{1} 0 \, (\frac{\partial u}{\partial R})^2 \, \bar{x} \, \bar{y} \, n \, r^{2n-1} \, dr \, d\eta$$

and gives

$$E \propto \int_{-1}^{1} \int_{0}^{1} 0 \, (r) \, dr \, d\eta \tag{9}$$

which is bounded for al n.

For quadrilateral elements, Hibbit (1977) shows that

$$E \propto \int_{0}^{1} \int_{0}^{1} 0 \, (\xi^{1-n}) \, d\xi \, d\eta \tag{10}$$

which is unbounded for $n \geq 2$.

Therefore collapsed quadrilateral (triangular) isoparametric elements are always to be preferred above quadrilateral elements.

The derivation of a singular cubic collapsed isoparametric element has been done by Pu, Hussain and Lorensen (1978). As they seeked a $r^{-1/2}$ singularity at the crack tip, they required the coefficient of ξ^3 etc. to be 0 for the co-ordinate interpolation, thus:

$$x = a_0 + a_1 \, \xi + a_2 \, \xi^2$$

for a one-dimensional equivalent but with

$$u = b_0 + b_1 \xi + b_2 \xi^2 + b_3 \xi^3$$

Now the leading term for the strain resorts back to $0 \, r^{-1/2}$.
For this element the mid-side nodes are shifted to the 1/9 and 4/9 positions measured outward from the crack tip.

Another interesting observation that Pu, Hussain and Lorensen (1978) have made is that curved sides opposite the crack tip are to be avoided as inferior results are obtained. Sides between adjacent crack-tip elements also have no need to be cur= ved and should be kept straight as the rays emanating from the crack tip containing the singularity, will be curved with curve sides.

REINFORCEMENT OF THE SINGULARITY

Lynn and Ingraffea (1978) have shown that the mid-side nodes of elements adjacent to crack-tip elements could be arranged such as to reinforce the singularity at the crack tip. The same approach is used as by Henshell and Shaw (1975). Recalling from eq. 3 that the Jacobian becomes singular as

$$1 - 8(1 - 2p)(p - x) = 0$$

it can be required that x be exterior to the element, for instance at the crack tip of a neighbouring element. Say $x = -q$ so that:

$$p = \frac{(1 - 2q) + 2\sqrt{q^2 + q}}{4} \tag{11}$$

Only the positive root has to be considered. Note also than when $q = 0$, $p = \frac{1}{4}$.
The arrangement used by Lynn and Ingraffea is shown in Fig. 5.

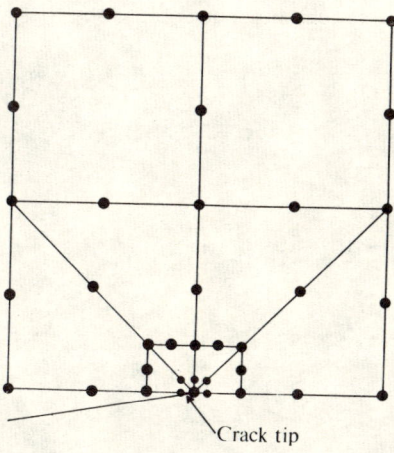

Fig. 5

Lynn and Ingraffea (1978) report that for the configuration used by them, the best results are obtained for a crack element to adjacent element size ratio of approximately 0,1.

OTHER ISOPARAMETRIC SPECIAL-ELEMENT FORMULATIONS

a) The Blackburn (1973) Element

Hellen (1977) pointed out that the special isoparametric element formulated by Blackburn (1973) was very similar to the isoparametric element as described above. The displacement variation for the collapsed quadrilateral element being:

$$u(\xi, \eta) = a_1 + \frac{(a_2 \xi + a_3 \eta)}{\sqrt{(\xi + \eta)}} + \frac{a_4 \xi \eta}{\xi + \eta} + a_5 \xi + a_6 \eta \tag{12}$$

while the displacement function used by the Blackburn element is:

$$u(\xi, \eta) = b_1 + \frac{(b_2 \xi + b_3 \eta + b_4 \xi \eta)}{\sqrt{(\xi + \eta)}} + b_5 \xi + b_6 \eta \tag{13}$$

b) The Stern and Becker (1978) Element

The Stern and Becker (1978) element is very similar to the Blackburn (1973) element and differs only in the $\xi \eta$ term so that

$$u(\xi, \eta) = c_1 + \frac{(c_2 \xi + c_3 \eta)}{\sqrt{(\xi + \eta)}} + b_4 \frac{\xi \eta}{(\xi + \eta)^{3/2}} + b_5 \xi + b_6 \eta \tag{14}$$

c) The Benzley (1974) Element

The element proposed by Benzley (1974) is an enriched linear quadrilateral isoparametric element. The displacement function is:

$$U_j = \sum_i^4 N_i U_{ij} + (Q_{Ij} - \sum_i^4 N_i \bar{Q}_{Iji}) K_I + (Q_{II} - \sum_i^4 N_i \bar{Q}_{IIji}) K_{II} \tag{15}$$

where

$$N_i = \tfrac{1}{4}(1 - \xi_0)(1 - \eta_0)$$

$$Q_{I1} = \frac{1}{G}\sqrt{\frac{\rho}{2\pi}} \cos \frac{\theta}{2} \left[\frac{\kappa - 1}{2} + \sin^2 \frac{\theta}{2} \right]$$

$$Q_{II1} = \frac{1}{G}\sqrt{\frac{\rho}{2}} \sin \frac{\theta}{2} \left[\frac{\kappa + 1}{2} + \cos^2 \frac{\theta}{2} \right]$$

for the 1 direction (16)

and

$$Q_{I2} = \frac{1}{G}\sqrt{\frac{\rho}{2\pi}} \sin \frac{\theta}{2} \left[\frac{\kappa + 1}{2} - \cos^2 \frac{\theta}{2} \right]$$

$$Q_{II2} = \frac{1}{G} \sqrt{\frac{\rho}{2\pi}} \cos \frac{\theta}{2} \left[\frac{\kappa - 1}{2} + \sin^2 \frac{\theta}{2} \right]$$

for the 2-direction,

and

$\bar{Q}_{Iji}, \bar{Q}_{IIji}$ are the respective Q^s evaluated at the ith node.

d) <u>The Gifford and Hilton (1978) Element</u>

This element uses a similar approach to that for the Benzley (1974) element, this time employing cubic shape functions as discussed before.

The displacement functions are:

$$u_j = \sum_i^{12} N_i u_{ij} + (Q_{Ij} - \sum_i^{12} N_i \bar{Q}_{Iji}) K_I$$
$$+ (Q_{IIj} - \sum_i^{12} N_i \bar{Q}_{IIji}) K_{II} \tag{17}$$

with

$$Q_{I1} = \frac{1}{4G} \sqrt{\frac{\rho}{2\pi}} \{ \cos \phi \left[(2\kappa - 1) \cos \theta/2 - \cos 3\theta/2 \right]$$
$$- \sin \phi \left[(2\kappa + 1) \sin \theta/2 - \sin 3\theta/2 \right] \}$$

$$Q_{II1} = \frac{1}{4G} \sqrt{\frac{\rho}{2\pi}} \{ \cos \phi \left[(2 + 3) \sin \theta/2 + \sin 3\theta/2 \right]$$
$$+ \sin \phi \left[(2\kappa - 3) \cos \theta/2 + \cos 3\theta/2 \right] \} \tag{18}$$

$$Q_{I2} = \frac{1}{4G} \sqrt{\frac{\rho}{2\pi}} \{ \sin \phi \left[(2\kappa - 1) \cos \theta/2 - \cos 3\theta/2 \right]$$
$$+ \cos \phi \left[(2\kappa + 1) \sin \theta/2 - \sin 3\theta/2 \right] \}$$

$$Q_{II2} = \frac{1}{4G} \sqrt{\frac{\rho}{2\pi}} \{ \sin \phi \left[(2\kappa + 3) \sin \theta/2 + \sin 3\theta/2 \right]$$
$$- \cos \phi \left[(2\kappa - 3) \cos \theta/2 + \cos 3\theta/2 \right] \}$$

The κ, $\bar{Q}^s$ are as before.

ϕ is the crack angle relative to the 1 axis.

DETERMINATION OF FRACTURE PARAMETERS

Fracture parameters for standard elements are often determined a posteriori, i.e. after displacements have been solved for, with isoparametric elements. Various methods of solution have been developed.

Substitution of displacements and stresses in classical solutions

The solution for the crack-tip displacements have been given by Williams (1957) as:

$$2 G u(\theta) = \sum_n \{(-1)^{n-1} r^{n-1/2} [d_{2n-1} D_{u1}(n, \theta) + a_{2n-1} A_{u1}(u, \theta)]$$

$$+ (-1)^n r^n [d_{2n} D_{u2}(n, \theta) + a_{2n} A_{u2}(u, \theta)]\} \quad (19)$$

$$2 G v(\theta) = \sum_n \{(-1)^{n-1} r^{n-\frac{1}{2}} [d_{2n-1} D_{v1}(n, \theta) + a_{2n-1} A_{v1}(n, \theta)]$$

$$+ (-1)^n r^n [d_{2n} D_{v2}(n, \theta) + a_{2n} A_{v2}(n, \theta)]\}$$

The stress-intensity factors K_I and K_{II} are then related to d_1 and a_1 by:

$$K_1 = -d_1 \sqrt{2\pi} \quad ; \quad K_2 = -a_1 \sqrt{2\pi} \quad (20)$$

$$D_{u1}(n, \theta) = (n - \tfrac{1}{2}) \cos(n - 5/2)\theta - (\kappa + n - 3/2) \cos(n - \tfrac{1}{2})\theta$$

$$D_{u2}(n, \theta) = n \cos(n - 2)\theta - (\kappa + n + 1) \cos n\theta$$

$$A_{u1}(n, \theta) = (n - \tfrac{1}{2}) \sin(n - 5/2)\theta - (\kappa + n + \tfrac{1}{2}) \cos n\theta$$

$$A_{u2}(n, \theta) = n \sin(n - 2)\theta - (\kappa + n - 1) \sin n\theta$$

$$D_{v1}(n, \theta) = (n - \tfrac{1}{2}) \sin(n - 5/2)\theta - (\kappa - n + 3/2) \sin(n - \tfrac{1}{2})\theta \quad (21)$$

$$D_{v2}(n, \theta) = n \sin(n - 2)\theta - (\kappa - n - 1) \sin n\theta$$

$$A_{v1}(n, \theta) = (n - \tfrac{1}{2}) \cos(n - 5/2)\theta + (\kappa - n - \tfrac{1}{2}) \cos(n - \tfrac{1}{2})\theta$$

$$A_{v2}(n, \theta) = n \cos(n - 2)\theta + (\kappa - n + 1) \cos n\theta$$

where

$\kappa = (3 + \nu)/(1 + \nu)$ for plane stress
$\quad (3 - 4\nu)$ for plane strain

Normally only few leading terms are required and the a^s and d^s can be solved for if as many displacements are available as are terms required:

$$\{u\} = [f(\theta)] \{\alpha\}$$

where $f(\theta)$ consists of the D^s and A^s and α of the d^s and a^s.

In a discussion of the paper of Barsoum (1976a), Tracey (1977) points out that for the collapsed parabolic triangular element the displacements vary as:

$$u_i = a_i + b_i \sqrt{r} + c_i \, r \text{ with } i = x, y$$

along rays emanating from the crack tip.

Considering only the $r^{\frac{1}{2}}$ term the stress intensity for the quadratic element becomes:

$$K_I = \{ -3U_i(0) + 4U_i(L/4) - U(L) \} / \sqrt{L} / f_i(\Theta)$$

and for the cubic element

$$K_I = \tfrac{1}{2} \{ -11U_i(0) + 18U_i(L/9) - 9U_i(4L/9) + 2U_i(L) \} / \sqrt{L} / f_i(\Theta)$$

with

$$f_x(\Theta) = \frac{\sin \Theta/2}{G\sqrt{8\pi}} (\kappa + 1 - 2\cos^2 \Theta/2)$$

$$f_y(\Theta) = \frac{\cos \Theta/2}{G\sqrt{8\pi}} (\kappa - 1 + 2\sin^2 \Theta/2) \quad (23)$$

Strain-energy release rate

The strain-energy release rate $\dfrac{\delta U}{\delta a}$ can be determined from a number of solutions of the same problem with slightly different crack lengths. With this method, proposed by Mowbray (1971), the strain-energy release rate, G, is determined as a function of a_1 and $\dfrac{\delta U}{\delta a}$ is then determined by numerical differentiation.

The method of virtual extensions/stiffness derivative technique

A method that is much more efficient than the above has been proposed by Parks (1974) and Hellen (1975), for determining the strain-energy release rate. This method is called: the method of virtual crack extensions or the :stiffness derivative technique'. Here the attributes of the frontal solution scheme, Hellen (1969), Irons (1970) are utilized, namely, in the frontal solution method, unknowns are solved for element by element. The method is as follows:

$$[K] \{u\} = \{b\} \quad (24)$$

is the equation system to be solved and the total potential energy of the system is:

$$\Pi = \tfrac{1}{2} u^T K u - u^T b$$

a small variation in Π is:

$$\delta \Pi = \tfrac{1}{2} \{u\}^T [\delta K] \{u\} + \delta u^T [K] \{u\} - \{\delta u\}^T b - \{u\}^T \delta b$$

substituting eq. 24 and requiring that $\delta b = 0$

$$\delta \Pi = \tfrac{1}{2} \{u\}^T [\delta K] \{u\} \quad (25)$$

Thus for a small variation in crack length, the variation in potential energy can be determined, this requiring only forward elimination, and if displacements and stresses are not required, only a number of back-substitutions are required for the crack-tip elements. A number of solutions can now be obtained by allowing small crack extensions emanating from the crack tip at various angles to determine also the direction of crack propagation if $\frac{\partial U}{\partial a}$ is assumed to be the determining parameter.

Parks (1974) also noted that this method effectively determines the J-integral for a contour surrounding the crack tip.

The J-integral Method

A path independent integral, termed the J-integral, was discovered by Rice (1968). It has been used successfully in many finite-element analyses which have proven its path independence for plane problems.

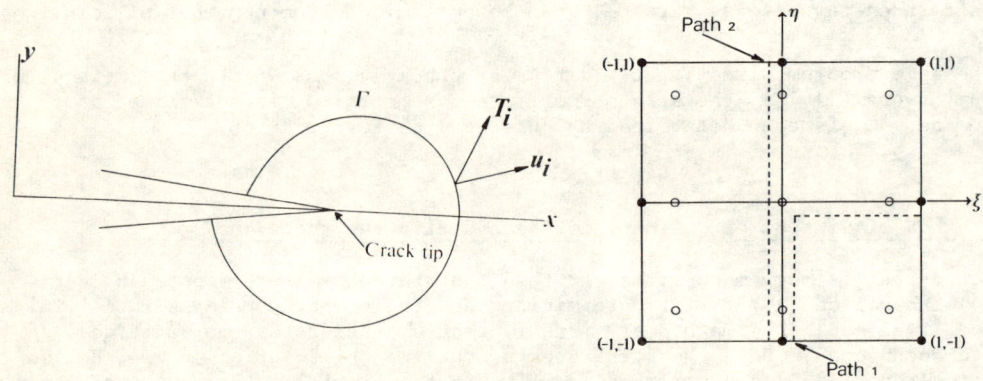

Fig. 6

$$J = \int_\Gamma W \, dy - T_i \frac{\partial u_i}{\partial x} \, ds$$

$W = \int_V \sigma \, d\varepsilon$ is the strain energy

T_i = The traction vector

U_i = The displacement vector

A pictorial representation of the integral is given in Fig. 6a.

De Lorenzi and Shih (1977) proposed a method for evaluating the J-integral with higher order elements. Two types of paths are considered, namely a path $\xi = 0$; $-1 \leq \eta \leq 1$ and a path around a corner: $\xi = 0$; $1 \leq \eta \leq 0$, and $\eta = 0$; $0 \leq \xi \leq 1$ (see Fig. 6b). De Lorenzi and Shih (1977), however, have simplified the last path by averaging for $\xi = 0$.

Considering the first path:

$$\Delta J = \sum_{i=1}^{3} H_i \{(W - \sigma_{11}\frac{\partial u_x}{\partial x} - \sigma_{21}\frac{\partial u_y}{\partial x})J_{11}$$

$$+ (\sigma_{12} U \frac{\partial \partial u_x}{\partial x} + \sigma_{12}\frac{\partial u_y}{\partial x}) J_{12}\}_i$$

Now $J = G$ for the linear elastic case, and H_i are the Gaussian weight functions.

OTHER ASPECTS

Three-dimensions

Barsoum (1976a) (1976b) used the same techniques outlined above, for the isoparametric solid and shell problems, in the latter case using Ahmads formulation for shell analysis.

Plasticity

As mentioned above, Hutchinson (1968) and Rice and Rosengren (1968) showed that the crack tip singularity changes from $r^{-1/2}$ for the linear elastic case to $r^{-1/(1+N)}$ where N is a hardening parameter. Barsoum (1977) has shown that the r singularity for perfect plasticity is obtained by allowing the crack-tip nodes of the collapsed quadrilateral to behave independently. Crack-tip blunting can then be modelled. Large plastic deformation has been treated by ADINA computer program developed by Bathe (1977).

CLOSING REMARKS

Isoparametric elements are available in most finite-element programs. The usefulness of the standard elements for fracture mechanics has been reviewed with most emphasis having been placed on linear elastic plane analysis. Accurate results accrue from the use of these elements, and it is the view of the author that no more effort should be made to develop crack-tip elements for linear analysis. However, the iterative methods used by Swedlow (1978) will still be used due to the inability of the above methods to account for the changing order of singularity surrounding the crack tip. An interesting alternative is to use high-order elements and to place nodes iteratively to obtain the correct order of elasto-plastic singularity.

REFERENCES

Akin, J.E. (1976). The generation of elements with singularities. *Int. J. Num. Meth. Engng.*, 10, 1249-1259.

Barsoum, R.S. (1976a). On the use of isoparametric finite elements in linear fracture mechanics. *Int. J. Num. Meth. Ingng.*, 10, 25-37.

Barsoum, R.S. (1976b). A degenerate solid element for linear analysis of plate bending and general shells. *Int. J. Num. Meth. Engng.*, 10, 551-564.

Barsoum, R.S. (1977). Triangular quarter-point elements as elastic and perfectly-plastic crack tip elements. *Int. J. Num. Meth. Engng.*, 11, 85-98.

Benzley, S.E. (1974). Representation of singularities with isoparametric finite elements. *Int. J. Num. Meth. Engng.*, 8, 537-545.

Bathe, K.J. (1977). ADINA - a finite element program for automatic, dynamic, incremental non-linear analysis. *Report 82448-1, Massachusetts Institute of Technology*.

Blackburn, W.S. (1972). Calculation of stress intensity factors at crack tips using special finite elements. *The mathematics of the finite element method, Brunell University*.

Byskov, E. (1970). The calculation of stress intensity factors using the finite element method with cracked bodies. *Int. J. Fracture Mech.*, 6, 159-167.

Chan, S.K., Tuba, I.S. and Wilson, W.K. (1970). On the finite element method in linear fracture mechanics. *Engng. Fracture Mech.*, 2, 1-17.

De Lorenzi, H.G. and Shih, C.F. (1977). Application of ADINA to elastic-plastic fracture problems. *Report 87448-6, Massachusetts Institute of Technology*.

Fawkes, A.J., Owen, D.R.J. and Luxmoore, A.R. (1979). An assessment of crack tip singularity models for use with isoparametric elements. *Engng. Fracture Mech.*, 11, 143-159.

Fawkes, A.J. (1978). An eight noded element for K_I and K_{II} determinations. *Numerical methods in fracture mechanics*, 374-384.

Gifford, L. and Hilton, P.D. (1978). Stress intensity factors by enriched finite elements. *Engng. Fracture Mech.*, 10, 485-496.

Hellen, T.K. (1969). A front solution for finite element techniques. *Report RD/B/N1459 - Central Electricity Board*.

Hellen, T.K. (1975). On the method of virtual crack extensions. *Int. J. Num. Meth. Engng.*, 9, 187-207.

Hellen, T.K. (1977). On special isoparametric elements for linear elastic fracture mechanics. *Int. J. Num. Meth. Engng.*, 11, 200-203.

Henshell, R.D. and Shaw, K.G. (1975). Crack tip finite elements are unnecessary. *Int. J. Num. Meth. Engng.*, 8, 495-507.

Hibbit, H.D. (1977). Some properties of singular isoparametric elements. *Int. J. Num. Meth. Engng.*, 11, 180-184.

Hutchinson, J.W. (1968). Singular behaviour at the end of a tensile crack in a hardening material. *J. Mech. Phys. Solids*, 16, 13-31.

Irons, B.R. (1970). A front solution program for finite element analysis. *Int. J. Num. Meth. Engng.*, 2, 5-32.

Lynn, P.P. and Ingraffea, A.R. (1978). Transition elements to be used with quarter-point crack-tip elements. *Int. J. Num. Meth. Engng.*, 12, 1031-1036.

Mowbray, D.F. (1970). A note on the finite element method in linear fracture mechanics. *Engng. Fracture Mech.*, 2, 173-176.

Parks, D.M. (1974). A stiffness derivative finite element technique for determination of crack tip stress intensity factors. *Int. J. Fracture Mech.*, 10, 487-502.

Pu, S.L., Hussain, M.A. and Lorensen, W.E. (1978). The collapsed cubic isoparametric element as a singular element for crack problems. *Int. J. Num. Meth. Engng.*, 12, 1727-1742.

Rice, J.R. (1968). A path independent integral and approximate analysis of strain concentration by notches and cracks. *Trans. ASME. J. Appl. Mech.*, 35, 379-386.

Rice, J.R. (1968). Fracture - an advanced treatise, 2, 191-311.
Rice, J.R. and Rosengren, G.F. (1968). Plane strain deformation near a crack tip in a power hardening law material. J. Mech. Phys. Solids, 16, 1-12.
Stern, M. and Becker, E.B. (1978). A conforming crack tip element with quadratic variation in the singular fields. Int. J. Num. Meth. Engng., 12, 279-288.
Swedlow, J.L. (1978). Singularity computations. Int. J. Num. Meth. Engng., 12, 1779-1798.
Taig, I.C. (1961). Structural analysis by the matrix displacement method. Report S017, English Electric Aviation.
Tong, P., Pian, T.H.H. and Lasry, S.J. (1973). A hybrid-element approach to crack problems in plane elasticity. Int. J. Num. Meth. Engng., 7, 297-308.
Tracey, D.M. (1971). Finite elements for determination of crack tip elastic stress intensity factors. Engng. Fracture Mech., 3, 255-265.
Tracey, D.M. (1977). Discussion of 'On the use of isoparametric finite elements in linear fracture mechanics' by R.S. Barsoum. Int. J. Num. Meth. Engng., 11, 401-402.
Williams, M.L. (1957). On the stress distribution at the base of a stationary crack. Trans. ASME. J. Appl. Mech.,
Wilson, W.K. and Thompson, D.G. (1971). On the finite element method for calculating stress intensity factors for cracked plates in bending. Engng. Fracture Mech., 3, 97-102.
Zienkiewicz, O.C. (1977). The finite element method - 3rd edition. McGraw-Hill.

FRACTURE AND PLASTIC DEFORMATION

P. J. Jackson and O. L. de Lange

*Physics Department, University of Natal, P.O. Box 375,
Pietermaritzburg 3200, R.S.A.*

ABSTRACT

The internal stresses produced by dislocations in a plastically deformed solid can nucleate cracks or cause plastic deformation. In this article the stresses of planar dislocation arrays are described, and it is pointed out that overlapping arrays of dislocations of opposite sign are particularly effective in producing large, localised internal stresses. Overlapping arrays can form when a group of gliding dislocations cross-slips along part of its length from one plane to another. Cross-slip can therefore play an important role in the production of the internal stresses which lead to fracture.

INTRODUCTION

The nucleation and propagation of cracks is the subject of the theory of fracture. There is an intimate association between these processes and the presence in a crystalline solid of the line defects called dislocations. The latter are mainly responsible for plastic deformation. Cracks may be nucleated by an accumulation of dislocations and crack propagation can be promoted or inhibited by dislocation motion. In a crack-free solid fracture is usually preceded by plastic deformation which can produce the stress concentrations required to nucleate a crack. Furthermore, slip can harden the material to the point where plastic deformation becomes a more difficult process than crack propagation: competition between plastic deformation and fracture will then be resolved in favour of fracture.

To appreciate the connection between crystal dislocations and fracture it is necessary to understand two aspects of dislocation theory. Firstly, how dislocations can produce stress concentrations which resemble those produced by cracks to the extent that cracks may usefully be represented by continuous distributions of dislocations, and secondly, how and why accumulating dislocations can increase the resistance to plastic flow.

In this article we describe these aspects of dislocation theory, and discuss a new dislocation mechanism which may be important in the fracture process.

THE STRESSES PRODUCED BY DISLOCATIONS

A dislocation is a line defect in the crystal structure which, when it moves, propagates a displacement (b) of one interatomic spacing. If the direction of displacement is perpendicular to the line, the dislocation is called an *edge dislocation*; if the displacement is along the line, the displacement is called a *screw dislocation*. For a detailed description of dislocations a text such as that of Kelly and Groves (1970) or the treatise by Nabarro (1967) should be consulted. We will describe neither the details of dislocation geometry, nor their complex interactions, but will focus attention instead on their role as sources of internal stress. Throughout this article we shall consider only infinitely long straight dislocations which are parallel to the z-axis of a cartesian coordinate system. The stress fields of these dislocations are two-dimensional.

The stresses[1] in an isotropic material are

$$p_{xz} = D_1 \frac{y}{r^2} \quad , \quad p_{yz} = D_1 \frac{x}{r^2}$$

for a screw dislocation, and

$$p_{xx} = -D_2 y(3x^2 + y^2)/r^4$$
$$p_{xy} = D_2 x(x^2 - y^2)/r^4$$
$$p_{yy} = D_2 y(x^2 - y^2)/r^4$$
$$p_{zz} = \nu(p_{xx} + p_{yy})$$

for an edge dislocation. Here $D_1 = \mu b/2\pi$, $D_2 = D_1/(1-\nu)$, where μ is the shear modulus. The stresses decrease inversely with distance, r ($= \sqrt{x^2 + y^2}$), from the z-axis. These expressions, which are based on the continuum theory of linear elasticity, are invalid closer to the dislocation than a few interatomic spacings, and therefore the divergence at $r = 0$ is unphysical. An important property of the stresses around a dislocation is their parity in the coordinates x or y. This is tabulated below.

TABLE 1 The Parity of the Cartesian Components of Stress of Single Dislocations

parity stress	x	y
p_{xz}	even*	odd
p_{yz}	odd	even
p_{xx}	even	odd
p_{xy}	odd	even
p_{yy}	even	odd
p_{zz}	even	odd

*even(odd) parity means the sign of stress is unchanged (changed) by the transformation $x \to -x$ or $y \to -y$.

---

[1] The stress p_{ij} is the force per unit area in the direction i across a plane perpendicular to the j axis.

Whereas the stresses of individual dislocations are too localised to nucleate cracks, the superimposed stresses of dislocations in a group are more effective in this regard. (Dislocation groups are, in fact, common in plastically deformed materials because slip is an inhomogeneous process.) Qualitative deductions of the results of this superposition may be made from a knowledge of the parity and shape of the stresses around single dislocations, which are illustrated in Fig. 1 below.

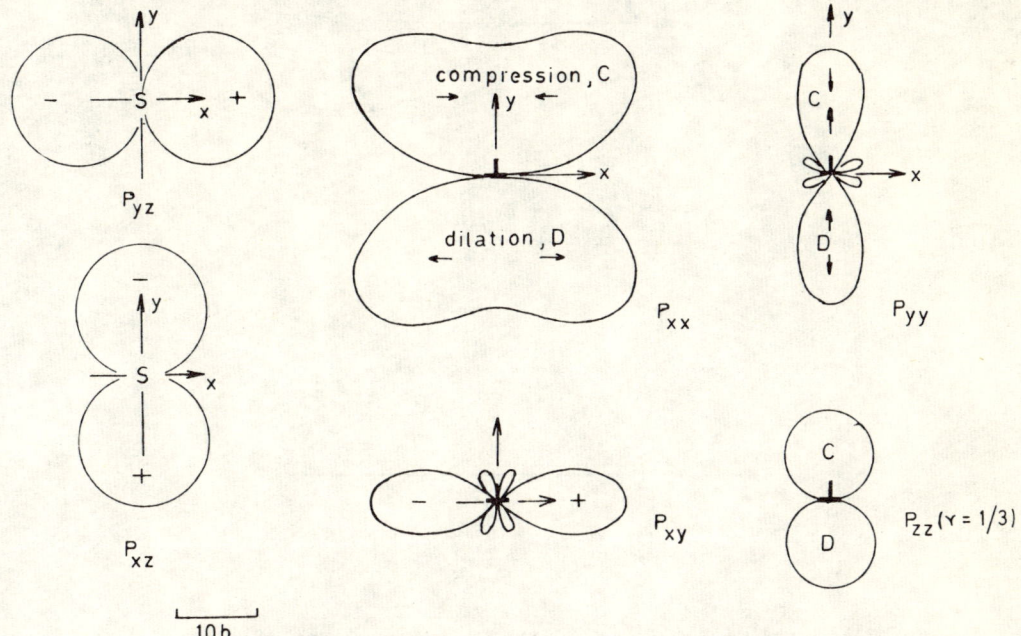

Fig. 1. Contours of equal stress around single dislocations. S = screw dislocation, ⊥ = edge dislocation. Contours are for a stress of $\mu/100$ for screw dislocations and $\mu/(1 - \nu)100$ for edge dislocations.

It is clear, for example, that above and below a planar group of dislocations there will be extended regions of compression and dilation, respectively (cf. the effect of superimposing stresses, such as p_{xx}, of neighbouring dislocations). These stresses can be large and reasonably constant within a region whose linear dimensions, on either side of the group, are of order the length of the group. One can also see that ahead of such a group the predominant stress will be p_{xy}. This is a stress which promotes propagation of the group on its glide plane.

A moving group which encounters an obstacle may pile up against that obstacle: the effect of this is to increase the stress p_{xy} at the head of the group, as shown in Fig. 2.

The stresses near a finite *wall*[2] of edge dislocations are different: on either side of the wall p_{xy} will be large and uniform, and at the ends of the wall large tensile concentrations will be present.

---

[2] A wall of edge dislocations is a planar array with the Burgers vector perpendicular to the plane of the array.

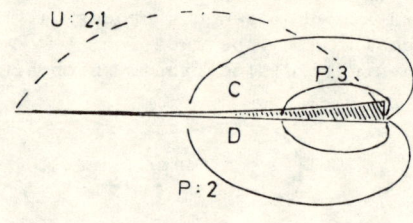

2(a)

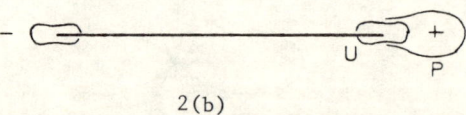

2(b)

Fig. 2.(a) The tensile stress p_{xx} of an array of edge dislocations. The contours P:2 and P:3 outline regions within which the stress is 2 and 3 times the applied stress, for an array piled up against an obstacle at the right hand end of the group. The density of dislocations increases to the right, as indicated schematically by the hatched region. The contour U:2.1 is for a uniform array with the same average spacing of dislocations as the pile-up.
(b) The shear stress p_{xy} of an array of edge dislocations. The contours are smaller than those in Fig. 2(a), even though they enclose regions within which the stress is only 1,25 times the applied stress. P: a piled up array. U: a uniform array with the same average spacing.

DISLOCATIONS AND CRACK NUCLEATION

The dislocations in a plastically deformed metal are often arranged in planar arrays which lie roughly parallel to the plane on which slip occurs (Fig. 3), and it has been proposed that the stress concentrations they generate may initiate fracture (Koehler 1952, Mott 1953). Cottrell (1959) and Petch (1959) have suggested that the superposition of the tensile stresses of individual dislocations above the head of a piled up array (cf. Fig. 2(a)) may be large enough to initiate a crack. Two arrays of edge dislocations on intersecting planes may pile up against each other at an obstacle such as a grain boundary. Near their point of intersection the tensile stresses of each array (Fig. 2(a)) are enhanced by superposition.

Cracks initiated in this manner have in fact been observed (Cottrell 1959).

Fig. 3. Etched texture on a plastically deformed copper crystal. The trace of the glide plane is GG. The criss-cross texture is due to dislocation arrays on planes roughly parallel and perpendicular to the glide plane.

Screw dislocations produce only shear stresses (Fig. 1) and therefore are not directly involved in the initiation of cracks.

CRACK PROPAGATION

The stresses near an array of dislocations may be relieved either by slip or by fracture, which in this sense are competitive processes. If a crack *is* nucleated, it is because slip is a difficult process in that region.

If a propagating crack moves into a region where slip is less inhibited, then the stresses around the crack (which are similar to those around a planar dislocation array) may cause localised slip (Gilman 1957, Tetelman and Robertson 1963). If this plastic flow is extensive it may be enough to relieve the stresses around the crack tip, and the crack will be arrested. Factors which inhibit slip will therefore promote crack propagation. Slip is inhibited in regions of high dislocation density, which can also provide the obstacles to slip responsible for crack nucleation. Alternatively, slip near a crack may be limited by the process of work hardening. Under some circumstances it is possible for plastic flow at the tip of a crack to aid crack growth by injecting into the tip dislocations which produce the same displacement as the crack itself (Tetelman and McEvily 1967).

We conclude that while plastic deformation can initiate fracture, it can also act to inhibit fracture once a crack has formed.

STRESS CONCENTRATIONS AND PLASTIC FLOW

It is useful to distinguish between two kinds of slip in a body which deforms plastically in response to an applied stress. Primary slip is caused directly by the applied stress: it is associated with the motion of primary dislocations. (Fig. 6 shows a simple example of primary slip in a single crystal.) Secondary slip is a result of the internal stresses of primary dislocations. Of these internal stresses p_{xy} (from edge dislocations) and p_{xz} (from screw dislocations) promote further primary slip, whereas other stresses (p_{xx}, p_{yy}, p_{zz} and p_{yz}) initiate secondary slip which, by interacting with primary slip, causes work hardening and inhibits plastic deformation. We see that concentrations of the tensile stresses p_{xx}, p_{yy} and p_{zz} can initiate fracture directly, or, if relieved by slip, can cause work hardening which may lead to fracture. (The stress concentration p_{yz} of screw dislocations affects fracture only indirectly through work hardening.) It is clear that primary slip which leads to configurations of dislocations in which tensile stresses are enhanced by superposition is important for fracture. In §2 we mentioned several such configurations: isolated planar groups, walls of dislocations and intersecting planar groups. In what follows we describe a simple configuration which may be as effective, and which has not been treated before.

THE STRESSES BETWEEN PLANAR ARRAYS

Consider two parallel planar arrays of dislocations in the planes $Y = \pm h$, each consisting of n equally spaced dislocations parallel to the z axis (Fig. 4), and suppose that the dislocations in one array have opposite sign to those in the other.

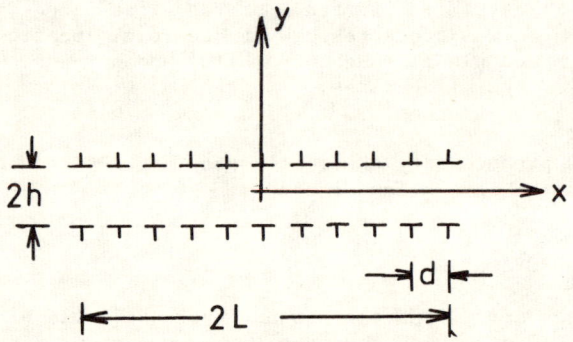

Fig. 4. Two overlapping groups of uniformly spaced dislocations.

An inspection of Fig. 1 and Table 1 makes it clear that the superposition of the stress fields of the individual dislocations in the arrays will have a simple result. Between arrays of edge dislocations the contributions to p_{xx} from each array will add. Outside the arrays they will cancel. The same is true of p_{yy}, whereas p_{xy}, because it is an odd function of x, will be small between the arrays

The result is that the region between the arrays is subjected to a uniform tensile stress. A recent analysis (de Lange, Jackson and Nathanson 1979) of the stresses between overlapping arrays shows that when the length of the arrays is greater than about 5 times their separation, the only appreciable stresses between edge arrays are p_{xx} and p_{zz}, with $p_{xx} \approx 2\mu b/[(1 - \nu)\bar{d}]$ and $p_{zz} \approx \nu p_{xx}$, where $\bar{d}$ is the average dislocation spacing in the arrays. These values of the stresses are very insensitive to the detailed arrangement of the dislocations in the arrays. Between screw arrays a similar situation prevails; only the uniform shear stress $p_{xz} \approx \mu b/\bar{d}$ is significant (see also Fig. 1). These results are illustrated in Fig. 5.

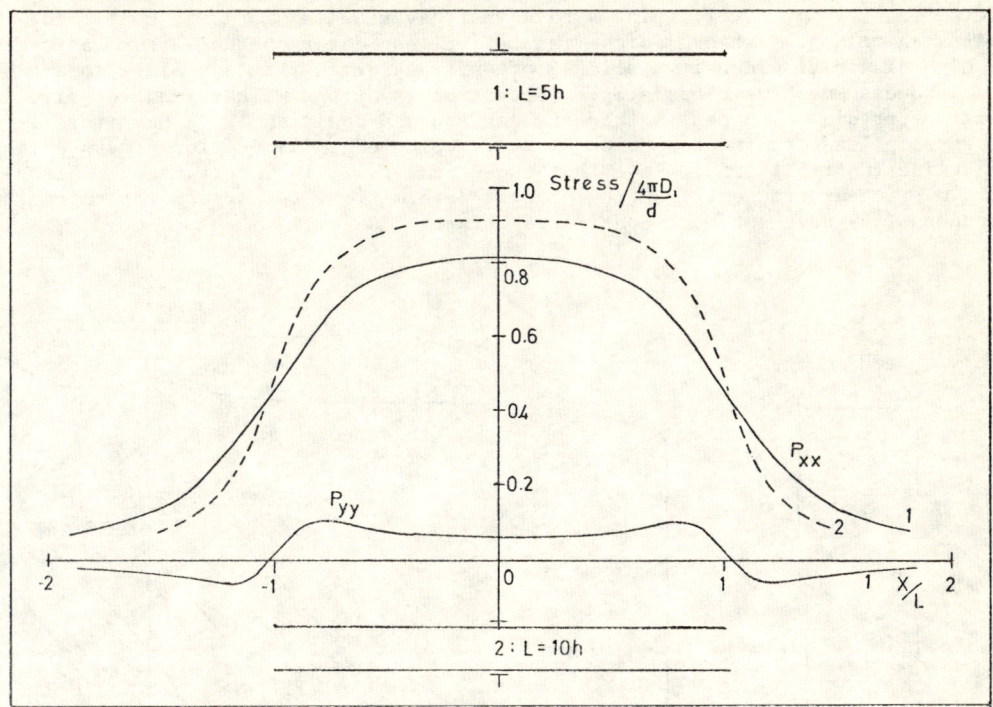

Fig. 5. The tensile stresses between arrays of continuous uniform distributions of edge dislocations. The upper curves show the variation of the predominant stress p_{xx} midway between the arrays. The lower curve shows the variation of p_{yy}, which is $\ll p_{xx}$ everywhere. The tensile stress $p_{zz} \approx \nu p_{xx}$ is not shown. The shear stress p_{xy} is near zero everywhere except near the edges, where it is less than for a single uniform array (Fig.2(b)).

The average dislocation spacing in a pure f.c.c. metal is related to the resolved shear stress, τ, at which it flows plastically (the flow stress) by the relation $h = \mu b/\omega\tau$, where ω is a constant in the range 2 to 7 (Nabarro, Basinski and Holt, 1964). Equating $\bar{d}$ and $\bar{h}$ leads to the conclusion that the uniform tensile stress between two edge arrays is $2\omega/(1 - \nu)$ times the flow stress. Since $\nu \approx 1/3$, this means a stress concentration factor of about 3ω; i.e. approximately 6 to 21. This factor does not depend on the size or separation of the arrays, provided that their length is $\gtrsim 5$ times their separation.

THE FORMATION OF OVERLAPPING ARRAYS

The overlapping arrays, whose stresses are described above, can form whenever there is a two-way traffic of dislocations of opposite sign on adjacent glide planes. A particularly simple example is that of a crystal deforming in single glide, as illustrated below (Fig. 6). In reality, however, slip is seldom as simple as depicted in Fig. 6. Overlapping arrays must nevertheless form frequently for the following reason: when gliding dislocations encounter obstacles such as regions of high dislocation density, slip is often transferred from one plane to another by a process known as cross-slip. Recent observations we have made of slip in neutron irradiated copper (Nathanson, Jackson and Spalding 1979) have revealed that cross-slip leads to the formation of overlapping edge dislocation arrays when pencil glide (Cottrell 1953), like that shown in Fig. 7, is interrupted. Although this must occur frequently during plastic deformation, the stress concentrations it produces (§6) have not been considered before.

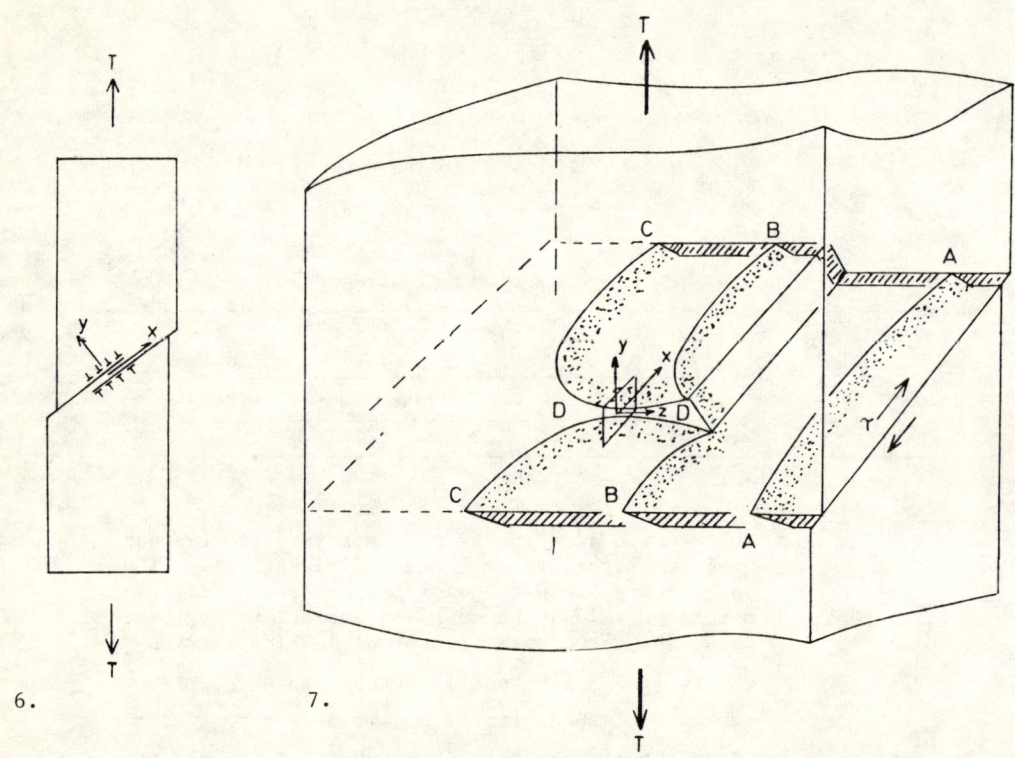

6.

7.

Fig. 6. Overlapping groups of edge dislocations produced by slip on adjacent primary glide planes (i.e. the xz planes) in a crystal deforming plastically in response to a tension T.

Fig. 7. Three stages in the formation of overlapping groups of edge dislocations by the cross-glide of a group of screw dislocations. A crystal is deforming by shear (pencil glide) which propagates from right to left in response to a tension T (shear stress τ). The line AA marks the boundary of slip in the first stage, when it is confined to a single plane. The stippled region contains a procession of screw dislocations, and the hatched areas are steps on the surface of the crystal. The line BB marks the boundary of slip in the second stage when slip has been partly transferred by cross-glide at an obstacle (not shown) to a second glide plane. The line CC is the slip boundary in the third stage. Slip spreading on the two adjacent planes now overlaps in the region D, forming processions of overlapping edge dislocations. The xy plane shown here is threaded by overlapping edge dislocations, like those shown in Fig. 4.

CONCLUSIONS

The formation of substantial stress concentrations must be a natural consequence of plastic deformation, since this proceeds by the motion of groups of dislocations. Those which can nucleate cracks are concentrations of tensile stress in the direction of glide. These occur next to dislocation arrays; near the heads of piled up arrays; and between overlapping arrays of opposite signs. Because the tensile stresses between overlapping arrays are enhanced by superposition, such regions will be particularly favourable sites for crack nucleation. These stress concentrations are also large enough to cause secondary glide (Nathanson et al., 1979) and work hardening, which is often a prerequisite for fracture. It is worth noting that when secondary slip occurs between overlapping dislocation groups, networks of sessile dislocations (Lomer-Cottrell locks) can be created. These networks may provide obstacles against which primary dislocations accumulate. This process can produce overlapping piled up groups of dislocations of opposite sign, between which the tensile stresses are large.

We have pointed out (§2) that the stresses of individual dislocations are not sufficiently extended to nucleate cracks, and that superposition of the stresses of many dislocations is required. Configurations in which this superposition is effectively accomplished involve, as we have seen, dislocations which move in groups on the glide plane. The nucleation of cracks by internal stress concentrations is a consequence of the inhomogeneous nature of plastic deformation.

REFERENCES

Cottrell, A.H. (1953). In *Dislocations and Plastic Flow in Crystals*, Clarendon Press, Oxford, p. 4.
Cottrell, A.H. (1959). In B.L. Averbach et al. (Eds.), *Fracture*, Wiley, New York. p. 20.
de Lange, O.L., Jackson, P.J. and Nathanson, P.D.K. (1979). To be published. See also Nathanson, P.D.K. (1979).
Gilman, J.J. (1957). *Trans. Met. Soc. AIME.*, 209, 449.
Johnson, T.L., Davies, R.G. and Stoloff, N.S. (1965). *Phil. Mag.*, 12, 305.
Kelly, A. and Groves, G.W. (1970). In *Crystallography and Crystal Defects*, Longman, London.
Koehler, J.S. (1952). *Phys. Rev.*, 85, 480.
Mott, N.F. (1953). *Proc. Roy. Soc.* A 220, 1.
Nabarro, F.R.N., Basinski, Z.S. and Holt, D.B. (1964). *Adv. Phys.*, 13, 193.
Nabarro, F.R.N. (1967). In *Theory of Crystal Dislocations*, Clarendon Press, Oxford.
Nathanson, P.D.K. (1979). Ph.D. Thesis, University of Natal.
Nathanson, P.D.K., Jackson, P.J. and Spalding, D.R. (1979). To be published.
Petch, N.J. (1959). In B.L. Averbach et al. (Eds.), *Fracture*, Wiley, New York, p.54.
Stokes, R.J., Johnston, T.L. and Li, C.H. (1959). *Phil. Mag.* 4, 920.
Tetelman, A.S. and Robertson, W.D. (1963). *Acta Met.*, 11, 415.
Tetelman, A.S. and McEvily, A.J. In *Fracture of Structural Materials*, Wiley, New York.

THE SIGNIFICANCE OF ROCK FRACTURING IN THE DESIGN AND SUPPORT OF MINE EXCAVATIONS

H. Wagner and N. Wiseman

Chamber of Mines Research Laboratories, P.O. Box 61809, Marshalltown 2107, R.S.A.

ABSTRACT

Due to the specific nature of the stress field around mining excavations the volume of rock fractured as a result of mining is generally confined to the immediate vicinity of these excavations. A narrow fracture zone around excavations is often a desirable feature since it tends to move the zones of high stress concentrations further into the rock mass and reduces the support requirements. Removal of the protective fracture zone can result in uncontrolled enlargement of underground excavations. The degree of rock fracturing around mine tunnels is best described by a strength criterion whereas an energy criterion is better for stoping excavations.

INTRODUCTION

In the South African mining industry close to 100 million cubic metres of rock are excavated annually in underground mining operations. On average about 80 per cent of the underground openings fall into the category of productive excavations whereas the remainder are classified as service excavations. The main difference between the two types of excavation is that the shape of productive excavations is determined by basic geological parameters such as the type of mineral deposit which could either be tabular, massive or pipe like, whereas the shape of service excavations is governed by technical parameters such as the tonnage of material to be transported or area served by these excavations.

In terms of the overall volume of rock affected by mining, the actual mining excavations comprise only a very small percentage. For example, if it is assumed that mining affects only the rockmass above the mining excavations then the ratio between gold mining excavations and the rockmass affected by these excavations is as low as 1:1 000 to 1:3 000. In other words, in gold mining less than one thousandth of the total volume of rock affected by mining is actually extracted. In the case of the more shallow coal mines this ratio could drop to a value of about one hundredth.

Since rock fracturing is usually confined to a small volume of rock in the immediate vicinity of the mining excavations, it follows that the effects of rock fracturing on the stability of a mine are significantly different from those in other engineering disciplines. A further significant difference exists between structural design in mining, civil engineering and mechanical engineering. In the latter two cases the engineer has a considerable degree of freedom in the choice of material whereas in the case of mining this choice is generally very restricted particularly as far as productive excavations are concerned. Furthermore, in most engineering disciplines, one of the basic principles is to design structures in such a way that fracturing does not take place. In mining with its very limited degree of flexibility in the choice of the material, the loading conditions and the shape of excavation, fracturing of the rock surrounding excavations is often unavoidable. The design of mine structures is, therefore, largely concerned with controlling the post-failure behaviour of rock, and the presence of fracturing is accepted as normal as long as it does not interfere with mining operations or lead to a structural collapse of a mine or a section of a mine.

Therefore, it is not surprising that the application of fracture mechanics in the design and support of mining structures has developed in a different direction from that in other engineering disciplines. One possible exception is the underground part of civil engineering although even in this particular field the emphasis is to design structures in such a way that rock fracturing is avoided wherever possible.

FAILURE THEORIES IN ROCK ENGINEERING

Name of Theory	Criterion	Remarks
Coulomb-Navier	$S_o = \frac{1}{2}(\sigma_1 - \sigma_3)$	Not applicable to rock in tension S_o = Shear strength
Modified Coulomb-Navier	$\lvert \tau_\theta \rvert = S_o + \mu \sigma_\theta$	τ_θ = Shear stress in failure plane σ_θ = normal stress acting on failure plane μ = coefficient of internal friction θ = angle of failure plane
Mohr (1900)	$\tau_\theta = f(\sigma_\theta)$ where failure envelope is straight $\tau_\theta = (S_o + \sigma_\theta \tan \emptyset)$ $\tan \emptyset = \mu$	$\emptyset$ = angle of internal friction
Plane Griffith (1924)	$(\sigma_1 - \sigma_3)^2 - 8 T_o(\sigma_1 + \sigma_3) = 0$ if $\sigma_1 + 3\sigma_3 > 0$ and $\sigma_3 = -T_o$ if $\sigma_1 + 3\sigma_3 < 0$	T_o = uniaxial tensile strength
Modified Griffith McClintock and Walsh (1962)	$\sigma_1[(\mu^2+1)^{\frac{1}{2}} - \mu] - \sigma_3[(\mu^2+1)^{\frac{1}{2}} + \mu] = 4T_o$	In this simplified form of the modified Griffith theory it is assumed that the stress σ_c which is required to close the Griffith cracks is negligibly small. μ = coefficient of friction between crack surfaces
Murrel (1965)	$\sigma_1 = C_o + b \sigma_3^m$	C_o .... uniaxial compressive strength b and m are constants. For quartzite Cook and Wagner found C_o 200$\pm$50 (MPa) $6 < b < 10$ and $m \simeq 1$

Table 1. Failure criteria in Rock Engineering.

Table 1 summarizes the most commonly-used failure theories in rock engineering. All except the original Coulomb-Navier criterion, have in common that the strength of rock tends to increase with increasing confining stress. However, most theories suggest that the strength of rock increases linearly with confining stress. Notable exceptions are the general MOHR failure theory and the original Griffith theory.

The failure theory that is most commonly employed in the design of underground structures in South African mines is

$$\sigma_1 = \sigma_c + k \sigma_3$$

where

σ_1 = maximum compressive stress a rock sample can withstand
σ_c = uniaxial compressive strength ($\sigma_3 = 0$)
k = a material property, and
σ_3 = confining stress

The strength properties of rock are normally determined in triaxial compression tests. Two different testing procedures are used, namely

i) to subject the rock specimens to a certain confining stress and to keep this stress constant throughout the tests, and

ii) to load the specimen so that the ratio between axial and confining stress is kept constant. Failure of the specimen will occur as long as the ratio $\sigma_1/\sigma_3 > k$ (Fig.1).

A critical examination of the in-situ loading conditions shows that neither the assumption of a constant confining stress nor that of a constant axial to lateral stress ratio are realistic. Hallbauer and others (1973) have suggested that it is far more realistic to determine the strength properties of rock in a loading system which has a certain lateral stiffness, k_1. In such a system the confining stress, σ_3, which acts on the specimen is a function of the lateral deformation of the rock sample, E_1, and the lateral stiffness, k_1.

$$\sigma_3 = k_1 E_1 \quad (2)$$

In the initial loading stages the build-up in confining stress is governed by the Poisson's ratio of the rock sample and the load-deformation characteristics of the confining system. However, as micro-cracks develop within the specimen and the latter dilates, the

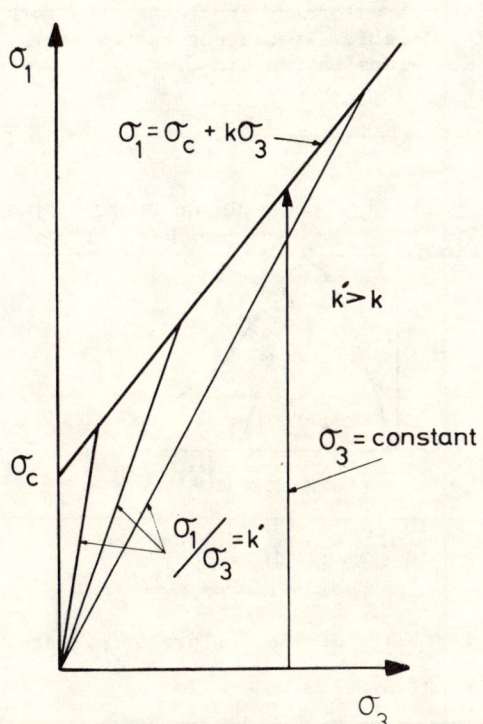

Fig. 1. Stress modes in triaxial compression tests (see text for explanation)

confining stress will increase rapidly. Clearly, this build-up in confining stress will be more pronounced in the case of rocks which display a considerable amount of dilatation. The creation of a system of localised confining stresses as a result of micro-fracturing in the rockmass is an important aspect in rock fracturing the significance of which has so far been neglected.

One of the most important aspects of the mechanical behaviour of rocks is their capability to provide resistance to deformation even after the material has been strained beyond the point of its ultimate load-bearing capability. Jaeger and Cook (1969) divide the load deformation behaviour of rocks into two distinctly different regimes, namely, into a work hardening and into a work softening or post-failure regime. In the past most attention was placed on the behaviour of rock prior to failure, and important rock properties such as Young's modulus and Poisson's ratio have been defined in this regime. Most recently, rock mechanics research has concentrated on the post-failure load deformation behaviour

Knowledge of the post-failure behaviour of rock is important for two reasons. Firstly, the progressive deterioration of the strength properties of rock with post-failure deformation, emphasises the need for controlling this deformation if the inherent strength properties of rock are to be preserved. This can often be achieved with suitable support measures. Second, the work softening characteristics of rock together with the overall structural response of the rockmass determine whether the mode of failure is stable or unstable. The criterion for stable failure is that the energy required for the rock fracturing process is greater than the energy which becomes available during the unloading process of the rock structure (Fig. 2a, b). This question is of particular importance as far as the design of pillar workings in relatively shallow mines is concerned.

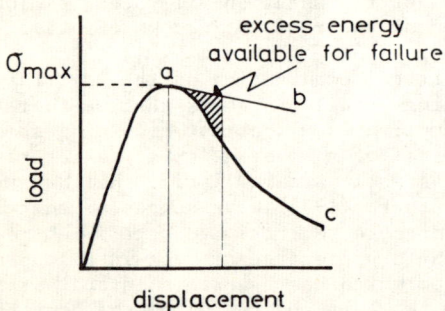

Fig. 2a. Unstable failure of pillars.

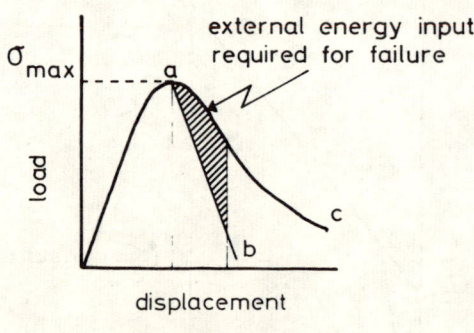

ab .... mine stiffness
ac   pillar stiffness

Fig. 2b. Stable failure of pillars.

DESIGN OF MINE STRUCTURES AND SUPPORT SYSTEMS

Because of the inherent difficulties in determining the mechanical and strength properties of the rockmass and in assessing the loading conditions on engineering structures in rock, the design principles in rock engineering differ greatly from those used in other engineering disciplines. Wherever possible, use is made of the concept of back analysis. This concept has proved to be particularly successful for designing structures in existing mines.

Two problem areas are encountered as far as the application of this concept is concerned. First, a methodology has to be developed to describe quantitatively the degree of fracturing in existing mine structures and second, realistic assumptions have to be made about the conditions that lead to either the partial or complete failure of a rock structure. A classical application of this concept is the design of bord and pillar workings in coal mines. Based on the statistical analysis of 125 failed and unfailed bord and pillar workings in South African collieries, Salamon and Munro (1967) developed an empirical relationship for the strength of coal pillars taking into account the effects of geometrical parameters such as the width, W, and the height, H, of the pillars. Salamon and Munro found that the strength of typical South African coal pillars can be described by the following relationship.

$$\sigma = 7{,}1 \, W^{0,46}/H^{0,66} \quad \text{MPa} \quad (3)$$

where the value of 7,1 MPa is a typical strength value for a one metre cube of coal.

The design of bord and pillar workings in collieries is simplified because the load acting on individual coal pillars in relatively shallow horizontal coal seams can be estimated with the aid of the tributary area concept. In terms of this concept each pillar has to carry the weight of the roof strata corresponding to the area supported by the pillar. Hoek (1971) similarly developed design procedures for rock slopes.

The design of the layout and support of tunnels in deep gold mines is a problem of particular importance both from an operational and an economic point of view. Unlike the design of bord and pillar workings in coal mines or the design of rock slopes, which are primarily dimensioning problems, the design of mine tunnels in gold mines is a problem of selecting the most suitable rock strata in which to site a mine tunnel taking into account the existing and future stress field, the strength differences of various rock strata and economic and operational constraints. Typically, the nature of the mining-induced stress field is such that zones of high stress concentrations develop in the vicinity of stope faces and rock pillars. These zones of high stress concentration extend for a distance which varies from tens to hundreds of metres ahead, above and below the stope face. The area behind the face is generally destressed. A second feature of the mining-induced stress field is that the magnitude of the induced stresses decreases with distance from the stope face.

Based on a quantitative system of describing rock conditions in existing mine tunnels in terms of degree of rock fracturing and rock deformation, Wiseman (1978, 1979) has shown that a

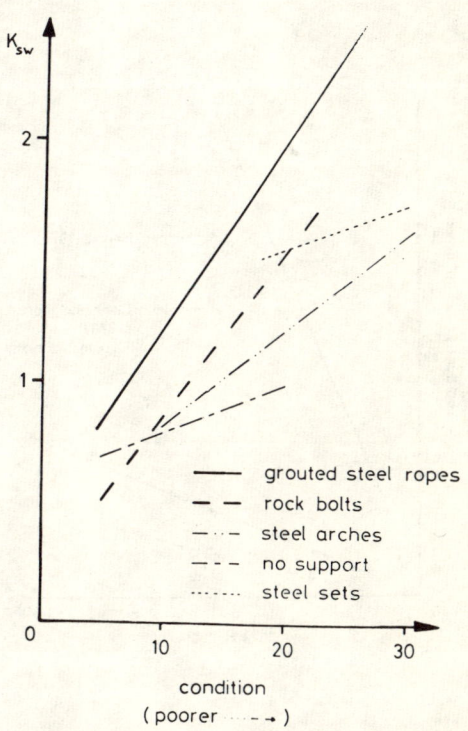

Fig. 3. Tunnel conditions as a function of sidewall stress and support.

well-defined relationship exists between the stress concentration factor in the sidewall of mine tunnels and the degree of fracturing and tunnel deterioration. This relationship is :

$$K_{sw} = (^3\sigma_1 - \sigma_3)/\sigma_c \quad (4)$$

where K_{sw} is the stress concentration factor. The studies by Wiseman have shown that in addition to the sidewall stress concentration factor, the type of support installed in the tunnel is an important parameter which determines tunnel conditions. (Fig. 3).

As far as the design of production excavations in deep tabular orebodies is concerned, extensive use is made of the concept of the spatial rate of energy release, ERR. The spatial rate of energy release (MJ/m^2) is a measure of the strain energy quantity which theoretically would be contained in a unit volume of rock, removed from underground during the process of enlarging the stoping excavation. Since the volume of rock that is being mined is already intensely fractured in most pratical situations, it does not contain any significant amount of strain energy. Salamon (1974), in a detailed study of the energy changes caused by the enlargement of an underground excavation, has shown that most of the missing energy quantity has been used in forming the mining induced rock fractures. Practical experience supports this argument and it has been found that in the absence of geological discontinuities such as faults and dykes, well-defined relationships exist between the spatial rate of energy release and mining-induced rock fractures. An example of this is given in Fig 4 which shows that a linear relationship exists between the number of reported rockbursts per 1 000 m^2 and the spatial rate of energy release.

Relationships of this kind are being used to design the layout and support of production excavations in such a way that the frequency of damaging rockbursts is being kept below a critical level. This can be achieved by carefully controlling the stoping geometry, by leaving strategically situated support pillars or by filling the mined-out area with waste rock. Analogue and digital computer models have been developed to assist mines with the planning of the layout of these excavations.

Controlling the post-failure deformation of fractured rock in the immediate vicinity of mining excavations is important for two reasons. First, if the movement of fractured rock into the excavation is not restrained by some form of support, then the inherent strength properties of rock are lost. Unrestrained rock movement can lead to a complete loss of coherence of the fractured rock due to fallout of individual rock fragments. This process can

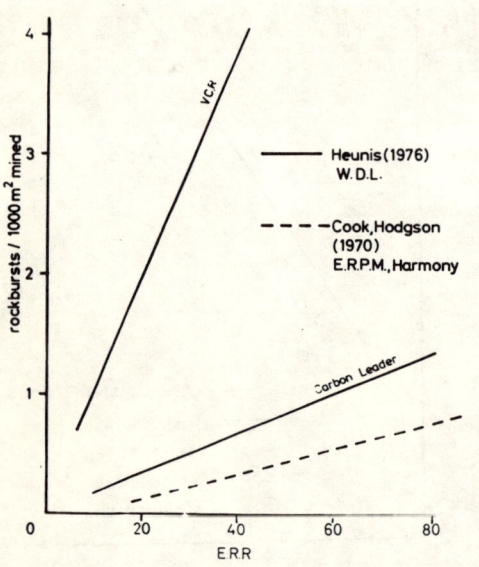

Fig. 4. The effect of energy release rate on rockburst incidence.

ultimately lead to a complete collapse of the underground structure. The maintenance of the excellent strength properties of fractured rock which is being constrained is of utmost economic importance if these properties are compared with those of conventional support materials (for example, concrete). Second, the development of a fracture zone of limited extent around an underground excavation is often a desirable feature because rock fracturing results in the zones of high stress concentrations migrating from the walls of the excavation deeper into the rockmass. An equilibrium is reached when the strengthening effect of the confinement provided by the fractured rock to the intact rockmass some distance away from the excavation equals the stresses acting on the rock. If the zone of fractured rock surrounding the excavation is removed, new fractures will form until a new equilibrium is reached. One of the main functions of support is, therefore, to preserve the stabilising effects of the fracture zone.

The efficiency of the support can be defined as its ability to constrain or prevent movement taking place in the zone of fractured rock around the excavation. This ability is termed the stiffness of the support and has become the principal criteria for the design of support systems, (Fig. 5). Ideally, a support system should have a high initial stiffness to keep the post-failure deformation of rock as small as possible but it should also have some yield properties to prevent overstressing of the support structure and to allow controlled development of the protective fracture zone. Pre-stressing of the support elements such as hydraulic props or rockbolts has the same effect as high initial stiffness. Supports which are installed with a certain pre-load are commonly known as active supports and are becoming increasingly important in rock engineering. Support structures of high stiffness but limited or no yield properties have restricted application since they tend to fail due to over loading. By contrast, support structures of low stiffness are unable to control post-failure deformation of rock and often result in structural failures due to excessive rock deformation.

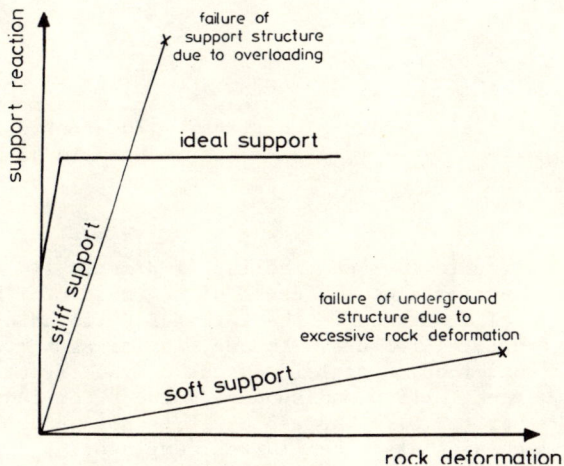

Fig. 5. Loading curves for various support types tunnel sidewall.

The application of the support principles discussed above is shown in Figure 6a, b which compare two mine tunnels under identical stress and geological conditions. The tunnel in Fig. 6a is supported by pre-tensioned rockbolts whereas that in Fig. 6b is supported by steel legs and caps. The latter form of support is typical of many of the old mine tunnels. The major principal stress is assumed to act in the vertical direction resulting in the formation of stress-induced rock fractures parallel to the sidewalls of the tunnels. The mode of tunnel failure in this instance would be by vertical slabbing of the sidewalls with subsequent buckling of the rock slabs. The latter process is assisted by dilatation of the fractured rock in the direction perpendicular to the orientation of the stress-induced fractures.

With rockbolting the movement due to dilatation is either prevented or restrained considerably, and consequently confining stresses are built up within the rock which induces a certain degree of self-support and resists further failure. The passive support system shown in Fig. 6b allows considerable movement of the sidewall slabs before any resistance to further movement is built up. Since the mode of failure of the tunnel is caused by the formation of rock slabs parallel to the sidewalls of the tunnel and not by vertical closure, it is obvious that the support system shown in Fig. 6b is totally unsuited to the prevailing conditions.

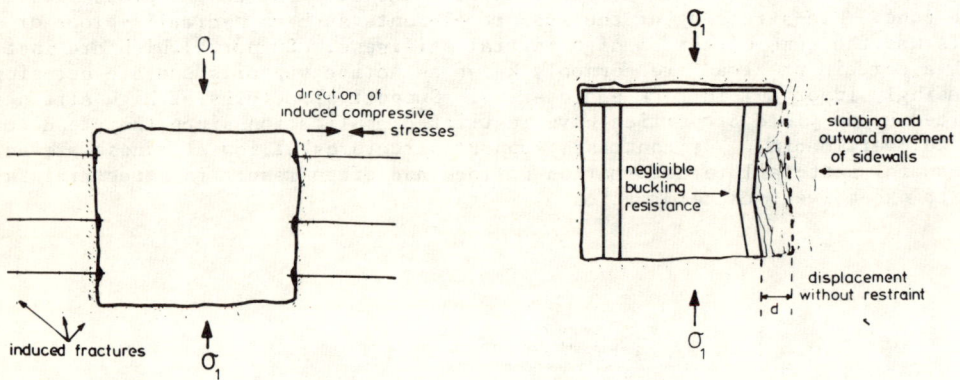

Fig. 6a. The control of failure by active support

Fig. 6b. The control of failure by passive support

CONCLUSIONS

The significance of fracturing in rock engineering is drastically different from that in other engineering applications. Because of the nature of most underground excavations the occurrence of fractures in the rockmass surrounding the excavation is not synonymous with structural failure. Indeed, as far as the economic design of support structures is concerned, a certain extent of rock fracturing is often a desirable feature and new construction and support methods for tunnels have been based on this concept, Wagner (1970). Therefore, it is not surprising that the direction that fracture mechanics has taken in rock engineering differs greatly from the classical approach of studying the conditions of propagation of individual fractures. In rock engineering the main emphasis is on the control of the post-failure behaviour of rock rather than on trying to prevent fractures from occurring.

Although great strides have been made in the past twenty years towards a better understanding of the fracture processes in rock and the application of sound engineering concepts in the design of rock structures, a number of questions remain to be solved. The most important task is the development of a theoretical model of the behaviour of fractured rock which takes into account the complete load deformation characteristics of rock, including the dilatation of rock material due to the formation of micro-fractures within the rockmass. To support this work further experimental data on the behaviour of fractured rock are essential. This may necessitate the development of new experimental techniques both in the laboratory and in the field. Only when this has been achieved will it be possible to replace some of the present design concepts which are based on statistical analysis of real life situations by more exact procedures.

Considering the complex nature of the rock fracturing process and the frequent lack of advance knowledge of structural defects in the rockmass, as well as the natural variation in rock conditions even within the same geological formation, the application of fracture mechanics to rock engineering problems will always remain a specific and difficult task.

REFERENCES

Hallbauer, D.K., Wagner, H. and Cook, N.G.W. (1973). Int. J. Rock Mech. Min Sci., 10, 713 - 726

Hodgson, K., and Cook, N.G.W. (1970). In D.A. Howells (Ed)., Dynamic Waves in Civil Engineering, Wiley Interscience, London pp 121 - 135

Hoek, E. (1971). Influence of structure on the stability of rock slopes. Proc. 1st Symp. Stability In Open Pit Mining, A.I.M.E., New York, pp 49 - 63.

Jaeger, J.C. and Cook, N.G.W. (1969). Fundamentals of Rock Mechanics, Methuen, London.

Salamon, M.D.G. (1974). Rock mechanics of underground excavations. Advances in Rock Mechanics. Vol 1, part B, National Academy of Sciences, Washington pp 951 - 1099.

Salamon, M.D.G. and Munro, A.H. (1967). A study of the strength of coal pillars. J.S. Afr. Inst. Min. Metall., 68 55 - 67.

Wagner, H. (1970). New Austrian tunelling method. TUNCON 70, Vol. 1. Ass. Sci. & Tech. Soc. S. Afr., Johannesburg pp 121 - 127

Wiseman, N. (1977). An evaluation of factors affecting mine tunnels. C.O.M. of S. Afr. Report 50/77, Johannesburg.

Wiseman, N. (1978). A study of the factors affecting the design and support of gold mine tunnels. C.O.M. of S. Afr. Report 50/78, Johannesburg.

ROCK FRACTURING PROCESSES IN DEEP MINES

N. C. Gay

Chamber of Mines Research Laboratories, P.O. Box 61809, Marshalltown 2107, R.S.A.

ABSTRACT

Around most excavations in deep mines there exists a zone of failed rock. In tunnels, this failure takes place by stable fracturing which causes slabs of rock to break away from the tunnel walls. In the areas of active mining, which have a tabular, slitlike form, the rock fractures in a stable manner in order to relax the high stresses developed in front of the excavations. Little or no movement occurs across these fracture surfaces. However, complicated shear surfaces are found occasionally; these contain relatively large amounts of comminuted material and displacements of several centimetres may occur across them. SEM observations show that the rock is essentially brittle during the formation of the first two types of fractures but that both brittle and ductile modes of deformation operate during the development of the shear zones.

INTRODUCTION

Gold mining in South Africa is now carried out at depths in excess of 3 km in some mines and the average mining depth is 1,5 to 2,0 km. The host rocks in which the gold conglomeratic reefs are found are quartzites, sandstones and shales. Depending on their mineralogical composition and structure, these rocks respond to the forces acting on them in an elastic or ductile manner. Particularly important is the quartz content in the rock (c.f. Price, 1966); the greater this is, the stronger the rock and the higher is its Young's modulus. By contrast, the more micaceous minerals there are in the rock, the more ductile is its behaviour.

The uniaxial compressive strengths of the rocks range on average from 150 to 300 MPa. These strength values increase with the application of confining pressure; nevertheless, the field stresses acting on the rocks surrounding mine excavations are sufficiently large to cause the rocks to fail. Thus around most mine excavations there is a zone of failed or fractured rock.

The nature of the fracture process can be deduced by examination of the fracture surface using SEM fractography. As in metals, four distinct types of failure, namely ductile failure, cleavage and intergranular brittle fracture, and fatigue failure, can be identified from fracture surface patterns. However, because the rocks are made up of various minerals, which may differ in their rheologic properties, the interpretation of the fractographs can be relatively complicated. Previous work using SEM for the study of deformation features in rocks, includes that of Dengler (1976) on microcracks and pores in crystalline rocks; that of Gay (1976) and Gay, Comins and Simpson (1978) on features seen on shear surfaces from a crypto-explosion structure; and that of Friedman and others (1974) on surface features induced during sliding friction experiments. The bulk of this work is concerned with documenting the appearance of the microscopic features, noting phase changes, such as the development of glass, and changes in chemical composition. Observations on material from mining-induced shear zones were described by Gay (1978) and Gay and Ortlepp (1979).

In this paper, three basic types of fracture surfaces which form around excavations in deep mines are described and the SEM fracture patterns seen on them are compared with those seen in metals and experimentally fractured rocks in an attempt to deduce the fracture mechanisms.

FRACTURES DEVELOPED AROUND EXCAVATIONS IN MINES

Excavations in mines are of two types: tunnels which have an approximately equidimensional cross-section, and tabular, slit-like excavations (stopes) with very

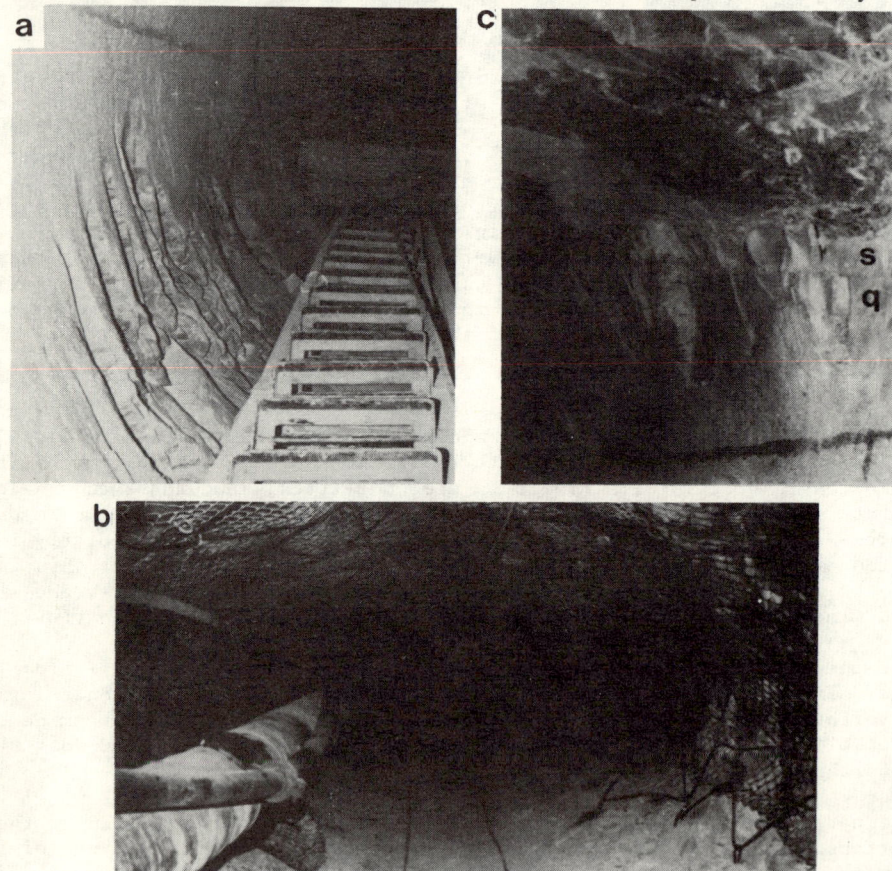

Fig. 1. a) Slabbing along the lower sidewall of a raise bored access way; b) Appearance of an originally equidimensional tunnel developed in a high stress region; c) Slabbing propagating from a shale layer(s) into the underlying quartzite (q).

large width to height ratios (≤ 500 : 1). In the tunnels, failure generally occurs by slabbing along the sidewalls (Fig. 1a); i.e. large slabs of rock fall away from the walls, resulting eventually in the modification of the tunnel shape to an elliptical cross-section (Fig. 1b). The fracture surfaces initiate on the weaker shale horizons and then propagate at a relatively controlled rate into the harder quartzites (Fig. 1c). The position at which slabbing initiates is also dependent on the orientation of the field stresses around the tunnel; generally it tends to be concentrated along those parts of the tunnel walls which are perpendicular to the least compressive principal stress. The width of the slabs is thin in relation to their area (1 - 10 sq.m.) and the surfaces are normally smooth with poorly developed rib marks and herring-bone structures.

Under the SEM, it is seen that the quartz grains are cleaved parallel to the macroscopically smooth surface (Fig. 2a) and individual grain surfaces show river patterns formed by intragranular cleavages at high angles to the surface (Fig. 2b). Conchoidal surfaces with curved striations are seen in some grains but were not common in the specimens examined. Micaceous minerals, form the matrix between the quartz grains; this has a relatively uneven dimpled appearance (Fig. 2c) and because it effectively encloses all the quartz grains, no intergranular fractures are seen. The observations indicate that the quartz grains fail by brittle fracture while the mica is more ductile. The fact that most quartz grains are cleaved parallel to the slab surface, is consistent with an extension or tensile origin for the slabbing, as a result of the maximum principal stress acting parallel to the sidewall at the locus of slab initiation.

Two types of fractures form in the rock surrounding the stopes (Kersten, 1964; McGarr, 1971 a, b). These are the areas of active mining, as a result of which their widths increase at a controlled rate with a resultant increase in the field stresses acting on the rocks surrounding the ends, or faces, of the stopes. As a result of these high stresses, the rock fails in a stable manner by fracturing along planes which are parallel to the stope face and which dip at moderate to high angles in the direction of face advance (cf. Fig. 3). The fractures are developed in a zone which extends for 5 to 10 metres in front of the face into the solid rock. They also propagate, in a discontinuous manner, for some distance (5 - 30 m) into the rock strata above and below the plane of the reef and individual fracture planes have surface areas

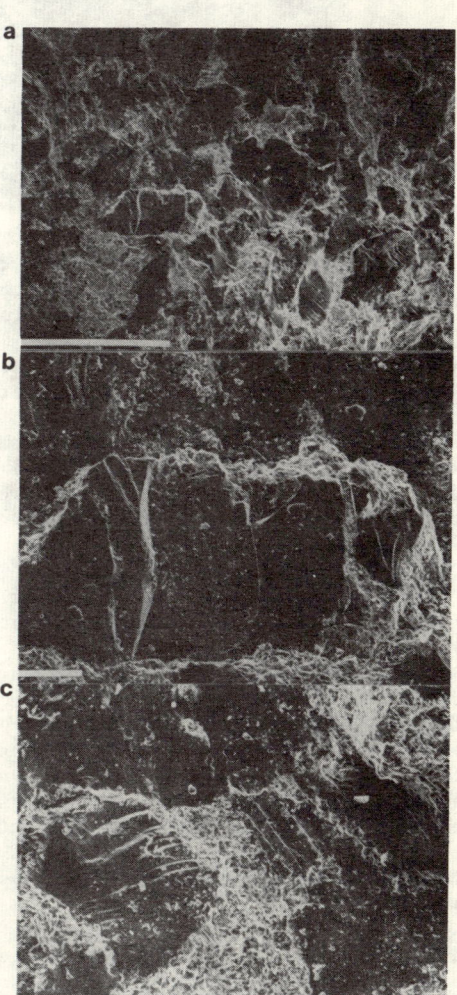

Fig. 2. SEM micrographs from a spalled slab; scale bars are: a -1mm; b,c - 0.1mm

Fig. 3. Stable fracturing in the hangingwall of a stope. Mining advanced from left to right. (Photograph by W. D. Ortlepp).

of tens of square metres. The fractures are generally closely spaced (3 - 50 cm), and are best developed in the deeper, more highly stressed environs. Very little, if any, comminuted material is seen along the surfaces in situ, which suggests that little movement occurs along them. Close to the mining faces, the slabs bounded by the fractures are packed tightly together but as the fractured rock is left behind in the mined-out area, cracks open up and permit the slabs to be extracted.

The fracture surfaces are much more irregular than those of the sidewall slabbing. Features seen in hand specimens include poorly developed rib-marks which may intersect each other to leave small lumps of whiteish, fractured grains. Plumrose structures are rare. The irregular surface persists to a microscopic scale and the SEM observations show regular, downward tear steps across the surface, (Fig. 4a). The steps are marked by comminuted quartz and/or micaceous material (Fig. 4b) and they separate plateau areas formed by grains cleaved parallel to the overall surface. Although there is not much crushing on these plateau areas, the damage is considerably more than on the sidewall slabs. Several orthogonal sets of intragranular cracks occur at high angles to the general surface (Fig. 4c) and where these intersect the grains are shattered into tongues of very much smaller particles (Fig. 4d). River patterns and some striations associated with conchoidal fracturing are also seen. These observations suggest that the rock is essentially brittle during the formation of the fractures.

A less common third type of fracture observed in deep mines has a much more complicated structure. These fractures have previously been correlated with violent, unstable failure of the rock (c.f. Pretorius, 1966; McGarr, 1971 b; Ortlepp, 1978; Gay & Ortlepp, 1979) but there is evidence for them forming in a stable manner as well (Roering, 1979, personal communication). The fractures are effectively shear zones which encompass several hundred square metres in area and across which displacements of several centimetres may occur. In detail, the shear zones are seen to be made up of smaller, en-echelon shear planes which are connected by subsidiary conjugate shear planes and extension fractures (Fig. 5). The entire shear zone tends to be parallel to the face of the advancing stope and forms several

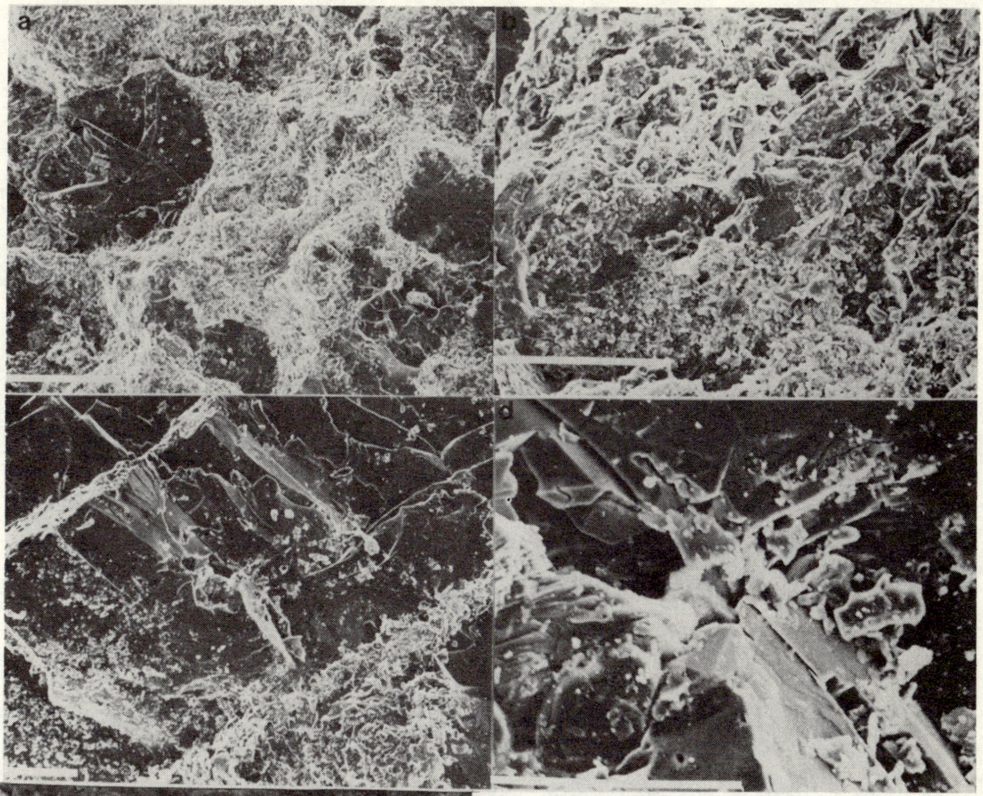

Fig. 4. SEM micrographs from a stable fracture in the vicinity of a stope; scale bars are a - 0.5mm b, c, d - 0.1 mm.

Fig. 5. Shear zone formed in the vicinity of a stope with well-developed gouge and secondary extension fractures. The width of the gouge zone is 50 mm.

metres in front of it. The zone dips at a high angle towards the advancing face and movement along it is such as to allow rock to move into the excavation.

The individual shear planes comprise a 10 - 30 mm wide zone of finely powdered, striated and compacted gouge (i.e. rock flour) in which larger fragments may be enclosed. Under the light microscope, the gouge material is seen to consist of finely comminuted quartz grains (diameter 5 - 10_μ m) and other particles, some of which are elongated and pulled apart to form fluxion structures reminiscent of turbulent flow. The adjacent wall rocks are shattered to form a microbreccia and are cut by subsidiary shear and extension fractures.

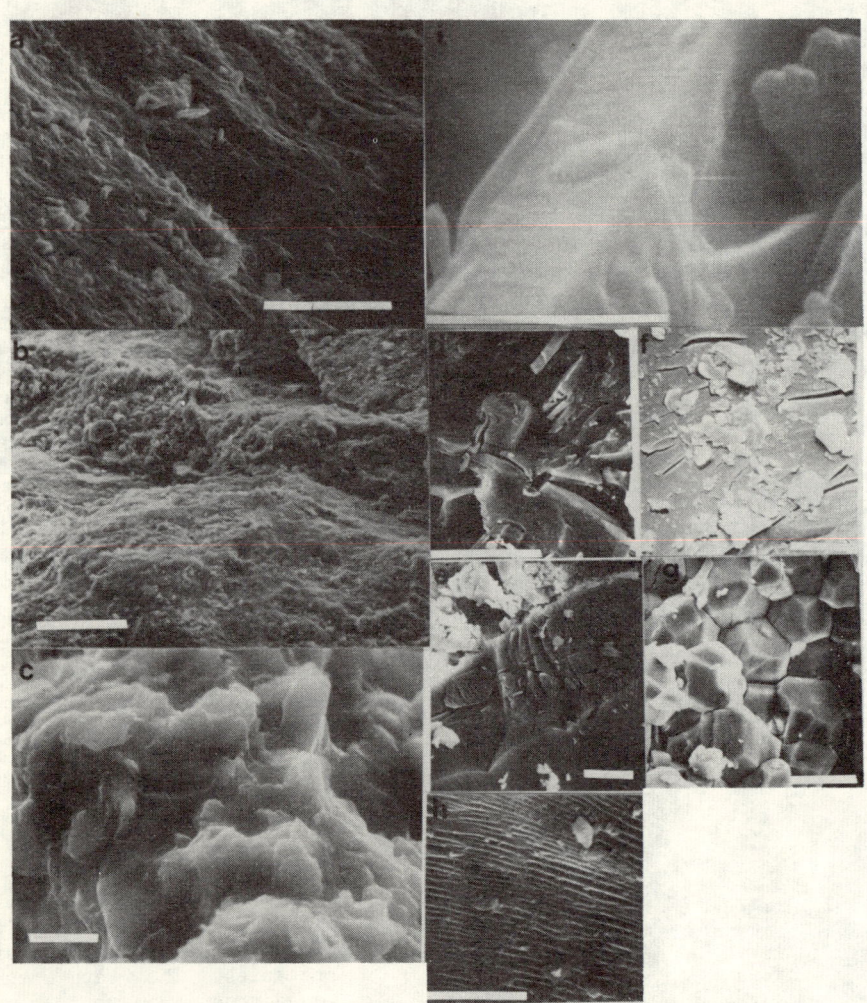

Fig. 6. SEM micrographs from shear zones: scale bars = 0.01 for all micrographs except a, b (0.1 mm) and g (0.02 mm).

Under the SEM, the shear surfaces are sometimes seen to be grooved or striated (Fig. 6a) with step like features similar to tear ridges (Fig. 6b). Irregular, possibly dimpled surfaces are also found (Fig. 6c). Transgranular cleavage (Fig. 6d), conchoidal fracturing (Fig. 6e) and intragranular cracking (Fig. 6f) are common, and a rhombohedral cleavage (possibly indicating rapid deformation rates) is seen occasionally (Fig. 6g). In addition, fatigue-like features such as striations (Fig. 6h) and lances (Fig. 6i) occur. These features suggest that both brittle and ductile modes of deformation operate during the formation of the shear zones.

DISCUSSION

The two main factors controlling the fracture processes operating in the rocks surrounding mine excavations are the strength and rheological properties of the rocks and the magnitudes of the field stresses acting on the excavation. The rock mass is normally heterogeneous and isotropic, consisting, as it does, of a sequence of layered strata with different properties. Thus, one finds that fracture initiates in the weaker rocks and propagates into the stronger ones. On a macroscopic scale, the mode of fracturing, i.e. shear or extension mechanisms, seems to be the same for all rock types, but on a microscopic scale it is found that the individual minerals which make up the rocks, respond differently and that brittle, ductile or mixed-mode mechanisms may operate.

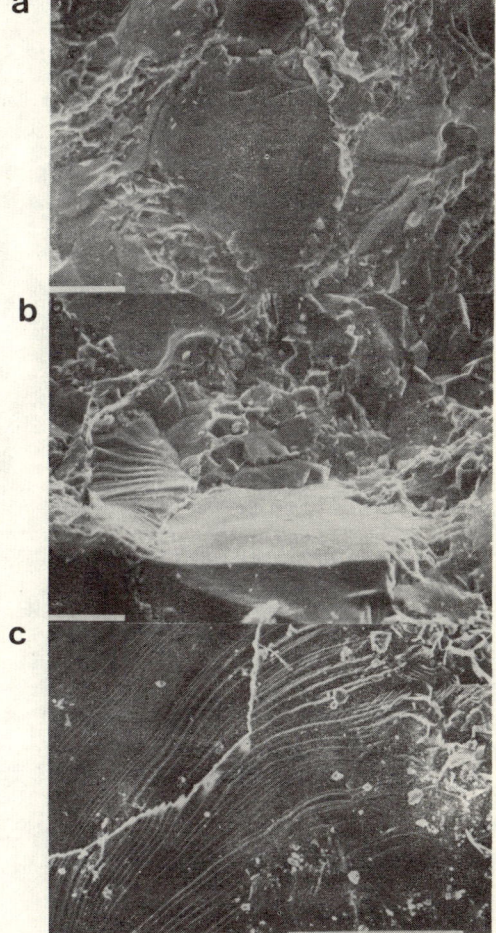

Fig. 7. SEM micrographs from quartzite specimens tested in a 3 point bending apparatus; scale bars = 0.1 mm.

The more important parameter in controlling fracture is the ambient stress state, the magnitude of which increases with depth and also varies with the geometry of the excavation. In deep level mines, the difference between the maximum and minimum principal stresses, is invariably large enough to cause the rocks forming the perimeter of an excavation to fall. Tunnels have a relatively fixed geometry and, unless they are sited in an area of active mining, the stresses acting on them are likely to remain constant. Thus, once the tunnel is constructed, failure occurs relatively instantaneously. Because the maximum principal stress is normally oriented near vertical in deep mines, this failure is exemplified by sidewall slabbing, with the fracture surfaces parallel to the plane containing the maximum and intermediate principal stresses. The mode of failure is extension fracturing and the SEM observations show that brittle fracture is the dominant mechanism. For comparison with the micrographs shown in Fig. 4, cleavage and conchoidal fractures induced in quartzites in a 3-point bending test are shown in Fig. 7 (a, b). These support the idea that failure occurs by tension.

The geometry of a stope is, however, not fixed, because its span, or width, increases at a relatively constant rate as the reef is extracted. The consequences of this are that the stresses acting on the rocks around the ends of the stope increase with each increment in stope width and that the roof and floor of the stope tend to converge, i.e. to close the excavation. To facilitate this and to relax the strain energy stored within it as a result of the high stress level, the rock fails in a stable manner by the development of fracture planes which dip in the direction of face advance. The spacing between these fractures decreases with increasing stress level, i.e. with increasing stope width, and the orientation the planes make with respect to the maximum principal stress trajectories suggests that they form by extension fracturing. This is supported by the fact that little or no shear displacement is observed across the fracture planes and that they tend to open up in the back areas of the stopes. Moreover, the SEM observations show that many grains are cleaved parallel to the fracture surface and, despite some evidence for ductile tearing, that the dominant fracture mode is a brittle one.

The complicated structure of the shear zones; namely their relatively wide zone of comminuted material, the en-echelon off setting of individual shear planes, the associated subsidiary fractures, and the large surface areas they encompass, suggests that they form over a relatively long period of time. On the basis of a detailed macroscopic and microscopic study, Gay and Ortlepp (1979) proposed the following sequence of events in their development.

The formation on a microscopic scale of closely spaced, extension cracks, sub-parallel to the maximum, ambient compressive stress and concentrated in zones sub-parallel to the planes of maximum shear stress; coalescence of these cracks to form conjugate shear planes; movement along the more favourably oriented of these shear planes to form the main gouge zones and associated secondary fractures. This movement may be accompanied by the release of seismic energy.

The SEM observations support, to some extent, their interpretation. Brittle cleavage is common and there is evidence for a rapid deformation rate during the development of the gouge in the sporadic occurrence of rhombohedral cleavage in the quartz. Tear ridges and dimple-like features suggest ductile deformation which would occur relatively slowly, presumably during the formation of the incipient shear planes. The significance of the fatigue like features, i.e. the striations and lances, is obscure. The striations in (Fig. 6h) are similar to fatigue striations formed in a quartzite specimen, which was subjected to a load equivalent to 70% of its instantaneous failure load at a rate of one cycle per minute and which failed after 950 cycles (Fig 7c). However, the process of mining does not involve the cyclic loading of the rock around the stope except for the stresses induced in the small volume of rock immediately in front of the face, during blasting each day. Rather, the rocks are subjected to a progressive build-up in stress prior to failure and it seems more likely that the features seen in the shear zone are ductile phenomena formed during this stress increase.

ACKNOWLEDGEMENTS

I am grateful to Dr. F.A. Koch for carrying out the fatigue tests and to Professor L.O. Nicolaysen for reviewing the manuscript.

REFERENCES

Dengler, L. (1976) Microcracks in crystalline rocks. In. H.R. Wenk (Ed.) <u>Electron Microscopy in Mineralogy</u>, Springer-Verlag, Berlin, pp 550 - 556.

Friedman, M., Logan, J.M. and Rigert, J.A. (1974) Glass-indurated quartz gouge in sliding-friction experiments on sandstone. <u>Bull. Geol. Soc. Amer. 85</u>, 937 - 942.

Gay, N.C. (1976) Spherules on shatter cone surfaces from the Vredefort structure, South Africa. <u>Science, 194</u>, 724 - 725.

Gay, N.C. (1978) SEM studies on fault gouge formed during rockbursts in mines. <u>Proc. Elec. Microscopy Soc. S.Africa, 8</u>, 139 - 140.

Gay, N.C., Comins, N.G.W. and Simpson, C. (1978) The composition of spherules and other features on shatter cone surfaces from the Vredefort structure, South Africa. <u>Earth and Plan. Sci. Letters, 41</u>, 372 - 380.

Gay, N.C. and Ortlepp, W.D. (1979) Anatomy of a mining induced fault zone. <u>Bull. Geol. Soc. Amer. 90</u>, 47 - 58.

Kersten, R.W.O. (1964) <u>Report on Chamber of Mines Project No. 111/64/10108</u>, Chamber of Mines of South Africa, Johannesburg.

McGarr, A. (1971a) Stable deformation of rock near deep-level tabular excavations. <u>J. Geophys. Res., 76</u>, 7088 - 7106.

McGarr, A. (1971b) Violent deformation of rock near deep-level tabular excavations. Bull. Seismol Soc. America 61, 1453 - 1466.

Ortlepp, W.D. (1978) The mechanism of a rockburst. <u>Proc. 19th U.S. Symp. Rock & Mech</u>., Stateline, Nevada, 476 - 483.

Pretorius, P.G.D. (1966) Discussion on paper by Cook and others. <u>J.S. Afr. Inst. Min. Metall. 66</u>, 705 - 713.

Price, N.J. (1966) <u>Fault and joint development in brittle and semi-brittle rock</u>, Pergamon Press, Oxford.

THE DEFORMATION AND FRACTURE OF QUARTZ

G. Glover and A. Ball

Department of Metallurgy and Materials Science, University of Cape Town, Private Bag, Rondebosch, Cape Town, R.S.A.

ABSTRACT

Experimental methods of obtaining information about the propagation of fractures in quartz under static and moving indentors are reviewed. Results obtained are discussed in relation to the methods of impacting and cutting quartz in the mining industry and the efficiencies of these methods considered. In addition the strain rate sensitivity of quartz is highlighted by the development of theories of plasticity at typical tectonic strain rates. In this regime the ability of diffusional processes to operate are found to be a controlling factor. The strain rate and temperature to which a geological area has been subjected is analysed by an interpretation of the dislocation structures observed in field samples.

INTRODUCTION

A large fraction of the earth's upper crust consists of the mineral quartzite. A study of the deformation of quartz is of obvious value to geological science, geophysics, mining and minerals engineering. Tectonic processes take place at low values of strain rate (10^{-10} -10^{-15} s^{-1}) while mining operations involve rapidly applied stresses, resulting in high strain rates ($\sim 10^2$ s^{-1}). The characteristic deformation of quartz in response to such a wide range of strain rates is expected to be varied. Hence, under tectonic conditions the plasticity of quartz enables large scale deformation to occur, while in the engineering field quartz is a true brittle material and useful fracture events take place. However, we could provide a unifying link between these two extremes by considering the phenomonon of earthquakes - in essence a large scale fracture, but which is obviously an end result of low strain rate deformation processes occurring in the earth's crust. This paper will demonstrate the wide area to which an understanding of the deformation of quartz can be applied.

BRITTLE RESPONSE

In the mining, comminution and grinding industries, the ease with which cracks can be nucleated and subsequently propagated is of major importance, in order that the best utilization is made of available input energy. We wish to know how cracks propagate in the various crystallographic directions through the quartz lattice, and in addition, how the microstructure influences crack geometry. For example, do the cracks propagate preferentially around the grain boundaries, through the grains, or do grain boundaries arrest the cracks? Mineralogically quartzite contains varying fractions of pure quartz, and it is known that the strength of the rock increases as the quartz content rises. At lower quartz contents (<65% quartz) the quartz grains are embedded in a weaker binding material, but as the quartz content rises the grains are strongly cemented and crack propagation takes place through the grains.

The processes of crack nucleation and propagation are well established. Griffith (1920) developed an energy balance approach to crack propagation. He also examined the basic crack nucleation event and discovered that microscopic surface flaws were the points of initiation. Inglis (1913) had previously shown that at the tip of a small elliptical crack, applied macroscopic stresses could be amplified. Hence, the local microscopic stress could rise well in excess of the bulk cohesive strength of the material, and a crack would propagate by the sequential rupture of atomic bonds. In a material displaying brittle behaviour crack geometries are likely to reflect closely the density and strength of atomic bonds across different crystal planes. A high bond density will be commensurate with a high fracture stress. Cleavage tendencies will also be a function of the crystal bonding structure. Quartz possesses no planes of easy cleavage, but it is reported that the rhombohedral planes do exhibit a weak cleavage tendency.

Ball and Payne (1976) examined in detail the tensile fracture of thin plates of single crystal quartz, as a function of orientation, environment and temperature. These authors found that the fracture stress required to propagate a crack was a minimum at approximately 500K. The decrease in fracture stress from 300K to 500K is probably due to a stress activated event which leads to crack tip sharpening. As the temperature was increased beyond 500K, the fracture stress increased, and this behaviour was explained by suggesting that local plastic flow could be effectively blunting the crack tip.

Several other experimental techniques have been applied to the quartz fracture problem by the present authors. These include hardness tests, Hertzian cone-crack generation, and scratching tests described below.

HARDNESS TESTS

The primary application of the Vickers hardness test has been on ductile materials. However, at low loads measurements have been made on brittle materials, although cracking and fracture around the indentation often precludes a straightforward interpretation of results. Quartz has a hardness value of ~ 1000Kg mm^{-2}, and there is some evidence to show that hardness is a function of crystal orientation. In the micro-hardness test (loads <1N) the hardness value often rises steeply as the load is reduced. In the case of quartz values of over 3000Kg mm^{-2} may be measured at loads of 0.05N. The reason for the increase is not well understood, but is thought to be connected with the probability that the indentor stress field interacts with stress relieving objects such as dislocations.

In a ductile material plasticity processes operate to relieve the high stresses generated below the sharp indentor tip. In the brittle case cracking occurs; median vents extending downwards are formed below the indentor point on loading, and lateral vents, running parallel to the surface, form on unloading. A fractographic examination of the crack system enables an estimate to be made of the likely extent of material removal under these high stress/point loading conditions. In particular the lateral vents from adjacent indentations may interact, enabling material to spall out from between the two craters.

A high hydrostatic pressure exists below the indentor point, which, in brittle solids, is known to prevent premature fracture, and generally assist plastic behaviour. Consequently, localised plastic flow could occur during the indentation of a brittle material such as quartz. A typical SEM micrograph of a Vickers indentation in quartz is shown in Fig. 1. Cracks (C), radiate from the diagonal corners, but the overall appearance is not one of brittle fracture but of plastic flow. The smooth, rounded features (S) within the indentation are reminiscent of slip line outcrops. This suggested that TEM could well reveal dislocations. However, a HVEM micrograph around the region of an indentation (Fig. 2) fails to resolve individual dislocations, and so strain relief by plastic flow remains unanswered.

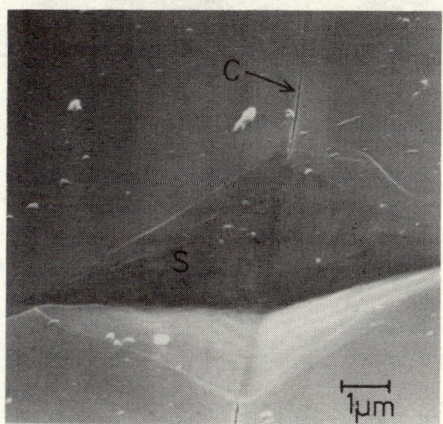 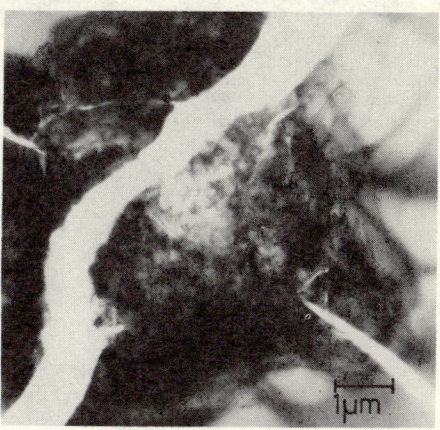

Fig. 1. Scanning electron micrograph of a Vickers indentation in quartz, showing the smooth 'plastic' appearance (Load = 1.0N).

Fig. 2. HVEM of a Vickers indentation. Individual dislocations cannot be resolved. The white areas are cracks that have propagated after thinning for microscopy. (Load = 1.0N).

The existence of dislocation processes is essential to resolving the effects of a sub-surface space charge on hardness. It is postulated that this type of charge can interact with charged dislocation lines near the surface, and either increase or decrease their mobility. A zero in the surface potential has been associated with a maximum in the hardness (and hence brittleness). These effects have been explored in laboratory drilling experiments, and clear correlations exist between the maximum drilling rate and hardness maximum. However, until dislocation generation beneath indentors is conclusively proved the origins of surface charge effects cannot be explained satisfactorily.

Finally, it is of interest to examine the indentation behaviour of a quartz grain near to a grain boundary in an argillaceous quartzite. A sequence of optical micrographs (Fig. 3) was obtained by indenting in exactly the same place and examining after each application of load. As the load increases the cracks extend in length towards the grain boundary region. At a load of 0.8N the region around the indent and also the grain boundary material shatters, demonstrating the weakness of this region.

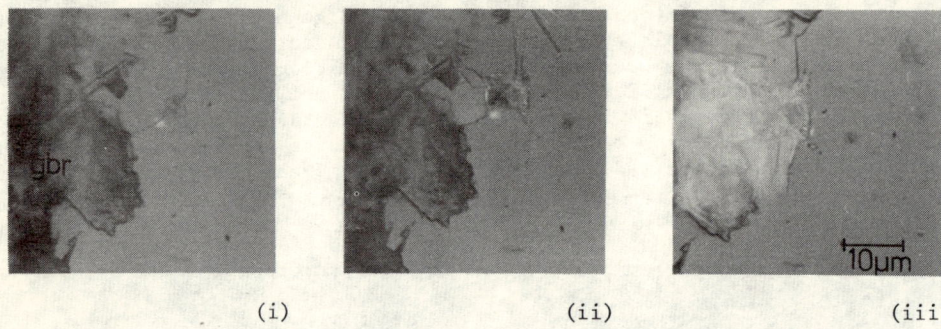

(i) (ii) (iii)

Fig. 3. A sequence of optical micrographs of an indentation in quartzite close to a grain boundary region (gbr). Note the eventual shattering of the boundary region (iii).
(i) 0.5N., (ii) 0.6N., (iii) 0.8N.

CONE-CRACK TEST

If a rigid sphere is pressed upon the surface of a sample, at some characteristic load a crack of conical section propagates downwards into the sample. The fracture load measures the fracture toughness of the test material. The test can be made as a function of such variables as specimen orientation, surface preparation, temperature and environment.

A systematic study of cone-cracking on a single crystal synthetic quartz was undertaken by Hartley and Wilshaw (1973). Tests were made in air and dry nitrogen as a function of temperature and crystal orientation. In dry conditions the fracture load was found to increase over values in ambient air, suggesting that water vapour assists fracture. As the temperature was raised the fracture load decreased, although a small increase which occurred over the range 600K - 700K could not be adequately explained.

The simplicity of the Hertzian test and its analysis, recommends it to a study of crack propagation paths in quartzite. A wide range of conditions could be tested, and the results would be of benefit for optimising impact processes in mining, and comminution facilities.

SCRATCHING TESTS

If a vertically loaded diamond pyramidal indentor has a horizontal component of force, F_T, applied to it, the resulting system of forces will enable the indentor to move forward at some critical value of F_T. A characteristic scratch track is produced, and corresponding deformation events include plastic and/or brittle behaviour. In the pure brittle case chips are formed, and extensive cracking occurs to the sides of the track. A median crack forms below the indentor, and runs along the complete length of the track.

Although we may expect quartz to behave in a brittle fashion when scratched, at fast scratching speeds (cm s^{-1}) high frictional temperatures may induce some degree of localised plasticity. The results are likely to be complicated by environmental effects; the presence of residual stress fields and incipient flaws could indicate that time-dependent cracking, due to water vapour interaction with crack tips, is possible.

By experimentally measuring the horizontal force F_T, it is possible to derive the specific energy of the cutting process. The specific energy is the amount of energy required to remove unit volume of material and is an index of efficiency. Note, however, that solely measuring the groove cross-section totally neglects the sub-surface crack system. On subsequent passes of a tool (and multiple passes are typical of industrial situations) cracks may interact and lead to material removal for relatively little expense of energy. Hence, laboratory measurements must be made on multiple and adjacent scratches that model the industrial sequence under consideration.

To assess the extent to which a small-scale scratching experiment could be scaled with a full size rock-cutter we have undertaken a series of experiments on single crystal quartz. Our primary aim was to examine the energetics of micro-scratching and link variations in specific energy with material properties and fracture events. By considering the real situation, for which the energetics are also known, it may well prove possible to utilize laboratory scale experiments in formulating the best techniques for cutting rock.

The scratch tool used in these experiments was a diamond Vickers pyramid, aligned with the diagonal parallel to the direction of scratching. The vertical force was in the range 0.1 - 1.0N. The scratch velocity could be varied between the limits 10 μms^{-1} and 100 μms^{-1}, but most experiments were undertaken at the lower speed. To exclude the effects of moisture, the scratching rig was placed within a sealed environmental chamber, and the relative humidity controlled. Samples were pre-annealed in vacuo at 800K and only exposed to the ambient environment within the chamber.

A typical SEM micrograph of a scratch on quartz made with a sharp tool is shown in Fig. 4. As the load increases the extent of chipping and cracking increases, and cracks radiate away from the groove. Material is removed in the form of chips, by multiple crack interaction. It has been our experience that the depth of the cut is deeper than may be expected from considerations of the depth of a static indentor under the same applied load. The scratch debris material was examined in the SEM, and observations such as Fig. 5 are typical: the indentor surface gooove is clearly visible but the cut has taken place well below this surface groove. This demonstrates the importance of fully understanding the stress field distribution beneath the indentor. We have also found a clear correlation between the specific energy and depth of cut: the specific energy increases as the depth of the scratch groove decreases.

The major effect of scratching at high temperatures (in the range 373K - 773K) was to increase the extent of cracking. Increasing brittleness as the temperature is raised has already been noted (Westbrook 1958). Consequently, to make use of this increased cracking tendency, there are advantages in working at higher temperatures and allowing the extensive crack propagation to aid material removal. For example, in rock cutting rather than allow the cutter to generate small chips, we desire a system of large cracks that by interacting with pre-existing cracks or with cracks generated at stope faces, allow large slabs of rock to be removed. In this way the specific energy is minimised.

Environmental effects were found to be of importance only when using a 'blunt' scratch tool. In the dry state the low load scratches were typical of Hertzian fracture tracks with little or no chipping, but with extensive microcracks running across the track. If the surface was then swamped with water and scratched the full cutting action appeared and grooves were formed (i.e. material was removed). In the dry case the value of specific energy is essentially infinite, as no material removal takes place until a vertical force of ~1.0N is applied. By contrast in water, even at the lowest loads cutting occurred. The horizontal force also decreased in water by ~25% over the full vertical load range. Thus the efficiency can be increased when scratching in a wet environment. Whether the water enhances crack propagation or if the effects are due to hydrodynamic lubrication between the specimen and tool, has yet to be clarified.

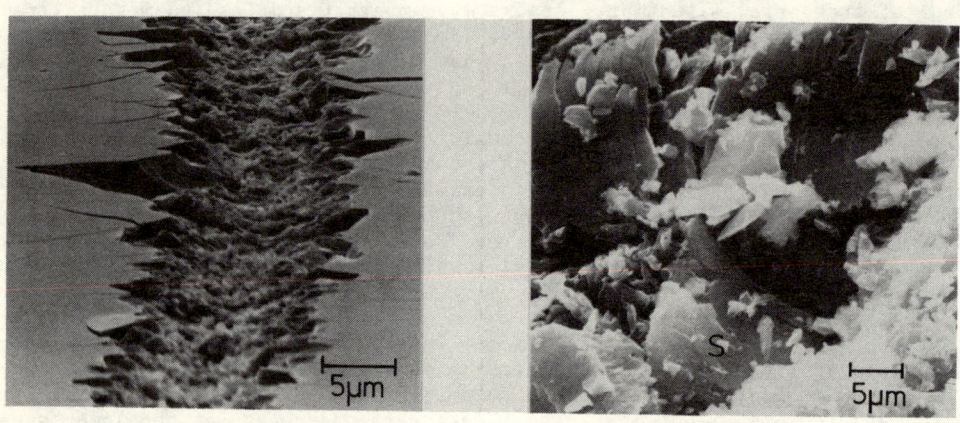

Fig. 4. Scanning electron micrograph of a scratch made on quartz with a sharp tool. Vertical load 0.8N.

Fig. 5. Scanning electron micrograph of scratch debris. Note the area (s) where the tool has passed over the surface, but the 'cut' has taken place at a much greater depth.

We conclude this section by suggesting that the increase in specific energy as the vertical force (F_N) decreases, linked with the occurrence of the "environmentally" controlled scratches at lower loads, demonstrates the need to clarify all relevant experimental and material parameters. It is our belief that some of the published results on environmentally assisted fracture (either in cutting or drilling) are from regions of high specific energy (low F_N). An increase in normal force would decrease the specific energy, but reduce the relative importance of environmental effects. However, an environment may assist in removing debris away from the bit, preventing choking, and helping to keep the bit cool; in other words, it prevents premature bit failure.

We feel that closer attention must be paid to the tool bit geometry, and its modification as wear proceeds. Accordingly, subtle changes in vertical force, speed and lubrication will maximise and maintain efficiency over the useful life of the tool and may ultimately increase it.

PLASTIC RESPONSE

The mechanism by which quartz deforms under typical tectonic conditions of stress ($\sim$1Kbar), strain rate ($\sim 10^{-15}$ s^{-1}) and temperature ($\sim$ 800K) can be extended to an understanding of such diverse phenomena as mountain building and earthquakes.

The geological history of a region may be obtained from an appraisal of dislocation structures in mineral samples. Typical tectonic sequences can involve plastic deformation, hot-working, cold-working and recrystallisation, and indeed, the remarkable similarity between dislocation structures in metals and those in minerals has often been mentioned in the literature. In consequence, the ideas of metallurgy can be used in examining deformed minerals, and deformation theories readily established. It should be noted that the use of high voltage electron microscopy (HVEM) has played a major role in the advancement of mineralogical deformation studies. Of interest to geophysicists are the stresses and/or strain rates to which a particular geological area has been subjected. To answer these questions we will demonstrate how quartz responds to various geological conditions of stress, strain rate and temperature. From field samples we may then link the observed dislocation features to good estimates of the deformation parameters.

At the outset a clear distinction must be drawn between the deformation of an isolated single crystal, and that of a single crystal grain embedded within a polycrystalline matrix. We emphasize this point, because studies of active slip systems are undertaken with orientated single crystals. This data must be carefully fitted to the general mechanisms of polycrystal deformation. One of the most important formulations on polycrystal deformation was deduced by von Mises (1928). His principle states that for a polycrystal to undergo a general homogeneous change of shape, without discontinuous phenomenon (such as cracking or pore formation), five independent slip systems must be available to the individual grains. However, Groves and Kelly (1963), showed that von Mises' principle could be relaxed if climb was added to slip as a mode of deformation. These points will be elaborated upon below.

The quartz crystal structure is based on a hexagonal lattice and slip may occur in the $\underline{a}$, $\underline{c}$ or ($\underline{a}$ + $\underline{c}$) directions. The $\underline{a}$ and $\underline{c}$ directions are relatively 'soft' i.e. they operate at low applied stresses and moderate temperatures. The ($\underline{a}$ + $\underline{c}$) system, however is 'hard' and considerable doubt surrounds its operation. If ($\underline{a}$ + $\underline{c}$) does not operate under typical tectonic conditions of stress and temperature, the remaining $\underline{a}$ or $\underline{c}$ systems do not satisfy von Mises' principle and a quartzite polycrystal in the earth's crust will not deform uniformly. HVEM studies of quartzite from areas of tectonic activity frequently show evidence for climb (Fig. 6a). That is, the dislocations are smooth and loops are often visible. Accordingly climb is probably overcoming the restriction of a lack of available slip systems.

We may now examine a model (Ball and White 1978) for quartz, based on dislocation climb and the Nabarro (1967) steady-state diffusional-creep theory.

Let a quartz crystal be subjected to an applied stress σ in the $\underline{c}$-axis direction (Fig. 7). The easy slip systems $\underline{a}$ or $\underline{c}$ do not experience any resolved shear stress, and these dislocations are essentially immobile. The hard ($\underline{a}$ + $\underline{c}$) system is orientated for the maximum resolved shear stress, but we assume that the stress is insufficient, or temperatures too low, to allow nucleation and slip of dislocations on these systems. The applied stress alters the chemical potential of vacancies in the vicinity of the dislocations, and chemical potential gradients are set up. Vacancies can diffuse down these gradients, and deformation proceeds by the climb of dislocations.

An essential feature of this model is the pre-existence of a stable dislocation network to provide the requisite sources and sinks of vacancies.

Babarro (1967) showed that the strain rate such a diffusional network could support was given by

$$\dot{\varepsilon} = \frac{Db\sigma 3}{\pi kTG^2} \bigg/ \ln\left(\frac{(4G)}{\pi\sigma}\right),$$

where D is the diffusion coefficient and the other symbols have the usual meanings. D is the only quantity that has an uncertain value. At typical crustal deformation temperatures (800K) the equation predicts a strain rate of

$$\dot{\varepsilon} \sim 1.6 \cdot 10^9 \times D \, s^{-1}.$$

D can be expressed in terms of an activation energy Q of the form

$$D = 5.10^{-18} \exp(-Q/kT) \quad \text{(White 1976)}.$$

A value for Q has been given as $84 \, KJmol^{-1}$, and hence $\dot{\varepsilon}$ has a realistic value of $2.5 \, 10^{-14} \, s^{-1}$.

It is also possible to estimate the strain rate from a knowledge of dislocation densities in a deformed polycrystal. In this case $\dot{\varepsilon} = bnv$, where n is the number of mobile dislocation and v the average dislocation climb velocity given by $D\sigma b^2/kT$. Hence $n = \dot{\varepsilon} \, kT/D\sigma b^3$. Thus, providing that we can determine n from TEM micrographs and make reasonable estimates for σ and D we can obtain the deformation strain rate.

Finally, to show conclusively that (a + c) did not operate at tectonic temperatures and slow strain rates Ball and Glover (1979) deformed synthetic quartz crystals along the c-axis. A confining pressure of 3Kbar was applied to prevent shattering. Experimental strain rates were 10^{-6} and $10^{-7} \, s^{-1}$ at temperatures of 800K and 1000K respectively. Thinned specimens were examined in the HVEM, and no evidence for (a + c) dislocation slip was found. The resulting dislocation structure was typical of climb, (Fig. 6b), and consistent with the above treatments.

Having determined the necessity for diffusional processes to assist the plastic response of quartz, this knowledge can be applied to an examination of likely deformation conditions. If temperatures are too low cataclastic behaviour will result, and the rock will be extensively fractured. An investigation of dislocation structures formed under these conditions may well show a cold worked appearance. A more accurate mathematical model of the quartz deformation problem is within reach, and we suggest that such an approach should be pursued.

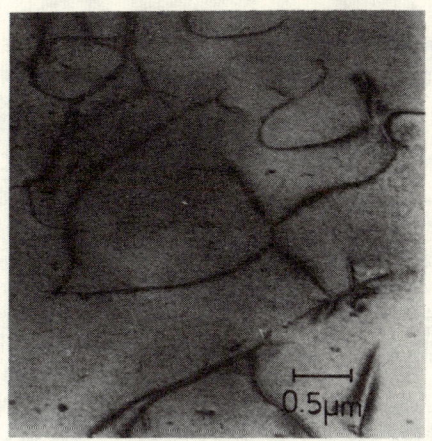

Fig. 6(a) HVEM of a naturally deformed quartzite, showing a typical climb dislocation structure.

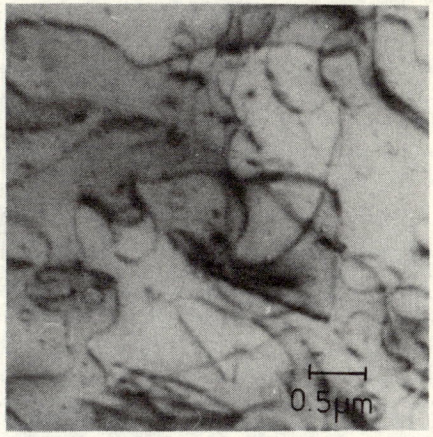

Fig. 6(b) HVEM of single crystal quartz deformed in the laboratory. Compare with Fig. 6(a).

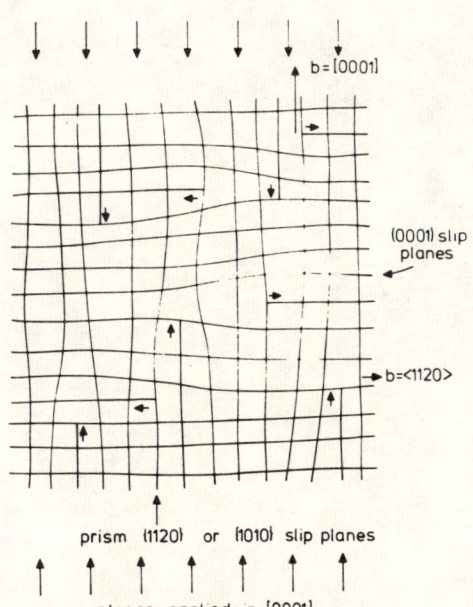

Fig. 7 A schematic representation of the application of the dislocation climb model of Nabarro (1967) to the deformation of a quartz crystal in the [0001] direction.

ACKNOWLEDGEMENT

This work was made possible by the generous financial support of the University of Cape Town and Chamber of Mines of South Africa. Laboratory assistance by Miss E. Lewis and Mr. B. Greeves is gratefully acknowledged.

REFERENCES

Ball, A., and Glover, G. (1979). Bull. Mineral., 102, 188-194.
Ball, A., and Payne, B.W. (1976). J.Mater. Sci., 11, 731-740.
Ball, A., and White, S. (1978). Phys. Chem. Minerals, 3, 163-172.
Griffith, A.A. (1920). Phil. Trans. Roy. Soc., A221, 163.
Groves, G.W., and Kelly, A. (1963). Phil. Mag., 8, 877.
Hartley, N.E.W., and Wilshaw, T.R. (1973). J.Mater. Sci., 8, 265-278.
Inglis, G.R. (1913). Trans. Inst. Naval Archit., 55, 219.
Nabarro, F.R.N. (1967). Phil.Mag., 16, 231-237.
von Mises, R. (1928). Z. ang. Math. Mech., 8, 161.
Westbrook, J.H. (1958). J. Amer. Ceram. Soc., 41, 433-440.

AUTHOR INDEX

Allen, C. 95
Anderson, R. P. G. 337

Ball, A. 95, 419

Campbell, J. R. 283

Davidson, D. 33
Dolby, R. E. 117, 313

Eccleston, P. J. 137

Fick, J. I. J. 327

Garrett, G. G. 79, 187, 337
Gay, N. C. 409
Glover, G. 419
Gurland, J. 63

Harrison, J. D. xix, 249
Heathcock, C. J. 95
Heymann, F. J. 371
Howat, D. D. 45

Jackson, P. J. 389
de G. Jones, D. 353

de Lange O. L. 389
Luyckx, S. B. 231

Marais, J. J. 203
Marriott, D. L. 219, 337

Nabarro, F. R. N. xvii

Persson, G. 3
Protheroe, B. E. 95

Taylor, R. N. 19
Twigg, D. 103

Van den Berg, W. J. 19
Van der Walt, W. 269
Van Rooyen, G. T. 295

Wagner, H. 399
Wiseman, N. 399